RENORMALISATION IN AREA-PRESERVING MAPS

ADVANCED SERIES IN NONLINEAR DYNAMICS

VOLUME 6

RENORMALISATION IN AREA-PRESERVING MAPS

R. S. MacKay

University of Warwick, UK

World Scientific

Singapore • New Jersey • London • Hong Kong

Published by

World Scientific Publishing Co. Pte. Ltd.
5 Toh Tuck Link, Singapore 596224
USA office: 27 Warren Street, Suite 401-402, Hackensack, NJ 07601
UK office: 57 Shelton Street, Covent Garden, London WC2H 9HE

British Library Cataloguing-in-Publication Data
A catalogue record for this book is available from the British Library.

RENORMALISATION IN AREA-PRESERVING MAPS

ISBN-13 978-981-02-1371-8
ISBN-10 981-02-1371-9

To Claude

Preface

It is 10 years since I wrote this thesis but many people still request copies. Despite frequent encouragements to publish it, I did not do so up to now, firstly because the style is sometimes informal, and secondly because there were many more investigations which I had intended to pursue to make it into a well-rounded story.

Although the central issues of existence of the critical fixed point for golden circles and their breakup on crossing its stable manifold remain open (from a rigorous point of view), there has been considerable progress on other aspects of the program the thesis initiated. So it seemed worthwhile to publish a new version bringing it up to date. I have chosen to do this by the addition of footnotes, so that it remains clear what was in the original thesis and what has been added. A supplementary bibliography has been added, and the list of problems extended to include some others which were implicit in the thesis. I have also taken the opportunity to correct several mistakes and the worst defects of style.

I thank Terri Moss for retyping the manuscript, thereby reducing the number of pages and making it easier to read, and Faridah Shahab and Tony Moore at World Scientific Publishing Company for arranging for its publication.

October, 1992

Abstract

Area preserving maps provide the simplest non-trivial class of conservative systems. They were introduced by Poincaré while studying the three body problem, and are relevant to many conservative problems. A few hours playing at a computer terminal is sufficient to convince one that almost all[1] area preserving maps have essentially the same features. Figure (i) is a typical picture. Firstly, one often sees stable periodic orbits. Their stability comes from being surrounded by closed invariant curves. Secondly, in between some of the larger invariant circles are "island chains", strings of alternating stable and unstable periodic points. Thirdly, when they are large enough, one can see that they are surrounded by a "sea" of stochastic orbits. These features are repeated on smaller and smaller scales. The stable periodic points of the island chains have their own invariant circles round them, and their own island chains, and so on! The aim of this thesis is to show that there are universal aspects of asymptotically exact self-similarity in area preserving maps. Renormalisation is the key towards understanding them. It means looking at the behaviour on successively longer time scales and smaller spatial scales.

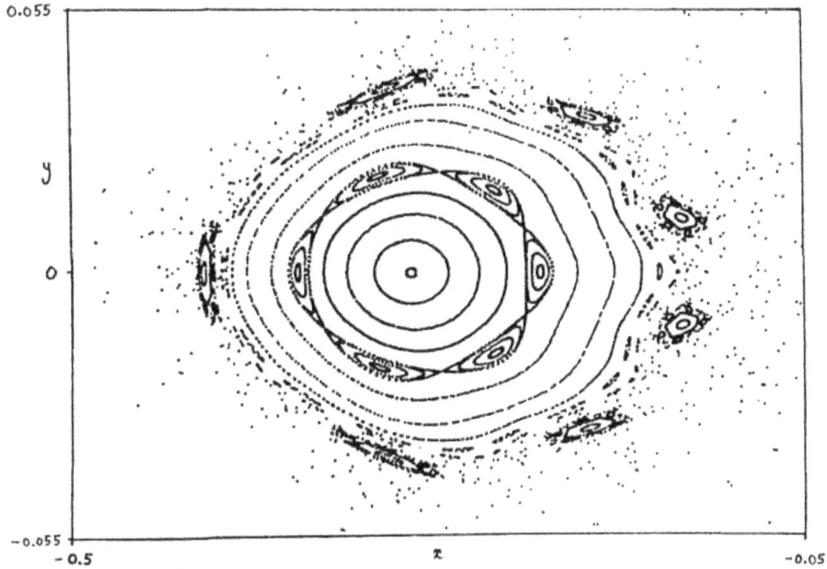

Figure (i): Some orbits of the area preserving quadratic de Vogelaere map at $p = -1.066$

[1] Leaving out the Anosov, integrable, and non-uniformly hyperbolic ones.

I discuss two particular phenomena exhibiting universal self-similarity in the above limit. They both involve consideration of how behaviour changes in a one parameter family of maps.

The first is the break-up of invariant circles. Heuristically, the island chains near to an invariant circle fatten up and squeeze gaps in it. Then stochastic orbits can leak through the gaps, so this transition is very significant for questions of confinement. In the critical case the nearby island chains repeat each other asymptotically exactly on smaller scales. This is shown in figure (ii), a blowup of the region near part of a critical circle. The behaviour of invariant circles depends strongly on the type of irrationality of its rotation number. For noble rotation number, it turns out that, close enough to a critical circle, all typical maps give asymptotically exactly the same picture as figure (ii). A universal one parameter family governs the transition from subcritical to supercritical in any one parameter family.

The other phenomenon is the disintegration of island chains, or more generally, of the trapped region associated with a stable periodic orbit. When a periodic orbit loses stability, it no longer has arbitrarily small invariant neighbourhoods. It typically gives birth, however, to a stable periodic orbit of twice the period, which has its own region of stability. This new orbit tends to be stable for a shorter interval of parameter, and then itself period doubles. Remarkably, one can generally find arbitrarily many successive period doublings,

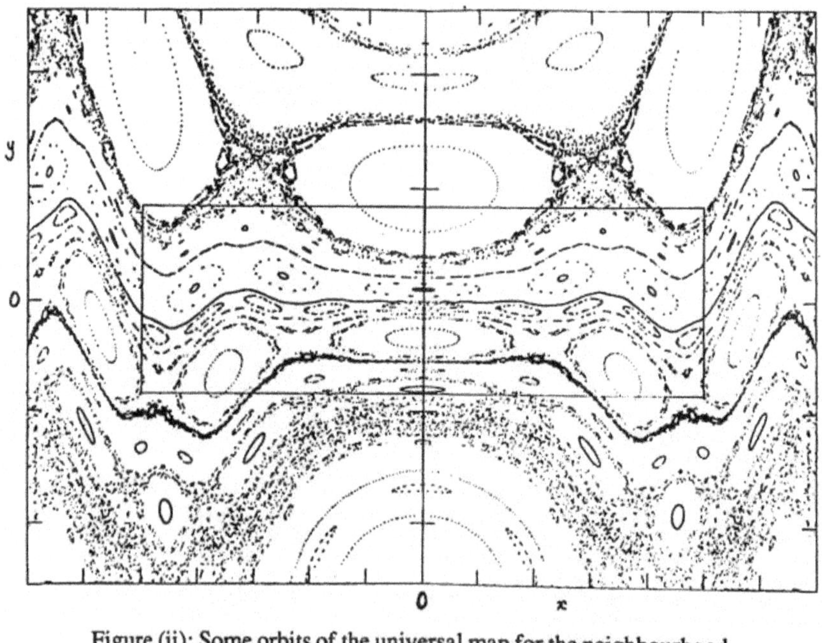

Figure (ii): Some orbits of the universal map for the neighbourhood of critical noble circles

and the local picture repeats itself asymptotically exactly as one follows this doubling sequence. Moreover, the local picture is the same for all typical cases. Figure (iii) is the universal picture for the disintegration of stable regions by period doubling as a parameter P varies. The solid lines represent the positions X of stable periodic orbits on a certain line, and the grey area represents a section of the region of stability along this line. Branches a, b, h, i are successive members of the period doubling sequence, and the picture can be seen to repeat itself exactly as one follows this sequence. At the accumulation parameter value and beyond, there appears to be practically no region of stability, locally. This transition is important because it replaces regular orbits by stochastic ones, and hence gives large regions of rapid transport.

Chapter 1 is an extensive introduction to area preserving maps. It is basically a review, but a few sections contain original work on topics that did not fit into the general framework of the thesis. Chapter 2 is an introduction to renormalisation, again mainly review, though it includes some original work too. In Chapter 3, I present numerical work on period doubling sequences in area preserving maps, leading to figure (iii), for example. A renormalisation explanation is given, and properties of the universal one parameter family described. In Chapter 4, I develop a renormalisation approach to invariant circles. Concentrating on noble rotation numbers, I show first that it could give a new proof of

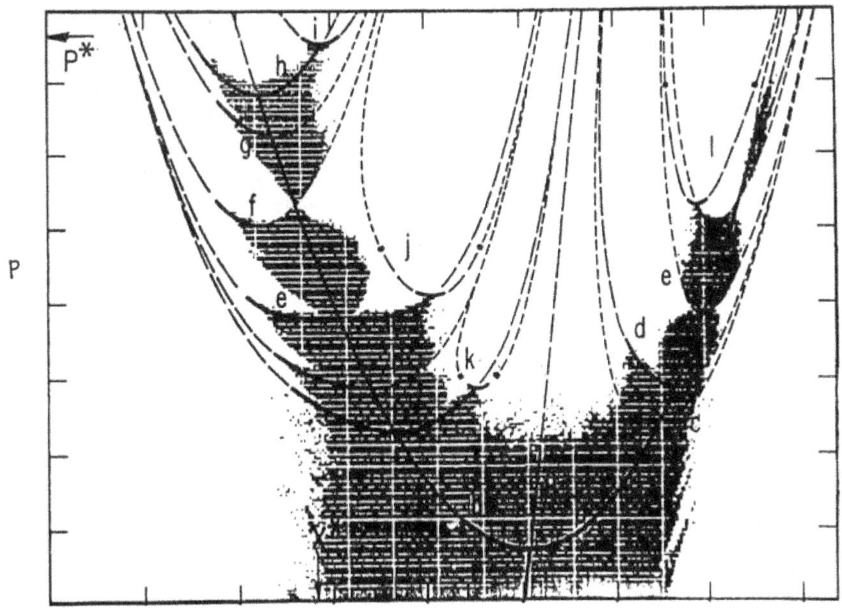

Figure (iii): The universal picture for disintegration of a region of stability by period doubling

persistence of invariant circles for small enough perturbation (the "KAM theorem").[2] Secondly, I describe numerical work on a critical fixed point of the renormalisation, corresponding to figure (ii). It looks as if its stable manifold gives at least part of the boundary of the set of maps with an invariant circle of given rotation number. Properties of the universal one parameter family describing the breakup of noble circles are given. This leads in particular to a criterion for existence of invariant circles. I also include a subsection on results for the analogous problem for one dimensional maps on a circle.

Acknowledgements

I would like to express my deep gratitude to John Greene for being such a wonderful advisor. He judged my level very accurately, and gave me problems which were challenging but just within my capability. He taught me to proceed in little steps. He gave me carefully selected things to read. He always had time for me, and we would meet together for discussion almost every day. When I asked him a question, however obscurely, he would understand what I was trying to say. He is very direct, so he would tell me exactly where I stood, what I had done well, what I had not, and what I needed to do next. He has great insight, and his suggestions have often been crucial. His criticism of my writing has been invaluable. The most fruitful times I have ever had have been working with him.[3] I consider myself extremely fortunate to have had the opportunity to work with such a great man.

I would also like to thank Martin Kruskal for his suggestions and interest in my work, and for being my advisor during the year 80/81, while John Greene was on sabbatical. He gave me so many ideas, that sometimes it was hard to work out which one to pursue, but I benefitted tremendously from discussions with him.

Thirdly, I would like to thank John Mather for teaching me a lot of the mathematics of the area in which I have been working. I have learnt a huge amount from him, and would like to thank him very much for putting up with a lot of ignorant questions, and for spending time explaining many things to me. I consider it a great privilege to have had him as a teacher.

There are many other people to whom I would like to express my gratitude. I mention in particular, John Krommes, who introduced me to the study of conservative systems, and Charles Karney, who has given me many useful suggestions. I thank Charles also for his useful comments as one of my thesis "readers". Conversations with Leo Kadanoff and Scott Shenker have been a great help. I would also like to thank three friends from whom I have learnt a lot in informal seminars and discussions: Dan Goroff, Rafael de la Llave, and Thea Pignataro. Studying with them has been invaluable. Other people who have influenced my work, and I would like to thank, are Serge Aubry, John Bagger, Jim Bialek, Tassos Bountis, William Browder, Alain Chenciner, Bob Dewar, Jean-Pierre Eckmann, Dominique Escande, Mark Evans, Mitchell Feigenbaum, Giovanni Gallavotti, Henry

[2] This goal has subsequently been achieved by Haydn in 1984 (published in 1990) and Khanin & Sinai (1986) (see also Sinai & Khanin (1988)).

[3] Perhaps this is still true!

Greenside, Robert Helleman[4], Michel Herman, Rick Jensen, Anatole Katok, Hans Koch, Oscar Lanford, Robert Littlejohn, Dean Montgomery, Phil Morrison, Sheldon Newhouse, Ian Percival, David Rand, Eliezer Rosengaus, George Schmidt, Harvey Segur, Jim Sethna, Eric Siggia, Philippe Similon, Bryan Taylor, Franco Vivaldi, Mike Widom, Arthur Wightman, and Albert Zisook.

Princeton University and the Plasma Physics Lab have been an excellent place to work. I would like to thank all those people who run the graduate program in plasma physics (especially Tom Stix), the people who have taught me, and my first and second year advisors Masaaki Yamada, Peter Barrett and Predhiman Kaw. I thank Steve Seiler for teaching an electronics course in Spring 1979, which extended what my father had taught me so that I could make the demonstrations described in §§2.2.3, 4.7.6. I am grateful to Joel Hosea for the use of figure 1.1.2.1.2. I thank the theory wing secretaries for their assistance and letting me use their word processor. The draftsmen, photolab and print shop have done excellent jobs for me. I have been really spoilt by the computer facilities. Finally, the libraries at PPL and Fine Hall have been a wonderful reference source.

I am grateful to the US-UK Educational Commission for maintenance support during 78/79 as a Fulbright-Hayes scholar. Also I acknowledge the support of the U.K. Science Research Council grant B/80/3001 for complete support for the years 80/81 and 81/82, and partial travel expenses to a summer-school at Les Houches (July 81). I acknowledge support of Princeton University and U.S. Department of Energy contract DE-AC02-76-CHO3073 for the year 79/80, the summers of 78, 79 and 80, tuition for 78/79, partial support for the summers of 81 and 82, and for attendance at the New York Academy of Sciences conference in Dec 79, the APS in 79 and 80, Dynamics Days, La Jolla (Jan. 82), the Sherwood meeting in April 1982, and partial support for Aspen (June 82). I am grateful to the University of Texas for partial support to attend their conference at Austin in March 81, to NATO and the University of Grenoble for maintenance support at Les Houches, to Los Alamos National Lab for the invitation to talk at their conference in May 82, and to the Aspen Centre for Physics for accommodation expenses and a good opportunity to talk with people and have ideas. Finally, I acknowledge the use of MACSYMA, a large symbolic manipulation program developed at the MIT Laboratory for Computer Science and supported by the National Aeronautics and Space Administration under grant NSG 1323, by the Office of Naval Research under grant N00014-77-C-0641, by the U.S. Department of Energy under grant ET-78-C-02-4687, and by the U.S. Air Force under grant F49620-79-C-020.

September, 1982

[4] I would like to add a special word of thanks to Robert Helleman, as it is through the conference that he organised in New York in December 1979 that I first became hooked on nonlinear dynamics, and Dynamics Day, La Jolla (January 1982) and Les Houches in July 1981, which he organised and co-organised were great opportunities for me to learn and discuss research. In addition, his encouragement to me has always been much appreciated.

Contents

Chapter 1
Introduction to area preserving maps

In this chapter, I discuss conservative systems, and show how their study can often be reduced to that of symplectic maps, the lowest dimensional case being area preserving maps. There are three important types of orbit in area preserving maps, to each of which I devote a section. They are the periodic orbits, which close on themselves, the quasiperiodic orbits, which densely fill an invariant circle, and the stochastic orbits, whose behaviour cannot be predicted for arbitrarily long time without precise knowledge of the initial conditions.

§1.1 CONSERVATIVE SYSTEMS AND AREA PRESERVING MAPS

In this section, I discuss the distinguishing features of conservative systems. I show how they can be reduced to maps by considering the return map on a surface of section. In the case of a Hamiltonian system with $1\frac{1}{2}$ or 2 degrees of freedom the return map is area preserving. I introduce two important classes of maps, the area preserving twist maps, and reversible maps.

§1.1.1 **Conservative systems.** This subsection gives an introduction to conservative systems. One of their distinguishing features is measure preservation, which leads to non-trivial long-time behaviour. I give a catalogue of types of conservative system, and conclude with some examples from physics, and a discussion of the possible behaviours.

§1.1.1.1 Introduction. Dynamical systems theory is the study of differential equations:

$$\frac{dx}{dt} = f(x) \qquad (1.1.1.1.1)$$

and their discrete time analogues:

$$x_{n+1} = F(x_n) \qquad (1.1.1.1.2)$$

with particular reference to features that are coordinate independent. The point x lies in $\mathbb{R}^n$, for some n, or a manifold (a topological space every point of which has a neighbourhood homeomorphic to $\mathbb{R}^n$), or an infinite dimensional space (e.g. a space of functions on some manifold).

Dynamical systems can be used to model many systems in physics, chemistry, biology, economics and other areas. A lot of the physical world can be modelled to a very good approximation by *conservative systems*. Prime examples are the systems described by Newtonian mechanics. In Newton's own words,

> "the laws which we have explained abundantly serve to account for all the motions of the celestial bodies, and of our sea."

Although it has required modifications at high velocities and small actions, Newtonian mechanics still holds a prominent place in physics. There have also been generalisations to other types of conservative system, some of which I will discuss in §1.1.1.2.

In common usage, conservative is usually understood to mean *energy conserving*. This means there is a (non-trivial) differentiable function on its state space, called the energy, which is invariant under its evolution. I will not use the word in this sense, however. Such a system can be reduced to a one parameter family of systems of one lower dimension, on the energy surfaces, but otherwise energy conservation is not really significant. To me, it is an incidental property of autonomous Hamiltonian systems.

The feature that really distinguishes conservative systems from general ones is preservation of a measure (see §1.1.1.2). Then the long-time behaviour cannot be trivial, by a famous theorem of Poincaré:

Theorem (Poincaré recurrence): For a measure preserving system with finite total measure, given $\varepsilon > 0$ and $T > 0$, almost every orbit comes back within ε of its initial position at some time $t > T$.

Thus the limit set for the system is the whole space. This should be contrasted with general dynamical systems in which there can be attractors of measure zero which attract all points in a neighbourhood, reducing the long-time behaviour of all orbits in its basin of attraction to that of the orbits on the attractor.

Long-time behaviour in conservative systems is of particular importance for questions of stability and confinement, on the one hand, and mixing and transport on the other. It is to such problems that this thesis is aimed. Even real systems with small dissipation or external noise may be treated as a perturbation of a nearby deterministic conservative system. So one should understand the unperturbed problem first.

§1.1.1.2 Types of conservative system. I begin by discussing a few of the types of conservative system. Apart from the reversible systems, these form a series of increasing specialisation.

(i) Measure preserving: A system is *measure preserving* if it has a measure such that the preimage of every measurable set has the same measure as the set. In a topological space one wants open sets to have positive measure.

(ii) Volume preserving: A system is *volume preserving* if it preserves a volume form (i.e., a non-degenerate n-form, for dimension n). A volume form induces a natural measure, namely, the integral of the absolute value of the volume form. So volume preserving implies measure preserving.

(iii) Symplectic: A system is *symplectic* if it preserves a symplectic form, i.e. a non-degenerate closed 2-form. For dimension $2n$, n is called the number of degrees of freedom. Darboux's theorem (Arnold, 1978) shows that locally coordinates p_i, q_i, $i = 1,\ldots,n$, called *canonical coordinates*, can be chosen to put any symplectic form ω^2 into the standard form:

$$\sum_{i=1}^{n} dp_i \wedge dq_i. \qquad (1.1.1.2.1)$$

A symplectic form induces a non-degenerate volume form:

$$(\omega^2)^n = \omega^2 \wedge \ldots \wedge \omega^2 \qquad (1.1.1.2.2)$$

on a finite dimensional space, so symplectic implies volume preserving (Liouville's theorem). A symplectic form also induces an isomorphism I between 1-forms ω^1 and tangent vectors ξ, by:

$$(I\,\omega^1(\xi))\,(\eta) = \omega^2(\xi, \eta) \qquad (1.1.1.2.3)$$

where η is also a tangent vector. This is analogous to the relation between covariant vectors ω^1 and contravariant vectors ξ in metric geometry, induced by the metric (e.g. Greene, 1981). Then it can be shown that every symplectic flow can be written as $I\,\omega^1$ for some closed 1-form ω^1 (i.e. $d\omega^1 = 0$). Such a flow is called *locally Hamiltonian.*

(iv) Hamiltonian: A system is *Hamiltonian* if it is symplectic and the 1-form ω^1 above is exact, i.e. $\omega^1 = dH$, for some function H, called the Hamiltonian. Locally, every closed 1-form is a gradient, but not necessarily globally, hence the distinction between symplectic and Hamiltonian. Thus the flow in a Hamiltonian system is $I\,dH$. In canonical coordinates, this means that:

$$\dot{q}_i = \frac{\partial H}{\partial p_i}, \quad \dot{p}_i = -\frac{\partial H}{\partial q_i} \qquad (1.1.1.2.4)$$

which looks more familiar. Autonomous Hamiltonian systems (i.e. systems having no explicit time dependence) conserve H, so they are energy conserving. Hamiltonian systems can be cast into a variational form, viz.

$$\delta \int p.dq - H\, dt = 0 \tag{1.1.1.2.5}$$

for variations with free end conditions, i.e.

$$p.dq = H\, dt = 0 \tag{1.1.1.2.6}$$

at the ends. In fact, this gives an alternative formulation of Hamiltonian mechanics (Arnold, 1978). Generalising from (1.1.1.2.5), one can say that a system is Hamiltonian if its solutions are given by a variational principle in an odd-dimensional space:

$$\delta \int \omega^1 = 0 \tag{1.1.1.2.7}$$

with respect to variations with free end conditions, where ω^1 is some 1-form.

(v) Lagrangian: A system is *Lagrangian* if it is given by a variational principle of the following form. The solutions are those paths for which the *action*:

$$\int dt\, L(q, \dot{q}, t) = 0 \tag{1.1.1.2.8}$$

is stationary with respect to variations with *fixed* end conditions. The function L is called the *Lagrangian*. Changing variables from $\dot{q}_i$ to p_i:

$$p_i = \frac{\partial L}{\partial \dot{q}_i} \tag{1.1.1.2.9}$$

shows that, provided the defining relations for p_i can be inverted, Lagrangian systems are Hamiltonian, with p_i, q_i as canonical coordinates, and Hamiltonian:

$$H(p, q) = p.\dot{q} - L(q, \dot{q}, t) \tag{1.1.1.2.10}$$

(vi) Reversible: A system is ***reversible*** if it is conjugate to its time-reverse, by a coordinate change whose square is the identity (called an *involution*). For example, reversing all the velocities of a potential system:

$$\begin{aligned} \dot{x} &= v \\ \dot{v} &= -\nabla V(x) \end{aligned} \tag{1.1.1.2.11}$$

reverses the flow. Similarly, all Hamiltonian systems with Hamiltonians even in the momenta (or positions) are reversible. But not all Hamiltonian systems are reversible. Neither are all reversible systems Hamiltonian, nor even measure preserving (see §1.1.4.5). Nevertheless, reversibility has significant consequences, and on its own leads to many properties similar to those of symplectic systems (see §1.2.1.3 and Devaney, 1976).

§1.1.1.3 Examples. Physically, the most important class of conservative systems is the class of Hamiltonian systems. This includes, for instance, frictionless mechanical systems, gravitational systems, electromagnetic systems, and semiclassical molecules. It also includes non-mechanical examples such as ray tracing for waves in the geometrical optics limit, and the configuration of magnetic field lines for a static field, or of flow lines for a steady fluid flow.

Hamiltonian systems are of particular relevance to plasma physics. As I have spent my time in Princeton at the Plasma Physics Lab, I will give more emphasis to examples in this field. The motion of charged particles in electromagnetic fields is Hamiltonian, and even various simplifying approximations. For example, for single particle motion in fields varying slowly on the time scale of a gyroperiod and the length scale of a gyroradius, the gyro-motion decouples, leaving *guiding centre* motion (Littlejohn, 1981a,b). This is a two degree of freedom system, parametrised by the magnetic moment. The case of zero magnetic moment and low energy reduces to field line flow, to be discussed below. Another approximation which is Hamiltonian is ideal magnetohydrodynamics (e.g. Morrison and Greene, 1980), in which the interaction of the particles is replaced by a smoothed field plus pressure.

Field line flow is a simple but very significant system which I shall use several times for illustration. It is of vital relevance to the design of devices for confining plasma magnetically, such as tokamaks and stellarators. Charged particles basically follow magnetic field lines, in tight helices. If the field lines remain confined then there is a chance that the particles will too. Of course, a better analysis of particle confinement would come from the guiding centre equations, but they can be treated in exactly the same way as field line flow.

To see that field line flow is Hamiltonian, consider magnetic systems with a coordinate ζ such that the flow is transverse to surfaces of constant ζ. Then flux preservation implies that the symplectic form:

$$B \cdot \xi \times \eta \qquad (1.1.1.3.1)$$

is preserved by the flow, where ξ, η are tangent vectors. So the flow can be regarded as a time-dependent one degree of freedom symplectic flow, ζ playing the role of time. Note that the same argument can be used to show that particle motion in a steady fluid flow is symplectic, replacing B by the mass flux ρv.

Lagrangian systems are also of considerable importance. For example, ray tracing has a natural formulation in terms of finding paths of stationary length with respect to variations with fixed end points. Ray tracing is important in plasma physics in wave heating and current drive, for example. Field line flow can be considered as a Lagrangian system, as it is given by the variational principle:

$$\delta \int A \cdot dx = 0 \tag{1.1.1.3.2}$$

with free end conditions, where A is a vector potential (see Littlejohn, 1981a,b). This enables one to see that the Hamiltonian for the previous formulation is $-A_\zeta$. Note that all autonomous two degree of freedom Lagrangian systems can be reduced to a single plasma physics problem, because Birkhoff (1927, §II.4) showed that on the energy surfaces they are equivalent to non-relativistic motion of a charged particle in some 2-D static electromagnetic field.

§1.1.1.4 Discussion. One degree of freedom systems can be solved exactly by quadrature, and some simple systems with more degrees of freedom have sufficient constants of the motion that they can be solved too. They are called *integrable*. Most conservative systems, however, remain intractible as far as their exact solution or even longtime behaviour is concerned, though perturbation theory and numerical methods yield good predictions for limited time. Even such simple systems as three bodies interacting under gravity, the periodically forced pendulum, non-axisymmetric field configurations, and an electron in a standing wave, have not been solved.

Contrary to Milton's verse:

> "In the beginning ... the heav'ns and earth
> Rose out of Chaos",

Poincaré (1899) showed the existence in the three body problem of very complicated orbits. Indeed, stochastic behaviour on some scale is typical for conservative systems. Chaotic behaviour is often attributed to external noise, but even very simple systems are capable of producing complicated behaviour on their own.

Chaotic behaviour of a deterministic system might appear to be a contradiction in terms. For example, Laplace wrote that:

> "If an intelligence, for one given instant, recognises all the forces which animate Nature, and the respective positions of the things which compose it, and if that intelligence is also sufficiently vast to subject these data to analysis, it will comprehend in one formula the movements of the largest bodies of the universe as well as those of the minutest atom: nothing will be uncertain to it, and the future as well as the past will be present to its vision."

If, however, one has only limited precision in one's observations, one will typically be able to predict to a desired accuracy for only a limited time ahead. Systems in which information is lost at least linearly in time, until it is all gone, are called stochastic.

Historically, the interest in longtime behaviour in conservative systems began with the question of the stability of the solar system. The problem of stability for long times is now one of considerable practical significance in the design of intersecting storage rings and magnetic fusion devices, for example, where particles are required to remain trapped for many millions of revolutions.

From the other direction, longtime behaviour has been a topic of hot debate in statistical mechanics, which describes the macroscopic behaviour of systems with many degrees of freedom by averages. The justification for averaging requires some degree of stochasticity (see Wightman, 1979), though exactly what form is still unclear to me. It was thought at one time that if a system was not integrable then it must be ergodic (§1.4.1) on its energy surfaces. KAM theory (§1.3.1.5) has shown this to be false. Indeed, Markus and Meyer (1974) show that generically Hamiltonian systems are not ergodic on every energy surface. This is slightly misleading, however, as they could still be ergodic on a lot of the energy surfaces.

Stochasticity (e.g. in the sense of existence of a large ergodic component) is also of considerable importance to applications like wave heating of plasma, where particles are required to be free to diffuse to large energies (though still confined spatially).

This is an appropriate place to mention other reviews on conservative systems. I recommend in particular, the articles by Berry, Moser and Treve in [1], and by Chirikov (1979), Helleman (1980), Lichtenberg and Liebermann (1983), Tabor (1980), Whiteman (1977), and Wightman (1979).[1]

§1.1.2 **Maps.** The study of continuous time systems can often be reduced to that of a discrete time system, or "map", the return map on a surface of section. For a symplectic system, the return map is also symplectic. For 2-D maps, symplectic is equivalent to area preserving. Lastly, I give a brief outline of the types of orbit one finds in area preserving maps.

§1.1.2.1 Surface of section. The first thing in the analysis of any dynamical system is to find its equilibria and their stability. I will not discuss this here, but proceed to the next step - finding periodic orbits and the behaviour in a neighbourhood. This problem can be reduced to the study of a related discrete time system of lower dimension, a great conceptual simplification. This is because in many examples, one can find a codimension 1 surface transverse to the flow, called a *surface of section*. Then the flow induces a map from this surface to itself, called the *return map*, defined by following the flow until the first return to the surface. Note that for a conservative system, Poincaré recurrence (§1.1.1.1) guarantees that orbits will return to the surface of section. Initial conditions for any orbit which crosses the section can be translated into initial conditions for the map, by

[1] For more recent reviews, see MacKay and Meiss (1987), Herman (1983), Sinai (1989), Arnol'd (1988), MacKay (1987, 1992a, 1992c), Meyer & Hall (1992).

taking the first intersection. The return map contains all the qualitative information and a lot of the quantitative information about the flow in the region of interest. In particular, they have the same long-time behaviour.

I illustrate the idea of surface of section with the example of field line flow in toroidal geometry. If the flow is transverse to a poloidal section, then following the field lines once around the device gives a return map, as sketched in figure 1.1.2.1.1. Figure 1.1.2.1.2 shows some orbits of such a return map for a real magnetic field (Sinclair et al., 1970). See also White et al. (1982) for orbits of a return map for an MHD simulation of PDX.

One can do the same for any time-periodic system (period T), taking the section $\{t = t_0 \pmod{T}\}$. For other systems, the choice of surface of section may not be so simple, but if one knows a periodic orbit, at any point of the orbit there will be locally a surface transverse to the flow which can be used as a local surface of section. For some of the issues involved in finding a suitable surface of section, see Hénon and Heiles (1964).

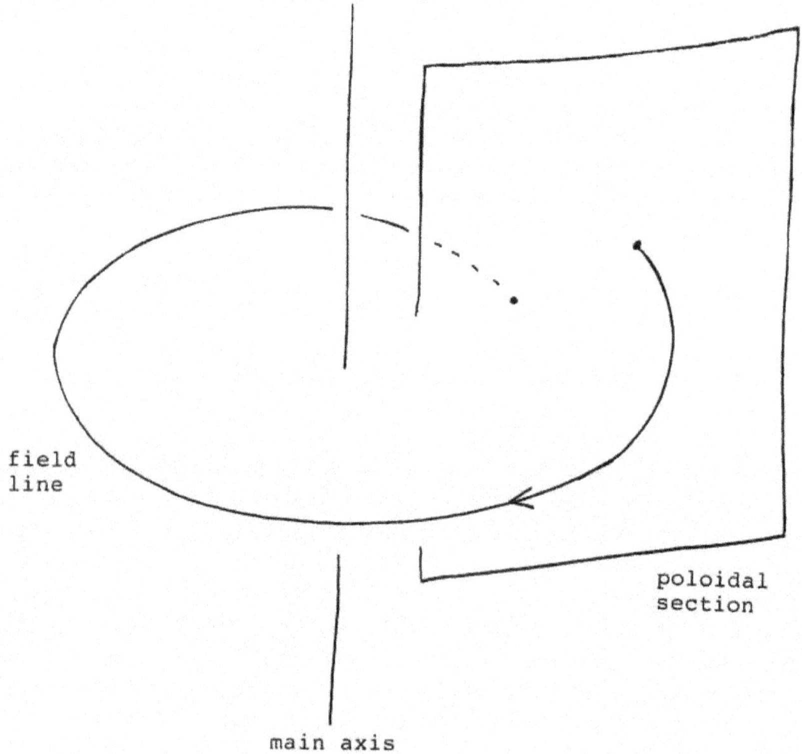

Figure 1.1.2.1.1: Construction of a return map for field line flow

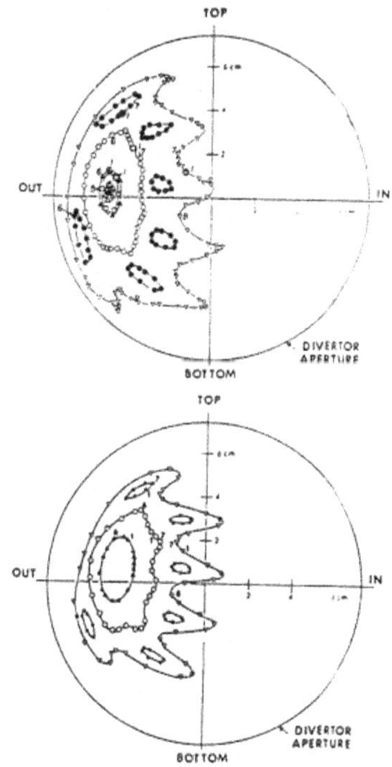

Figure 1.1.2.1.2: Some orbits of a return map for a stellarator field (from Sinclair et al., (1970). The upper picture was produced by following a low energy electron beam injected parallel to the field, and the lower one by integrating the field numerically.

§1.1.2.2 Symplectic maps. For a conservative system, the return map is also conservative. For example, for field line flow, since magnetic flux is conserved, the return map preserves the area form $B\cdot(\xi \times \eta)$, where ξ, η are tangent vectors to the poloidal section, so it is symplectic. In general, for a time-periodic Hamiltonian system, the return map on the above surface of section will be symplectic with the same symplectic form.

For an autonomous Hamiltonian system with symplectic form ω^2, the Hamiltonian is conserved, so one can regard the system as a one parameter family of flows on manifolds of one lower dimension, the energy surfaces. Thus one can take a surface of section within each energy surface, with two fewer dimensions. The return map is symplectic with symplectic form:

$$\omega^2 = d\delta \wedge dH \qquad (1.1.2.2.1)$$

where δ is the time to return (Abraham and Marsden, 1978). Alternatively, choosing coordinate q_1 to define the section by q_1 = constant, H = constant, and using canonical coordinates (note that Darboux's theorem leaves one the freedom to choose q_1 arbitrarily), then the map preserves the symplectic form:

$$\sum_{i=2}^{n} dp_i \wedge dq_i \qquad (1.1.2.2.2)$$

A one degree of freedom symplectic map T is called *area preserving* when expressed in canonical coordinates, because in that case the volume form (1.1.1.2.2) it induces is precisely the (oriented) area element. Equivalently,

$$\text{Det}\, DT = 1 \qquad (1.1.2.2.3)$$

everywhere. Here D stands for differentiation, so DT is the 2×2 matrix of partial derivatives. Note that in two dimensions, area preserving (in which I include orientation preserving) implies symplectic. As this is the simplest non-trivial form of conservative system, I will devote my attention almost entirely to area preserving maps. They exhibit many of the features of higher dimensional and continuous time conservative systems, though there are some phenomena only possible in higher dimensions, such as Arnold diffusion (e.g. Chirikov, 1979, Tennyson et al., 1980) and the Krein crunch (§1.2.1.6). In any case, many conservative systems give area preserving maps, specifically, all autonomous (Hamiltonian or Lagrangian) systems with two degrees of freedom, and one degree of freedom systems with periodic time dependence. This includes other examples from plasma physics, such as the motion of a charged particle in a 2-D field, guiding centre motion in 3-D, and ray tracing for waves in 2-D. They also have applications in other fields, such as celestial mechanics (Moser, 1973) and solid-state physics (Aubry, 1983).

Other types of conservative system lead to related types of map. In particular I will discuss discrete time Lagrangian systems in §1.1.3.2, and reversible maps in §1.1.4.

§1.1.2.3 Types of orbit. In this section, I give a brief outline of the most important types of orbits in area preserving maps. They will be discussed in detail in §§1.2, 1.3, and 1.4. This should not be regarded as a complete classification of the orbits. For instance, unbounded systems can also have escape orbits.

I have already mentioned periodic orbits in continuous time systems. They have their analogue in discrete time. A point x is said to be a *fixed point* of a map T, if $Tx = x$. A point x is said to be *periodic* if it is a fixed point of some iterate of T. The smallest positive integer q such that $T^q x = x$ is called its *period*, and its orbit is called a *periodic orbit*. Periodic points are important because they govern the behaviour in a neighbourhood. The type of the nearby behaviour is given almost completely by the eigenvalues of the linearisation of the map about the periodic point, i.e. of the derivative $D(T^q)$. This gives a classification of periodic orbits into linearly stable and unstable.

The importance of periodic orbits is not limited to area preserving maps. The second type of orbit I will discuss, however, is much more common in conservative systems than non-conservative ones (see §4.7, nevertheless). These are quasiperiodic orbits which densely fill an invariant circle. Any typical linearly stable fixed point in an area preserving map has invariant circles arbitrarily close to it, so they are common. In fact, these invariant circles make the fixed point non-linearly stable. So invariant circles are important for questions of stability and confinement.

Lastly, there are orbits whose behaviour is as random as a sequence of coin tosses. They wander over large regions, and predictive power about the future, given incomplete knowledge of the initial conditions, however accurate, decays exponentially in time. These are called *stochastic orbits*. They cause mixing and rapid transport.

These types of orbit can all be seen in figure (i), for example. The stochastic looking orbit on the outside eventually escapes to infinity, but the separatrices round island chains are typically not curves, but stochastic layers, as shown in figure 1.1.2.3.1. Before I discuss them in detail, I will describe in §§1.1.3, 1.1.4 two particularly important classes of area preserving map.

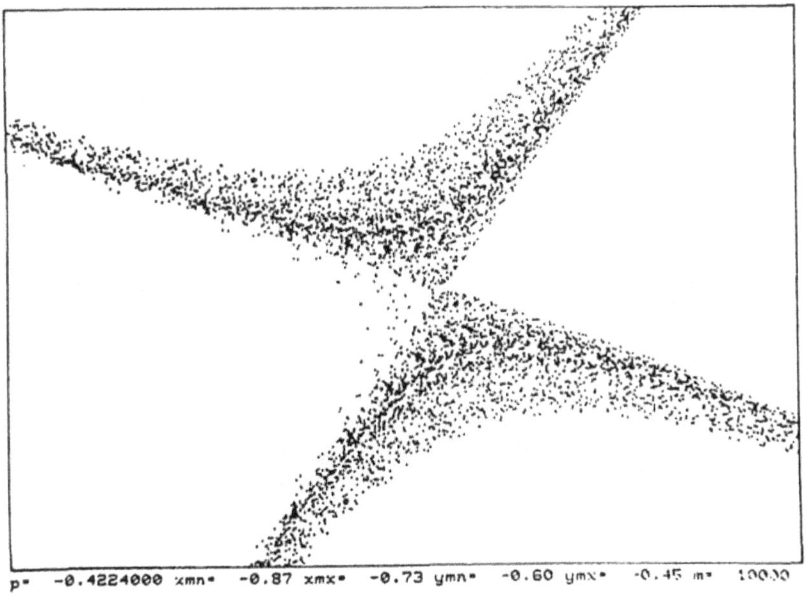

Figure 1.1.2.3.1: Stochastic layers

§1.1.3 **Twist maps.** A particularly important class of area preserving maps is the class of area preserving twist maps. They arise naturally in two settings. Firstly, near a fixed point a map typically has twist, giving a periodic twist map. Secondly, Lagrangian systems give rise to twist maps (not necessarily periodic). It turns out that any area preserving twist map can be expressed as a discrete time Lagrangian system, giving a useful representation for them, called the action representation. I conclude with an appendix on other generating functions for area preserving maps.

§1.1.3.1 Periodic twist maps. I motivate this section with the example of field line flow again. As discussed in §1.1.2.1, the flow can be reduced to a return map on a poloidal section, and coordinates can be chosen in which this map is area preserving. Typically, a tokamak or stellarator has a magnetic axis, that is, a field line which closes after one revolution. This corresponds to a fixed point of the return map. Also, the field is designed to have *rotational transform*, i.e. the other field lines twist around this axis. In the poloidal section orbits rotate around the fixed point. Finally, tokamaks generally have *magnetic shear*, that is, the rotation rate varies with distance from the magnetic axis. This is sketched in figure 1.1.3.1.1.

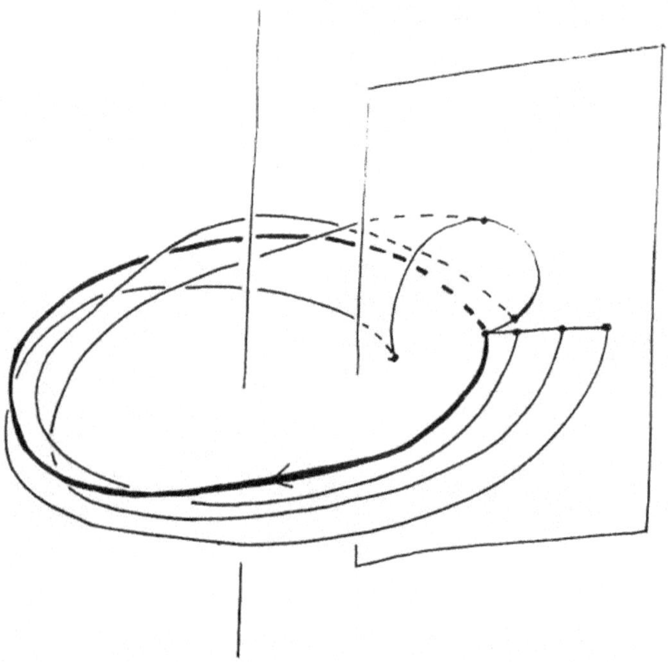

Figure 1.1.3.1.1: Rotational transform and twist

This leads us to consider *area preserving twist maps*. These are maps T for which there exist coordinates (θ, z), θ an angle variable, such that:

$$\frac{\partial \theta'}{\partial z} \text{ has constant sign} \tag{1.1.3.1.1}$$

where

$$(\theta', z') = T(\theta, z). \tag{1.1.3.1.2}$$

I also require that $\text{Det}\, DT = 1$ in these coordinates.

For the field line problem (and many others), (θ, z) are some sort of polar coordinates centred on a fixed point. Note, incidentally, that expressing an area preserving map in regular polar coordinates makes the conserved area form position dependent. But the transformation to "canonical polar coordinates" (ρ, θ):

$$\begin{aligned} x &= (2\rho)^{\frac{1}{2}} \cos\theta \qquad & \rho &= \frac{x^2 + y^2}{2} \\ y &= (2\rho)^{\frac{1}{2}} \sin\theta \qquad & \theta &= \tan^{-1}\left(\frac{y}{x}\right) \end{aligned} \tag{1.1.3.1.3}$$

is area preserving. So:

$$d\rho \wedge d\theta = dx \wedge dy: \tag{1.1.3.1.4}$$

There is a coordinate singularity at the fixed point. It is convenient to remove the singularity, by removing the fixed point. This gives a map which can be regarded as acting on part of a cylinder.

Maps of the cylinder derived in this way, however, have some special properties. Firstly, they are end-preserving, meaning that points sufficiently far up/down the cylinder remain far up/down. Secondly, given an area-preserving map of a cylinder, and a measurable set U containing all points below some level z_1 and no points above some other level z_2, the difference in the area of U and $T(U)$:

$$C = \text{Area}\,(T(U) \setminus U) - \text{Area}\,(U \setminus T(U)) \tag{1.1.3.1.5}$$

is independent of U, and is called the *Calabi invariant* (or net flux) of T. It measures the average upward drift. For example, the map:

$$\theta' = \theta + z' \\ z' = z + f(\theta) \tag{1.1.3.1.6}$$

has Calabi invariant $\int d\theta f(\theta)$. For maps of the cylinder derived from a map of the plane by removing a fixed point, the Calabi invariant is clearly zero. Thus I will restrict attention to the *class A* of end preserving, area preserving twist maps of a cylinder with zero Calabi invariant.

Often I will want to consider a *lift* of T, rather than the map T itself. This means that θ is regarded as a coordinate on a line rather than a circle, so we get a *periodic* map (of period 1, say), i.e.:

$$\theta'(\theta + 1, z) = \theta'(\theta, z) + 1. \tag{1.1.3.1.7}$$

Maps of class A are relevant to many other conservative problems. Typically any elliptic fixed point has a neighbourhood in which there are coordinates for which the map is a periodic twist map. We will also see (§1.1.3.2) that discrete time Lagrangian systems with a periodic coordinate give maps in class A.

§1.1.3.2 Discrete time Lagrangian systems. Twist maps also arise from return maps for Lagrangian systems. The return map can be given in a variational form derived from that for the original system. Let the section be defined by a coordinate surface for an angular coordinate, say $\varphi = \varphi_0 \pmod 1$. Then, for a given energy, one can define the *action* $\tau(x, x')$ from position coordinates x to x' on the section to be the value of the action:

$$\tau(x, x') = \text{sta} \int L(x, \dot{x}, \varphi, \dot{\varphi}) \tag{1.1.3.2.1}$$

for the orbit of that energy from (x, φ_0) to $(x', \varphi_0 + 1)$ ("sta" stands for the stationary value, assuming there is a locally unique one). Then:

$$\text{sta} \int_{x_0, \varphi_0}^{x_n, \varphi_0 + n} L(x, \dot{x}, \varphi, \dot{\varphi}) = \text{sta} \sum_{i=0}^{n-1} \tau(x_i, x_{i+1}) \tag{1.1.3.2.2}$$

so the orbits of the discrete time system are given by sequences x_n for which the total action is stationary for all finite segments, with respect to variations with fixed ends. Discrete time Lagrangian systems turn up naturally in some contexts, for example, 1-D crystal models (e.g. Aubry, 1983). In these models, the lattice position plays the role of time, and the free energy that of the Lagrangian.

Discrete time Lagrangian systems give rise to symplectic maps, much as for continuous time. I restrict attention to the case of one degree of freedom. Subscript i will refer to the derivative with respect to the ith argument. Given an action $\tau(x, x')$, with $\tau_{12}(x, x')$ of constant sign, the relations:

$$\begin{aligned} y' &= \tau_2(x, x') \\ y &= -\tau_1(x, x') \end{aligned} \quad (1.1.3.2.3)$$

can be inverted to generate a map $T : (x, y) \to (x', y')$ (note that Mather uses the opposite sign convention). It is area preserving, since the Lagrangian yields canonical coordinates. Alternatively, the derivative is:

$$DT = \begin{vmatrix} -\dfrac{\tau_{11}}{\tau_{12}} & -\dfrac{1}{\tau_{12}} \\ \tau_{21} - \dfrac{\tau_{22}\tau_{11}}{\tau_{12}} & -\dfrac{\tau_{22}}{\tau_{12}} \end{vmatrix} \quad (1.1.3.2.4)$$

which has unit Jacobian. Furthermore, τ_{12} of constant sign implies that:

$$\frac{\partial x'}{\partial y} \text{ has constant sign.} \quad (1.1.3.2.5)$$

This *twist condition* is the generalisation of (1.1.3.1.1) to non-periodic maps.

To see that orbits of T correspond to paths of stationary action:

$$\sum_i \tau(x_i, x_{i+1}) \quad (1.1.3.2.6)$$

note that stationarity with respect to x_i is equivalent to:

$$\tau_2(x_{i-1}, x_i) + \tau_1(x_i, x_{i+1}) = 0. \quad (1.1.3.2.7)$$

But from (1.1.3.2.3) this can be expressed as

$$y_i - y_i = 0 \quad (1.1.3.2.8)$$

which justifies the correspondence.

§1.1.3.3 Action representation. The significance of discrete time Lagrangian systems for my purposes, is that any area preserving map satisfying the twist condition (1.1.3.2.3) can be generated in the above way, at least locally (Spivak, 1965). Non-trivial de Rham cohomology may prevent existence of a global (i.e. single-valued) generating function, but this is discussed later in the section. In a given coordinate system, the *generating function* $\tau(x, x')$ is unique up to addition of a constant, which can easily be seen to have no dynamical effect. Thus generating functions with τ_{12} of constant sign provide a representation for area preserving twist maps which automatically restricts them to be area preserving and to satisfy the twist condition. I call it the *action representation*, in view of its interpretation as a discrete time Lagrangian system. Examples will be given in §1.1.4.3. There are other representations by generating functions which can be useful, some of which will be discussed in §1.1.3.4 for area preserving maps and in §1.1.4.4 for reversible maps.

I will want to compose different maps. The rule for composition of action generating functions, where defined, is that of stationarity, just like the rule for the orbits. But note that the composition of two twist maps need not be a twist map, even the composition of a twist map with itself. Twist means that the image of the vertical line x = constant is the graph of a single-valued function of x. But the second image need not have this property, as is sketched below.

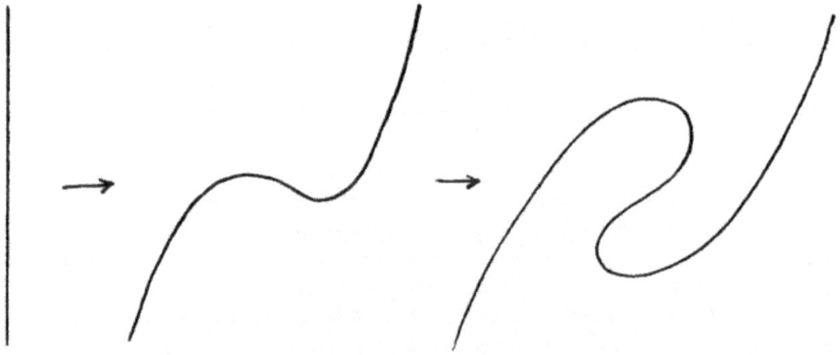

Assuming that the composition is a twist map, I will spell out the composition law for generating functions. The generating function for the composition TU of two maps T, U with generating functions τ, υ, is:

$$\upsilon \oplus \tau(x, x'') = \upsilon(x, x') + \tau(x', x'') \quad (1.1.3.3.1)$$

where $x'(x, x'')$ is chosen to make the sum stationary with respect to variations in x', i.e.

$$0 = \upsilon_2(x, x') + \tau_1(x', x''). \quad (1.1.3.3.2)$$

That $\nu \oplus \tau$ generates TU can be seen immediately from the generating relations (1.1.3.2.1).

The de Rham cohomology of a manifold measures its connectedness (Arnold, 1978). Any lift of a periodic area preserving twist map has a global action generating function $\tau(\theta, \theta')$, because the de Rham cohomology of the plane is trivial. On the cylinder, however, it need not have a global generating function. It does, however, in the case of zero Calabi invariant. Since $\tau(\theta + 1, \theta' + 1)$ generates the same map as $\tau(\theta, \theta')$, they can differ only by a constant. This constant is the Calabi invariant. To see this, take a curve γ joining (θ, θ') to $(\theta + 1, \theta' + 1)$. Then γ defines a circle and its image, on the cylinder, by:

$$z = -\tau_1(\theta, \theta')$$
$$z' = \tau_2(\theta, \theta') \quad (1.1.3.3.3)$$

The Calabi invariant can be evaluated as the difference in the area under these circles. Thus,

$$\tau(\theta + 1, \theta' + 1) - \tau(\theta, \theta') = \int_\gamma \tau_1\, d\theta + \tau_2\, d\theta' = \int -z\, d\theta + z'\, d\theta' \quad (1.1.3.3.4)$$

is the Calabi invariant.

§1.1.3.4 Other generating functions. There are several other representations for area preserving maps by generating functions. The best known is the *mixed generating function*:

$$y' = F_1(x', y)$$
$$x = F_2(x', y). \quad (1.1.3.4.1)$$

These relations can be inverted to give an area preserving map $T : (x, y) \to (x', y')$, provided F_{21} has constant sign. This restricts T to have $\partial x'/\partial x$ of constant sign. $F(x', y) = x'y$ gives the identity, so this is useful in perturbation theory for maps near the identity (e.g. Gallavotti, 1982).

Poincaré's generating function is sometimes useful (e.g. Meyer, 1970). It generates an area preserving map by:

$$y' - y = G_1(x' + x, y' + y)$$
$$x' - x = -G_2(x' + x, y' + y) \quad (1.1.3.4.2)$$

There is also a related generating function with the sums and differences interchanged (Rimmer, 1979).

Lastly, I mention *Lie generating functions*. These are useful in many applications, but as I have not used them I will just give some references (Dewar, 1976, Dragt and Finn, 1976, Cary, 1981).

§1.1.4 **Reversible maps.** Reversible systems give rise to reversible maps, defined by a symmetry property. The symmetries give rise to symmetry lines. I give some examples of reversible area preserving maps, and discuss the effect of symmetry on generating functions. Lastly, I discuss how general reversibility is likely to be in the class of area preserving maps.

§1.1.4.1 Definition. A map T is said to have *symmetry* S if S is orientation reversing and

$$S^2 = (TS)^2 = \text{identity}. \qquad (1.1.4.1.1)$$

Then STS^{-1} is the inverse of T, so T is invertible and conjugate to its inverse by an involution. Thus possession of a symmetry is equivalent to reversibility (§1.1.1.2) (Birkhoff, 1927, DeVogelaere, 1958, 1962, Moser, 1973, Devaney, 1976). If S is a symmetry for T, then so are T^nS, $n \in Z$. I call them a *family* of symmetries. In particular, I call TS a *complementary* symmetry to S, because T factorises as their product $TS.S$. One could also regard ST as a complementary symmetry, as $T = S.ST$. In fact, factorisation as a product of two involutions is equivalent to reversibility, because either involution can easily be seen to be a symmetry. Note that reversible maps need not be area preserving.

§1.1.4.2 Symmetry lines. Particularly important to the study of reversible maps are the fixed points of its symmetries. I will use the notation Fix(S) for the fixed points of S. Their existence will be discussed later in this section.

A fixed point of an orientation reversing involution S has a neighbourhood in which there exist coordinates (X, Y) such that

$$S(X, Y) = (X, -Y) \qquad (1.1.4.2.1)$$

Thus the fixed points form curves, called *symmetry lines* if S is a symmetry for some map. This was shown by Finn (1974), and as it is a constructive and useful proof, I reproduce it here. A more general case was shown by Montgomery and Zippin (1955). They proved local conjugacy of any involution in any dimension to its derivative at a fixed point.[2]

[2] Also Meyer (1981) showed that if S is antisymplectic then (1.1.4.2.1) can be achieved with symplectic coordinates.

$S^2 = 1$ plus orientation reversal implies that DS has the normal form:

$$\begin{vmatrix} 1 & 0 \\ 0 & -1 \end{vmatrix} \tag{1.1.4.2.2}$$

at any fixed point of S. Thus write

$$\begin{aligned} S : x' &= x + f(x, y) \\ y' &= -y + g(x, y) \end{aligned} \tag{1.1.4.2.3}$$

where f, g are of higher order. Then $S^2 = 1$ implies that

$$\begin{aligned} f(x', y') &= -f(x, y) \\ g(x', y') &= g(x, y). \end{aligned} \tag{1.1.4.2.4}$$

Define new coordinates by:

$$\begin{aligned} X &= x + \tfrac{1}{2} f(x, y) \\ Y &= y - \tfrac{1}{2} g(x, y). \end{aligned} \tag{1.1.4.2.5}$$

This is an invertible transformation in a neighbourhood of 0, because f and g are of higher order. Then S reduces to the above form, as claimed. If S is a symmetry of some map, I call such coordinates *symmetry coordinates*. Note that they are not unique, e.g. $\xi = X + h(Y)$, $\eta = Y$, are also symmetry coordinates for any even h.

The openness or closedness of symmetry lines, and their number depends on the topology of the manifold (Bredon, 1972). Smith proved that on an n-sphere, the fixed point set of an involution has the homology of an m-sphere for some $-1 \le m \le n$. All cases can occur, though orientation reversal limits the possibilities. Note that $m = -1$ means the empty set. For $m \neq 3$, all simply connected manifolds having the homology of an m-sphere are m-spheres, the case $m = 3$ remaining open, but I don't know to what extent the fixed point set will be simply connected. Poincaré gave an example to show that simple connectedness is necessary for $m = 7$. In any case, on the 2-sphere, the fixed point set of an orientation reversing involution is either one circle or empty. For the plane, one can add a point at infinity to get a 2-sphere. This point will have to be a fixed point of the involution, since there is no other point for it to interchange with. Thus the fixed point set is non-empty, so it is a circle going through the point at infinity, giving an open line on the plane. On the cylinder it is easy to construct examples with one circle, or two lines or the empty set, and on the torus, with two circles, one circle or none.

Note that:

$$T^{2n}Sx = x \text{ iff } T^n ST^{-n} x = x \text{ iff } S(T^{-n}x) = T^{-n}x. \qquad (1.1.4.2.6)$$

So

$$\text{Fix}\,(T^{2n}S) = T^n \,\text{Fix}\,(S). \qquad (1.1.4.2.7)$$

Thus a family of symmetries separates into two *half-families*, $\{T^{2n}S\}$ and $\{T^{2n+1}S\}$.

§1.1.4.3 Examples. Now I give some important examples of reversible maps. They will also be area preserving.

The DeVogelaere maps (1958):

$$\begin{aligned} x' &= f(x) - y \\ y' &= x - f(x') \end{aligned} \qquad (1.1.4.3.1)$$

have the following factorisation $S_2.S_1$ into involutions:

$$\begin{aligned} S_1 : x' &= x & \qquad S_2 : x' &= y + f(x) \\ y' &= -y & y' &= x - f(x'). \end{aligned} \qquad (1.1.4.3.2)$$

DeVogelaere maps can also be written in McMillan form (1971):

$$\begin{aligned} x' &= 2f(x) - Y \\ Y' &= x \end{aligned} \qquad (1.1.4.3.3)$$

where

$$Y = y + f(x). \qquad (1.1.4.3.4)$$

Then S_1, S_2 have the forms:

$$\begin{aligned} S_1 : x' &= x & \qquad S_2 : x' &= Y \\ Y' &= 2f(x) - Y & Y' &= x. \end{aligned} \qquad (1.1.4.3.5)$$

Another important class of reversible maps is the *generalised standard maps*:

$$y' = y + h(x)$$
$$x' = x + g(y'). \tag{1.1.4.3.6}$$

The *standard*, or *Taylor-Chirikov map* (Taylor, unpublished, Chirikov, 1979) has:

$$h(x) = -k \sin x, \;\; g(y) = y \tag{1.1.4.3.7}$$

If h is odd, the generalised standard maps have the factorisation $R_1.S_1$ into involutions:

$$R_1: \;\; y' = y \qquad S_1 : y' = y + h(x)$$
$$x' = -x + g(y) \qquad x' = -x \tag{1.1.4.3.8}$$

and if g is odd, the factorisation $R_2 S_2$ into involutions:

$$R_2 : y' = -y \qquad S_2 : y' = -y - h(x)$$
$$x' = x - g(y) \qquad x' = x. \tag{1.1.4.3.9}$$

Note that the standard map possesses both factorisations, so I say it is *doubly reversible*. The fixed lines of these symmetries are shown in figure 1.1.4.3.1, regarding the map as acting on the torus. The double factorisation for maps with both h and g odd can be removed by identifying points under reflection through the origin. This also removes some anomalous behaviour of such maps (see §1.2.4.7).

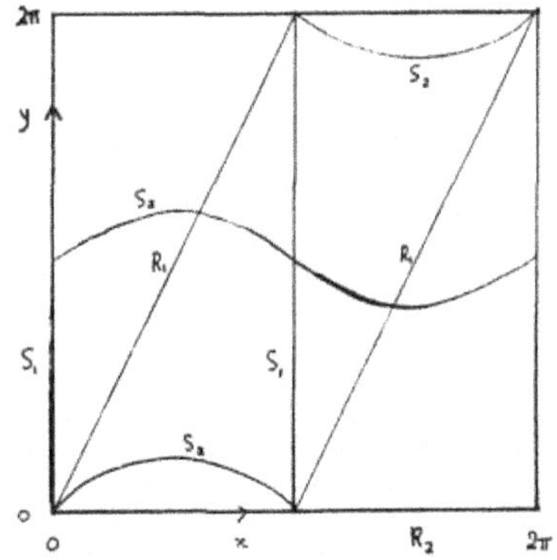

Figure 1.1.4.3.1: Symmetry lines of the standard map

Note that McMillan maps (and so DeVogelaere maps) can be put into generalised standard form with $g(y) = y$, and $h(x) = 2f(x) - 2x$, by the coordinate change:

$$y = x - Y. \tag{1.1.4.3.10}$$

Reversibility appears to very common in area preserving maps (see §1.1.4.5). An example of an area preserving map which appears to have no symmetry, however, is Rannou's map (1974). It (or actually, its inverse!) is of generalised standard form with:

$$\begin{aligned} h(x) &= \lambda(\sin x + 1 - \cos x) \\ g(y) &= -y - 1 + \cos y. \end{aligned} \tag{1.1.4.3.11}$$

§1.1.4.4 Reversibility and generating functions. It is easy to restrict to reversible maps in the action representation. I consider two special symmetries. Any symmetry can be reduced to either form, of course, by §1.1.4.2, but the distinction is important for actions. Symmetry with respect to:

$$\begin{aligned} x' &= x \\ y' &= -y \quad \text{is equivalent to} \quad \tau(x', x) = \tau(x, x') \end{aligned} \tag{1.1.4.4.1}$$

and symmetry with respect to:

$$\begin{aligned} x' &= -x \\ y' &= y \quad \text{is equivalent to} \quad \tau(-x', -x) = \tau(x, x'). \end{aligned} \tag{1.1.4.4.2}$$

As an example, the action for the DeVogelaere maps, which have the first symmetry, is:

$$\tau(x, x') = xx' - F(x) - F(x') \tag{1.1.4.4.3}$$

where F is an indefinite integral of f. The generalised standard maps with g = identity, have action:

$$\tau(x, x') = \frac{1}{2}(x - x')^2 - H(x) \tag{1.1.4.4.4}$$

where H is an indefinite integral of h. This can be made symmetric with respect to the second symmetry by using symmetry coordinates for R_1 or S_1. For example, the coordinate change:

$$Y = y - \frac{1}{2}h(x) \tag{1.1.4.4.5}$$

converts them to the form:

$$\begin{aligned} Y' &= Y + \frac{1}{2}h(x) + \frac{1}{2}h(x') \\ x' &= x + Y - \frac{1}{2}h(x) \end{aligned} \tag{1.1.4.4.6}$$

which has reflection symmetry about $x = 0$, and has generating function:

$$\tau(x, x') = \frac{1}{2}(x - x')^2 - \frac{1}{2}H(x) - \frac{1}{2}H(x'). \tag{1.1.4.4.7}$$

There are closely related generating functions to the action generating function which restrict to symmetric maps alone, without imposing area preservation. Extending an idea of Eckmann et al. (1982), these work as follows:

$$\begin{aligned} y' &= h(x, x') \qquad & y' &= k(x, x') \\ y &= -h(x', x) \qquad & y &= k(-x', x). \end{aligned} \tag{1.1.4.4.8}$$

If h_1, respectively k_1, is of constant sign, these relations can be inverted to give a map with the first, respectively second, symmetry above, in both cases satisfying the twist condition (1.1.3.2.3). Conversely, every map with one of the above symmetries and satisfying the twist condition can be generated in this way. I call these *symmetric representations*. Area preservation can be imposed by requiring that:

$$\begin{aligned} h_1(x, x') &= h_1(x', x) \\ k_1(x, x') &= k_1(-x', -x) \end{aligned} \tag{1.1.4.4.9}$$

respectively. Composition in the symmetric representations is not so easy, in view of the fact that symmetry is not necessarily preserved under composition. As I ended up not using the symmetric representation, I won't pursue it here.

Symmetry is also easy to impose on Poincaré's generating function (Rimmer, 1979).

§1.1.4.5 Generality of reversibility in area preserving maps. Most area preserving maps given in the literature are reversible, though no-one has found a symmetry for Rannou's map (1.1.4.3.11). This leads one to ask whether they all are. The answer is no. Mather (private communication) gave me a counterexample.[3] The flow of figure 1.1.4.5.1 can be chosen to be area preserving, so its time-1 map is an area preserving map. But the flow has no symmetry. A symmetry would have to extend to the quotient space formed by identifying points of the same trajectory. There are three special trajectories which can not be "housed-off" from each other by open sets of trajectories. They must map into each other under the symmetry, with reversal of directions, but there is no way this can be done.

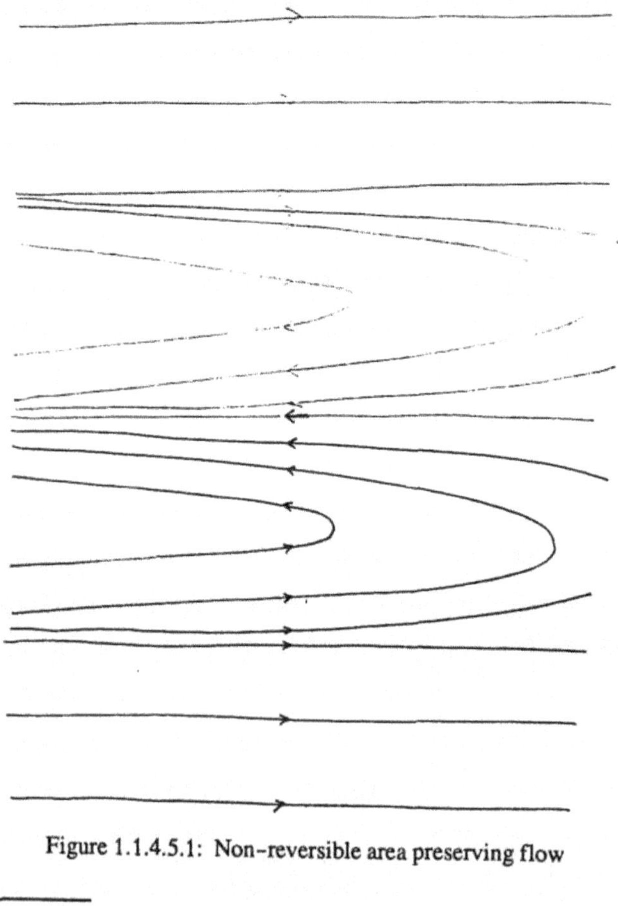

Figure 1.1.4.5.1: Non-reversible area preserving flow

[3] Roberts and Capel (1992) have given another, based on an idea in the next paragraph (see also problem (8) in the section "Problems for the future").

This example leads one to consider just *local reversibility*, i.e. reversibility restricted to some invariant set. There are two considerations, however, which make local reversibility unlikely to be general. The first is that the classifications of generic bifurcations at multipliers $+1$ are different for one parameter families of area preserving maps and for one parameter families of reversible area preserving maps (see §1.2.4.6). So we at least have to exclude periodic orbits with multipliers $+1$. The second is that Greene (private communication) found area preserving perturbations of the standard map (of generalised standard form) which appeared not to have any symmetries near to the unperturbed ones. He did this by seeing how the symmetric periodic points (see §1.2.3) of the standard map moved with the perturbation. For the standard map they lay on a smooth curve (a symmetry line), but the perturbed points looked as if they would not admit any smooth curve through them.

On the other hand, every hyperbolic fixed point has a neighbourhood on which the map has symmetry. This is because Moser (1956) showed that every hyperbolic fixed point has a neighbourhood N in which there is a convergent coordinate change to the normal form:

$$\begin{aligned} x' &= \sigma\, x\, e^{w(xy)} \\ y' &= \sigma\, y\, e^{-w(xy)} \end{aligned} \tag{1.1.4.5.1}$$

where $\sigma = \pm 1$ and w is a convergent power series. The normal form has the obvious symmetry:

$$x' = y, \;\; y' = x \tag{1.1.4.5.2}$$

proving the claim. Presumably this symmetry can be extended to the invariant set $\bigcup_n T^n(N)$, by

$$ST^n x = T^{-n}Sx \;\text{ for }\; x \in N \tag{1.1.4.5.3}$$

though it might not necessarily be continuous, let alone differentiable. If it were, then ergodic maps with a hyperbolic fixed point would be reversible.

The Birkhoff normal form for an elliptic fixed point does not in general converge, but it may well be that some of the symmetries that it possesses formally are real.

There is a conjecture, known as the Jacobian conjecture (Wright, 1981), that every polynomial map with constant Jacobian has a polynomial inverse. Engel (1955) claimed to have proved it, but Vitushkin pointed out a mistake (Campbell, 1973). If true, however, it would lead to some useful statements with respect to reversibility (Engel, 1958, Helleman

1983). Namely, it would imply that any polynomial area preserving map of prime degree, or whose polynomials have coprime degree, can be put into McMillan form, and so are reversible.

I conclude this section by discussing the converse question, whether reversibility implies area preservation. The answer is that it does not even imply preservation of any measure equivalent to Lebesgue. This is shown by the counterexample of figure 1.1.4.5.2. As a second counter-example,

$$\ddot{x} = F(x), \quad \text{curl } F \neq 0 \tag{1.1.4.5.4}$$

can not be expressed in Hamiltonian form, but is reversible.[4]

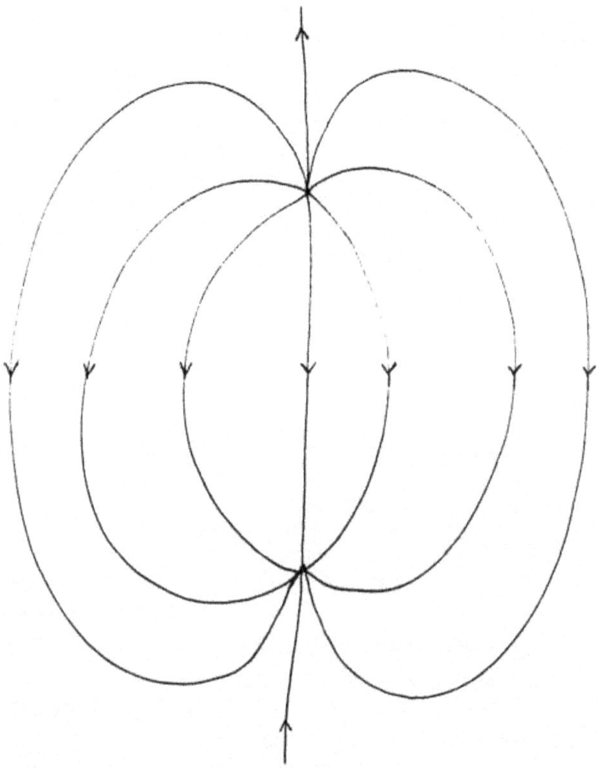

Figure 1.1.4.5.2: Reversible non-area preserving flow

[4] Roberts and Quispel (1992) discuss further the differences between area-preserving and reversible.

§1.2 PERIODIC ORBITS

In this section, I discuss the first type of orbits, the periodic orbits. Their linear stability is analysed. Then I state the Poincaré-Birkhoff theorem, which proves existence of many periodic orbits in area preserving twist maps. Next I discuss symmetric periodic orbits of reversible maps. I conclude with a section on bifurcations of periodic orbits, by which yet more periodic orbits are produced.

§1.2.1 **Stability**. I discuss the stability analysis of periodic orbits in a general map, and introduce the multipliers of a periodic orbit. A periodic orbit is shown to be isolated from other periodic orbits of the same period if it has no multiplier +1. Restricting attention to conservative systems introduces special features. I give details for the case of area preserving maps. Then the idea of Poincaré index is introduced. Finally, I analyse stability of periodic orbits in 4-D symplectic maps.

§1.2.1.1 Linear stability analysis. The definition and significance of periodic orbits were given in §1.1.2.3. I begin by discussing the stability of periodic orbits in general maps, restricting to conservative maps in §1.2.1.3. As a periodic point of period n of a map T can be considered as a fixed point of T^n, I will consider only fixed points, without loss of generality (wlog).

A fixed point x of a map T is said to be *attracting* if it has a neighbourhood U such that:

$$TU \subset U, \quad \bigcap_{n>0} T^n U = \{x\} \tag{1.2.1.1.1}$$

It is *stable* if every neighbourhood U has a subneighbourhood V, such that:

$$T^k(V) \subset U, \forall k \in \mathbb{Z}^+ \tag{1.2.1.1.2}$$

where $\mathbb{Z}^+$ is the non-negative integers. So an attracting fixed point is stable, but not necessarily vice versa. A fixed point is said to be *unstable* if it is not stable. The definitions of stable and unstable will be strengthened in §1.2.1.3 for conservative systems.

The first step in stability analysis is to examine the stability of the linearisation of the map (i.e. the derivative DT) at the fixed point. Its stability is called the *linear stability* of the fixed point. This is given almost completely by the eigenvalues of DT, which are called the *multipliers* of the fixed point. If all the multipliers lie inside the unit circle, the fixed point is attracting under DT. If there are any multipliers outside the unit circle, it is unstable under DT. In fact, in these two cases the fixed point always has the same stability under T as under its linearisation, so the multipliers are sufficient to determine stability. In the case that some of the multipliers are on the unit circle, the rest being inside, one needs to do further work, even for linear stability.

Matrices are classified up to similarity transformation, i.e. coordinate change (though not to symplectic coordinate change - see §§1.2.1.4, 1.2.1.6) by their *Jordan normal form*:

$$\begin{vmatrix} \Lambda_1 & & & & \\ & \Lambda_2 & & & \\ & & \Lambda_3 & & \\ & & & \cdot & \\ & & & & \cdot \end{vmatrix} \qquad (1.2.1.1.3)$$

where the Λ_j are *Jordan blocks* of the form:

$$\Lambda = \begin{vmatrix} \lambda & 1 & & \\ & \lambda & \cdot & \\ & & \cdot & 1 \\ & & & \lambda \end{vmatrix}. \qquad (1.2.1.1.4)$$

The λ_j are eigenvalues of the matrix. If DT has a non-trivial Jordan block (i.e. of dimension greater than 1) with eigenvalue on the unit circle, then the fixed point is unstable for the linear map and the nonlinear one. The remaining case, when all the multipliers lie on or inside the unit circle, and those on the unit circle have diagonal Jordan blocks, is linearly stable, but could be attracting, stable but not attracting, or unstable.

§1.2.1.2 Isolation. If x_0 is a fixed point of a map T then, provided that it has no multipliers +1, it is isolated, i.e. it has a neighbourhood in which there are no other fixed points. This is because *Newton's method*:

$$x \mapsto x - (DT - 1)^{-1}. (T(x) - x) \qquad (1.2.1.2.1)$$

has the same fixed points as T, and is a contraction on a neighbourhood of x_0, so it has a unique fixed point in that neighbourhood. Similarly, periodic orbits with no multiplier equal to +1, are isolated from periodic orbits of the same period, or a submultiple.

§1.2.1.3 Stability in conservative systems. In conservative systems there are restrictions which lead to special behaviour of periodic orbits. For one, volume preservation implies that there are no attracting periodic orbits. Since symplectic maps are volume preserving, the same goes for them.

Furthermore, a linear symplectic map (e.g. the derivative at a fixed point) and its inverse have the same eigenvalues (Arnold, 1978). To see this, a linear map L is symplectic iff:

$$L\,J\,L^tJ = I \text{ where } J = \begin{vmatrix} 0 & -I \\ I & 0 \end{vmatrix} \qquad (1.2.1.3.1)$$

in canonical coordinates, I is the identity and t refers to the transpose. Thus:

$$\det(L - \lambda I) = \det(L^t - \lambda I) = \det J(L^t - \lambda I)J = \det(L^{-1} - \lambda I). \qquad (1.2.1.3.2)$$

The characteristic polynomial is said to be *reflexive*. So the eigenvalues of L come in reciprocal pairs λ, $1/\lambda$ (counted according to multiplicity). This follows directly from (1.2.1.3.2), except for eigenvalues ± 1. For the case of -1, L symplectic implies that $\det L = +1$, and the product of the eigenvalues is always $\det L$, so the eigenvalue -1 must have even multiplicity. Since symplectic spaces have even dimension, $+1$ has even multiplicity too, by elimination. Thus the multipliers of any periodic orbit come in reciprocal pairs.

The same applies for *symmetric periodic orbits* of reversible maps. These are orbits which are their own reflection under one of the symmetries (and actually all the symmetries of the same family - see §1.1.4.2). Firstly, a symmetric periodic orbit can not be attracting, because if U were a neighbourhood of a symmetric fixed point x, satisfying (1.2.1.1.1), then SU would be a neighbourhood of x too, so:

$$\exists\, n_0 \text{ s.t. } T^nU \subset SU \text{ for } n \geq n_0. \qquad (1.2.1.3.3)$$

Applying T^nS to this shows that:

$$T^nU \supset SU \text{ for } n \geq n_0 \qquad (1.2.1.3.4)$$

which contradicts T^nU decreasing to $\{x\}$.

Secondly, for a symmetric periodic orbit of a reversible map, the multipliers come in reciprocal pairs (Devaney, 1976). Write the derivative as the product RS of two linear involutions. So its inverse is SR. Then:

$$\det(RS - \lambda I) = \det R(RS - \lambda I)R = \det(SR - \lambda I) \qquad (1.2.1.3.5)$$

since $R^2 = I$. Again, eigenvalues ± 1 require special consideration. As linear involutions have determinant ± 1, there are two possibilities for $\det RS$. The multiplicity of eigenvalue -1 must be even/odd according as $\det RS = \pm 1$. The multiplicity of $+1$ is determined by elimination, depending on $\det RS$ and the parity of the dimension of the space. Note, however, that reversible and symplectic are not equivalent. Devaney (1976) gives an example of a 4-D linear reversible map which is not symplectic. The 2-D case was discussed in §1.1.4.5.

Reflexiveness motivates stronger notions of stability and instability for conservative systems, because linear stability/instability in one direction of time implies the same in the other direction. So a fixed point is said to be *stable* if every neighbourhood U has a subneighbourhood V such that:

$$T^n V \subset U, \ \forall n \in Z \qquad (1.2.1.3.6)$$

where Z is all the integers. Also, I strengthen the notion of instability to the following. A fixed point is said to be *unstable* if it has a neighbourhood such that the only point whose orbit stays inside it for both directions in time is the fixed point. Although this is stronger in general than "not stable", no-one knows any examples of a conservative map with a point which is neither stable nor unstable (Siegel and Moser, 1971).

Reality implies that complex eigenvalues always come in complex conjugate pairs. A particular consequence of the reflexive property of the characteristic polynomial is that possession of a complex conjugate pair of simple eigenvalues on the unit circle is stable to symplectic or reversible perturbation, unlike the general case. This is because a complex conjugate pair whose product is 1 must have modulus 1. So the only way a pair can get off the unit circle is at ± 1, splitting into a reciprocal real pair, or by collision with another pair on the unit circle (Krein crunch - see §1.2.1.6). Even then there are restrictions (at least in the symplectic case) which might not permit them to leave. This makes the conservative analysis much more involved.

§1.2.1.4 Stability in area preserving maps. For a 2-D system the multipliers λ are the roots of:

$$\lambda^2 - TrDT\,\lambda + \text{Det}\,DT = 0. \qquad (1.2.1.4.1)$$

Area preservation implies that Det $DT = 1$, so the product of the multipliers of a periodic orbit must be 1. Together with reality of DT, this restricts them to be a reciprocal pair of reals, or a complex conjugate pair on the unit circle. In fact, they are:

$$\lambda = \frac{t}{2} \pm \left(\frac{t^2}{4} - 1 \right)^{1/2} \qquad (1.2.1.4.2)$$

where

$$t = \text{Trace}\,DT. \qquad (1.2.1.4.3)$$

Thus the linear stability is determined by the trace of the derivative, as in the following table, or in terms of a derived quantity, the *residue*, defined by:

$$R = \frac{2 - Tr\, DT}{4}. \tag{1.2.1.4.4}$$

The reason for introducing the residue is that multipliers +1 (residue 0) is a special case, as we saw in §1.2.1.1.

Trace	Residue	Classification	Multipliers
$t > 2$	$R < 0$	regular hyperbolic	reciprocal pair of positive reals
2	0	regular parabolic	pair at +1
$-2 < t < 2$	$0 < R < 1$	elliptic	complex conjugate pair on the unit circle
-2	1	inversion parabolic	pair at -1
$t < -2$	$R > 1$	inversion hyperbolic	reciprocal pair of negative reals.

These cases are sketched in figure 1.2.1.4.1, for linear maps. Elliptic points are linearly stable, hyperbolic points linearly unstable, and parabolic points linearly stable or unstable according as their Jordan normal form is diagonal or not. Note that in the elliptic case the multipliers are $e^{\pm 2\pi i \nu}$ with:

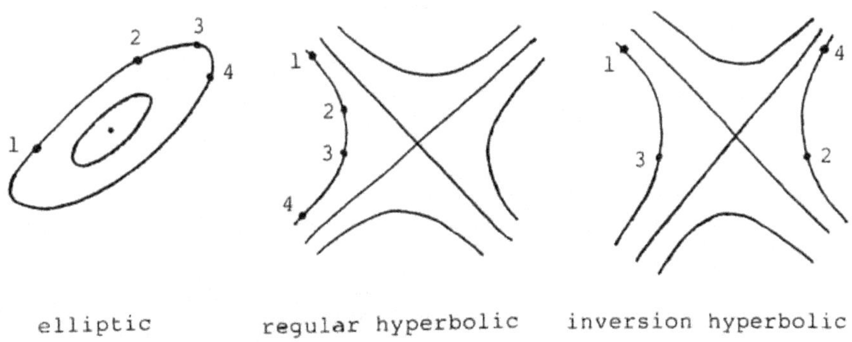

Figure 1.2.1.4.1: Orbits of linear area preserving maps

$$t = 2\cos 2\pi\nu$$
$$R = \sin^2 \pi\nu. \tag{1.2.1.4.5}$$

Apart from the question of non-diagonal Jordan normal forms, the eigenvalues of an area preserving matrix do not give a complete classification up to area preserving coordinate change. There is an additional invariant under area-preserving coordinate change in the case that the eigenvalues are on the unit circle, called the *Krein signature* σ (Arnold and Avez, 1968, Moser, 1958). For 2-D maps this can be evaluated from the off-diagonal elements of DT:

$$DT = \begin{vmatrix} A & B \\ C & D \end{vmatrix} \tag{1.2.1.4.6}$$

as follows:

$$\sigma = \begin{matrix} + \\ 0 \\ - \end{matrix} \text{ according as } B - C \begin{matrix} > \\ = \\ < \end{matrix} 0. \tag{1.2.1.4.7}$$

The case $\sigma = 0$ can occur only at multipliers ± 1. To see that the Krein signature is invariant under area preserving conjugacy, simply evaluate it for a general area preserving coordinate transform of DT. Essentially it measures the direction of rotation of DT. For 2-D maps it doesn't have great significance, but for higher dimensional symplectic maps it is very significant (see §1.2.1.6).

Now I mention what linear stability can tell one about actual stability. In fact, further analysis shows that parabolic points with diagonal Jordan normal form are typically unstable, and elliptic points are typically stable, apart from the cases $R = 3/4,\ 1/2$, corresponding to third and fourth order resonance, $\nu = k/m$, k, m coprime, $m = 3,4$. The case $m = 3$ is typically unstable, and $m = 4$ can be either. This follows from normal form analysis, to be discussed in §1.2.4.4, and the Moser twist theorem (§1.3.1.5). It will be discussed a little in §1.3.1.5 (see also Meyer, 1971).

§1.2.1.5 Poincaré index. For isolated fixed points of a map on a surface, a useful quantity can be defined, called the *index*. In fact I won't use it explicitly in this thesis, but I use it a lot for my own intuition.

For a map T of a surface, the index of a closed curve C that passes through no fixed points is defined as the number of times that the displacement vector:

$$v(x) = T(x) - x \tag{1.2.1.5.1}$$

encircles 0 as x traverses C, taken to be positive or negative according as the encirclement and traversal are in the same or opposite directions. Provided C crosses no fixed points, the index is a continuous function of C, so it is constant, being integer-valued. Thus the index of a closed curve containing no fixed points is zero, provided it can be shrunk to a point. This is because it can be shrunk to a curve small enough that the direction of $v(x)$ is restricted to some angle, so no revolutions are possible. Also, we can speak of the index of an isolated fixed point, meaning the index of any closed curve surrounding it and no other fixed points. The index of a fixed point of an area preserving map with non-zero residue can easily be seen to be the sign of its residue. Note that while a fixed point of T is a fixed point of T^n, its index can depend on n.

The index is summable, i.e. the index of a curve formed by traversing first one and then another is the sum of the separate indices. Thus the index of a curve is the sum of the indices of the fixed points that it contains. Index can be used to locate periodic orbits numerically (Bialek, private communication). If one finds a closed curve of non-zero index, then dividing it in two, one half must also have non-zero index. In this way one can shrink down on a fixed point. Index is also useful in bifurcation analysis (see §1.2.4.1).

§1.2.1.6 Stability in 4-D symplectic maps. In this section I evaluate stability criteria for linear 4-D symplectic maps. I also briefly discuss Krein's theorem.

The characteristic polynomial $P(\lambda)$ of a symplectic matrix L is reflexive (§1.2.1.3), so the eigenvalues come in reciprocal pairs. Write them as $r,\ r^{-1}, s, s^{-1}$. Then:

$$P(\lambda) = (\lambda^2 - \rho\lambda + 1)(\lambda^2 - \sigma\lambda + 1) \qquad (1.2.1.6.1)$$

where

$$\rho = r + r^{-1},\ \ \sigma = s + s^{-1}. \qquad (1.2.1.6.2)$$

Thus

$$\rho + \sigma = A \equiv \text{trace } L. \qquad (1.2.1.6.3)$$

and

$$2 + \rho\sigma = B \equiv \sum_{i<j} \begin{vmatrix} L_{ii} & L_{ij} \\ L_{ji} & L_{jj} \end{vmatrix} = \frac{1}{2}(A^2 - \text{trace } (L^2)) \qquad (1.2.1.6.4)$$

So ρ, σ are the roots of:

$$\tau^2 - A\tau + B - 2 = 0 \tag{1.2.1.6.5}$$

i.e.

$$\tau = \frac{A}{2} \pm \left(\frac{A^2}{4} - B + 2 \right)^{1/2} . \tag{1.2.1.6.6}$$

In the case:

$$\frac{A^2}{4} - B + 2 \geq 0 \tag{1.2.1.6.7}$$

ρ, σ are real. The eigenvalues corresponding to $\tau = \rho$ or σ are a complex conjugate pair on the unit circle, or reciprocal pair on the positive/negative real axis according as $|\tau| < 2$, $\tau > 2, \tau < -2$. They are given by (1.2.1.4.2) with τ replacing t.

If

$$\frac{A^2}{4} - B + 2 < 0 \tag{1.2.1.6.8}$$

then write the roots as a quadruplet $z\,e^{\pm i\theta}, z^{-1}\,e^{\pm i\theta}$, with $z > 0$. Then setting:

$$\zeta = z + z^{-1} \tag{1.2.1.6.9}$$

with inverse:

$$z = \frac{\zeta}{2} + \left(\frac{\zeta^2}{4} - 1 \right)^{1/2} . \tag{1.2.1.6.10}$$

I obtain:

$$2\zeta \cos\theta = A \tag{1.2.1.6.11}$$

and

$$\zeta^2 - 4(1 - \cos^2\theta) = B - 2. \tag{1.2.1.6.12}$$

Thus

$$\zeta^2 - (B+2) + \frac{A^2}{\zeta^2} = 0 \qquad (1.2.1.6.13)$$

i.e.

$$\zeta^2 = \frac{B}{2} + 1 + \left(\left(\frac{B}{2} + 1 \right)^2 - A^2 \right)^{1/2} . \qquad (1.2.1.6.14)$$

Note that this exceeds 4, by the hypothesis (1.2.1.6.8), so $\zeta > 2$, and we get a real value of z, as desired.

This completes the linear stability analysis, apart from cases with multiple eigenvalues. It is summarised in the stability diagram of figure 1.2.1.6.1. Linear instability of a fixed point implies nonlinear instability, but the same is not true in general for stability. In the case, however, that all the multipliers are on the unit circle and do not satisfy any conditions of strong resonance, and the nonlinearities are non-degenerate, stability does hold in the weak sense that given $\varepsilon > 0$, there exists a neighbourhood V of the fixed point such that the fraction of points in V which do not remain within V for ever is less than ε (Moser, 1973).

The stability diagram gives the following possibilities for a periodic orbit to lose stability as parameters are varied:

(i) tangent bifurcations: For $A^2/4 - B + 2 \geq 0$, a pair of eigenvalues can leave the unit circle at $+1$ and split along the positive real axis as $S = B + 2 - 2A$ decreases through 0. Note that, since persistence is not guaranteed at multipliers $+1$, this typically only happens in the generalised sense of a tangent bifurcation (cf. §1.2.4).

(ii) period doubling: For $A^2/4 - B + 2 \geq 0$, a pair of eigenvalues can leave the unit circle at -1 and split along the negative real axis as $R = B + 2 + 2A$ decreases through 0. This is accompanied by period doubling.

(iii) Krein crunch[5]: For $A^2/4 - B + 2 < 0$, the eigenvalues on the unit circle can collide in two pairs and split off the circle as a quadruplet.

[5] Maybe better termed "complex instability".

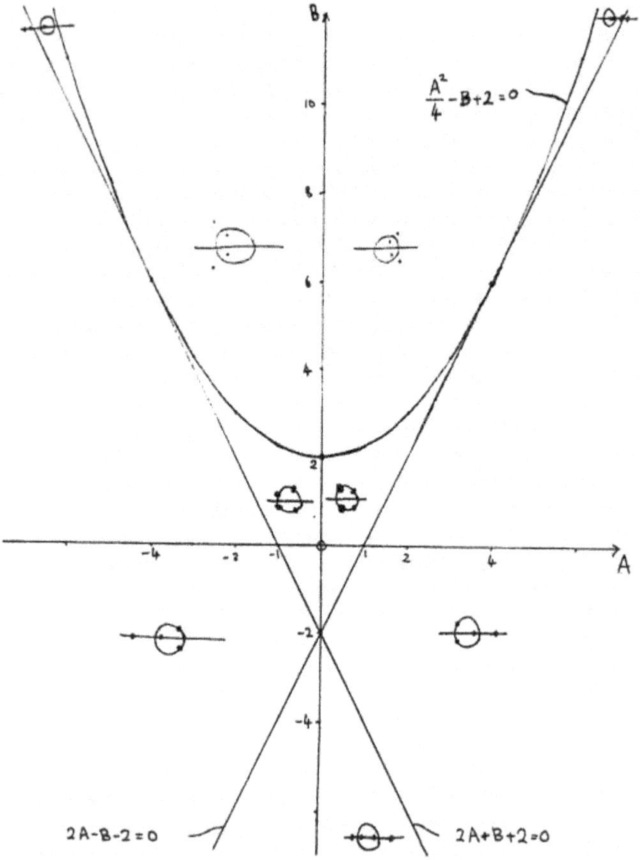

Figure 1.2.1.6.1: Stability diagram for periodic orbits of 4–D symplectic maps

The last possibility is something I would love to investigate more. It could lead to all sorts of interesting behaviour in a neighbourhood.[6] A caution, however, is that collision of the eigenvalues does not always lead to their splitting off the unit circle. They can pass through each other, and this can even be stable to perturbation. This is because although the stability of a symplectic matrix is classified by its Jordan normal form, it does not provide a complete classification up to symplectic conjugacy. For each pair of eigenvalues on the unit circle there is an additional invariant, called the *Krein signature* of the pair (e.g. see Arnold and Avez, 1968, App. 29). It can take the values ± 1 (and also 0, in the case

[6] Bridges, Cushman and I (1993) now have some results on this.

of a pair at ± 1), and measures the direction of rotation in the eigenspace associated with the pair. Krein's theorem, which is most clearly explained by Moser (1958) who discovered it at the same time, states that two pairs of eigenvalues on the unit circle with the same signature cannot leave the circle on collision. The typical case for opposite signatures is that they fall off. Thus one should think of the stability diagram as having two sheets in the regions 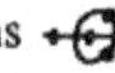 and , and four sheets in region 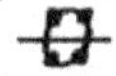. labelled by the Krein signatures of the pairs on the unit circle. Only two of the sheets of 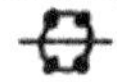 connect to 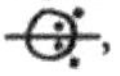, the other two joining to each other by a fold.

§1.2.2 **Periodic orbits in twist maps.** The definition and stability of periodic orbits are considered in the action representation. Then I discuss the Poincaré-Birkhoff theorem which proves existence in periodic area preserving twist maps of at least two periodic orbits for every rational rotation number in the range of the twist.

§1.2.2.1 Periodic orbits in the action representation. In the action representation, periodic orbits are given by sequences $x_0, \ldots, x_q = x_0$, for which the action:

$$\sum_{i=0}^{q-1} \tau(x_i, x_{i+1}) \tag{1.2.2.1.1}$$

is stationary with respect to all variations. Considering the case $q = 1$, a fixed point is determined by:

$$\frac{d}{dx}\tau(x, x) = 0 \tag{1.2.2.1.2}$$

and its residue is:

$$R = \frac{\dfrac{d^2}{dx^2}\tau(x, x)}{4\,\tau_{12}(x, x)}. \tag{1.2.2.1.3}$$

Thus the sign of the residue depends on whether the orbit maximizes or minimizes the action, and the sign of the twist.[7]

§1.2.2.2 Poincaré-Birkhoff theorem. A point (θ_0, z_0) is said to be a *periodic point of type (p, q)* of a periodic map T if:

$$R^p T^q(\theta, z) = (\theta, z) \tag{1.2.2.2.1}$$

[7] This was generalised to $q > 1$ by MacKay and Meiss (1983).

where

$$R(\theta, z) = (\theta - 1, z) \qquad (1.2.2.2.2)$$

and p, q are the smallest integers ($q > 0$) satisfying (1.2.2.2.1). Mather (1982b) and Katok (1982) require also an order preserving condition, viz. that the θ_i come in the same order on the circle as for uniform rotation at rate p/q, but Aubry (1982) derives this as a property of the orbits he finds. In the action representation a periodic orbit of type (p, q) corresponds to a sequence $\theta = \theta_0, \theta_1, \ldots, \theta_q = \theta_0 + p$ for which the *action*:

$$W(\theta) = \sum_{i=0}^{q-1} \tau(\theta_i, \theta_{i+1}) \qquad (1.2.2.2.3)$$

is stationary with respect to variations in θ.

Birkhoff (1927) showed that every map in class A has at least two periodic orbits of type (p, q) for each rational p/q in lowest terms, in an appropriate interval, which I will call the *range of the twist*. If $\tau_{12} < 0$, the periodic orbits are obtained by respectively minimizing and minimaximizing the action $W(\theta)$ over an appropriate set of θ. This is explained also by Mather (1982c). For $\tau_{12} > 0$ (the case Mather considers), interchange "max" and "min". I shall restrict attention without loss of generality to the former case.

The stability analysis of §1.2.2.1 shows that the minimizing periodic orbits found by Birkhoff have non-positive residue R^-, and the minimaximizing orbits have non-negative residue R^+. They give rise to *island chains*. Some island chains can be seen in figure (i), for example. Also figure 1.3.2.1.1 shows some island chains for the quadratic map:

$$x' = p - y - x^2$$
$$y' = x \qquad (1.2.2.2.4)$$

for parameter $p = 2.38216325159$. For purposes of orientation, thin island chains, as when they are born by bifurcation from a periodic orbit, have residue close to zero.

There is an alternative proof of the Poincaré-Birkhoff theorem which gives at least two periodic orbits (of opposite index in the case that they are isolated) of type (p, q), but only for maps close enough to an integrable twist (see §1.3.1.2), and the proximity depends on (p, q). It works by finding a circle whose points are only radially translated under $F^q R^p$. Area preservation plus zero Calabi invariant imply that it is intersected by its image, the intersections giving periodic points of type (p, q). For details, see Arnold and Avez (1968).[8]

[8] Note that the original form of the Poincaré-Birkhoff theorem, conjectured in 1912 by Poincaré and proved in 1913 by Birkhoff, required different hypotheses from those used here.

§1.2.3 **Symmetric periodic orbits of reversible maps.** Symmetric periodic orbits play a special role in reversible maps. In this section I describe their connection with symmetry lines, and discuss their stability. For reversible area preserving twist maps, I discuss actions for symmetric periodic orbits, and the dominant symmetry for Birkhoff orbits. I conclude with an appendix on counting periodic orbits in the quadratic map.

§1.2.3.1 Symmetric periodic orbits and symmetry lines. Reversibility implies that the reflection of a periodic orbit by any symmetry is the time reverse of a periodic orbit with the same multipliers and same Krein signature. If it is its own reflection, a periodic orbit is called *symmetric*. This definition is independent of the symmetry considered within a given family. Note that not all periodic orbits of reversible systems are symmetric (e.g. see §1.2.3.5). The ideas of this section are due to DeVogelaere (1958, 1962).

If x is a point of a symmetric periodic orbit of a reversible map T, then:

$$Sx = T^k x \qquad (1.2.3.1.1)$$

for some k. If k is even, then

$$S.T^{k/2}x = T^{k/2}x \qquad (1.2.3.1.2)$$

so the orbit has a point $T^{k/2}x$ on the fixed line of S. If k is odd, then

$$TS.T^{\frac{k+1}{2}}x = T^{\frac{k+1}{2}}x \qquad (1.2.3.1.3)$$

so the orbit has a point $T^{\frac{k+1}{2}}x$ on the fixed line of TS.

Conversely, a periodic orbit with a point on some symmetry line is symmetric. Note that an intersection of the fixed lines of T^mS and T^nS is a fixed point of T^{m-n}, and hence gives a symmetric periodic orbit. Conversely, a point of period q on Fix(T^nS) lies at an intersection with Fix($T^{m\pm kq}S$), for all $k \in Z$.

If x lies on Fix(S) and is periodic of even period $2n$, then $T^n x$ also lies on Fix(S), because:

$$S.T^n x = T^{-n}Sx = T^{-n}x = T^{-n}T^{2n}x = T^n x. \qquad (1.2.3.1.4)$$

Similarly, if x lies on Fix(S) and has odd period $2n+1$ or $2n-1$, then $T^n x$ lies on Fix(ST) or Fix(TS), respectively, because:

$$\begin{aligned} ST.T^n x &= T^{-1-n}Sx = T^{-1-n}x = T^{-1-n}T^{2n+1}x = T^n x \\ TS.T^n x &= T^{1-n}Sx = T^{1-n}x = T^{1-n}T^{2n-1}x = T^n x. \end{aligned} \qquad (1.2.3.1.5)$$

Conversely, if x lies on Fix(S) and $T^n x$ on Fix(S), or Fix(ST), or Fix(TS), then x is periodic with period dividing $2n$, $2n+1$, $2n-1$, respectively, because:

$$\begin{aligned} Sx = x,\ S.T^n x = T^n x &\Rightarrow T^{-n} x = T^n x \\ Sx = x,\ ST.T^n x = T^n x &\Rightarrow T^{-1-n} x = T^n x \\ Sx = x,\ TS.T^n x = T^n x &\Rightarrow T^{1-n} x = T^n x. \end{aligned} \tag{1.2.3.1.6}$$

Thus, given a complementary pair of symmetries, a periodic orbit of even period has precisely two points on one line and none on the other, and an orbit of odd period has precisely one on each. I say an orbit ***belongs*** to a symmetry (actually, its whole half-family - §1.1.4.2) if it has a point on its fixed line.

The results of the previous paragraph provide an easy way to find a symmetric periodic orbit. If you know a symmetry to which it belongs, then it is necessary only to search that line for a point whose appropriate image falls on the appropriate line. This is only a 1-D root search, rather than 2-D as would be the case for a general periodic orbit. Furthermore, it requires only half the number of iterations for each error evaluation. I use the secant method, as it is the fastest simple method, with errors ε_n decreasing asymptotically like:

$$\varepsilon_{n+1} \sim \varepsilon_n^{\gamma},\ \gamma = \frac{\sqrt{5}+1}{2}. \tag{1.2.3.1.7}$$

I stop iterating when round-off errors make the steps stop decreasing.

§1.2.3.2 Stability of symmetric periodic orbits. I show that symmetric periodic orbits of reversible maps have the same classification for linear stability as periodic orbits of area preserving maps. Let us use symmetry coordinates (1.1.4.2.1) for a symmetry to which the periodic point of interest (wlog, fixed point) belongs, so that:

$$DS = \begin{vmatrix} 1 & 0 \\ 0 & -1 \end{vmatrix}. \tag{1.2.3.2.1}$$

Write

$$DT = \begin{vmatrix} A & B \\ C & D \end{vmatrix}. \tag{1.2.3.2.2}$$

Then symmetry implies that $(DS.DT)^2 = I$, i.e.

$$(DS.DT)^2 = \begin{vmatrix} A^2 - BC & B(A-D) \\ C(D-A) & D^2 - BC \end{vmatrix} = \begin{vmatrix} 1 & 0 \\ 0 & 1 \end{vmatrix}. \tag{1.2.3.2.3}$$

So,

$$\text{either } A = -D = \pm 1, \ B = C = 0$$

$$\text{or} \quad A = D, \quad AD - BC = 1. \tag{1.2.3.2.4}$$

The first case is pathological when $A = +1$ because it can have no nonlinear terms. To see this, write:

$$T = S + \delta T. \tag{1.2.3.2.5}$$

Then symmetry implies that:

$$I = (S + \delta T)S(S + \delta T)S = I + \delta T\, S + \delta T\, S. \tag{1.2.3.2.6}$$

So

$$\delta T(x, -y) = 0. \tag{1.2.3.2.7}$$

When $A = -1$, there is no such pathology, but I will ignore it as it is orientation reversing. The second case is area preserving, as claimed.

The derivative DT^q for a symmetric periodic point x of period q can be evaluated by going only halfway round the orbit. If x lies on $\mathrm{Fix}(S^i)$ and $T^m x$ on $\mathrm{Fix}(S^f)$, then the tangent map at x round the orbit is given in terms of the derivative halfway round by:

$$DT^q_x = DS^i_x \cdot (DT^m_x)^{-1} \cdot DS^f_{T^m x} \cdot DT^m_x \tag{1.2.3.2.8}$$

as can be easily checked. This saves a factor of two on evaluation of the derivative. It may be more convenient to evaluate DT^m in coordinates other than symmetry coordinates, e.g. to speed up the matrix multiplication, but generally I convert the final result to symmetry coordinates.[9]

---

[9] i.e if $S^{i/f} = C^{i/f}\, \mathcal{S}\, {C^{i/f}}^{-1}$ with $\mathcal{S}(x, y) = (x, -y)$ then ${DC^i}^{-1} DT^q_x DC^i = D\mathcal{S}\, M^{-1} D\mathcal{S}\, M$, and

${DC^f}^{-1} DT^q_{T^m x} DC^f = M\, D\mathcal{S}\, M^{-1} D\mathcal{S}$, where $M = {DC^f}^{-1} DT^m_x DC^i$. In particular, if the coordinate

changes $C^{i/f}$ are chosen area preserving, so $M = \begin{bmatrix} a & b \\ c & d \end{bmatrix}$ satisfies $ad - bc = 1$, then ${DC^i}^{-1} DT^q_x DC^i =$

$\begin{bmatrix} ad+bc & 2bd \\ 2ac & ad+bc \end{bmatrix}$ and ${DC^f}^{-1} DT^q_{T^m x} DC^f = \begin{bmatrix} ad+bc & 2ab \\ 2cd & ad+bc \end{bmatrix}$ and the residue $= -bc$.

Often I want to find a parameter value for which there is a periodic orbit of some specified residue. This can be done by two nested 1-D root finders, but it is more efficient to use a 2-D root finder. I used a method of Broyden (Dahlquist and Björck, 1974), modified to use a well-adapted inner product, which worked well on the whole.

§1.2.3.3 Actions for symmetric periodic orbits. Actions for symmetric periodic orbits can also be evaluated by going only halfway round. For example, with the symmetry of (1.1.4.4.2):

$$\tau(-x', -x) = \tau(x, x'). \tag{1.2.3.3.1}$$

If x_0 is a point of period q belonging to that symmetry, i.e. $x_0 = 0$, then:

$$\sum_{i=0}^{q-1} \tau(x_i, x_{i+1}) = 2 \sum_{i=0}^{m-1} \tau(x_i, x_{i+1}) + \sigma.\tau(x_m, -x_m) \tag{1.2.3.3.3}$$

where

$$q = 2m + \sigma, \;\; \sigma = -1, 0, 1. \tag{1.2.3.3.4}$$

For specific cases like the standard map, I treated the other possibilities too, but I don't know how to do it in general.

§1.2.3.4 Dominant symmetry for Birkhoff orbits. In this section I consider symmetries for periodic twist maps. I begin with the setting of a periodic twist map derived by removing a fixed point.

If T has symmetry S, then an intersection of the fixed lines of S and TS gives a fixed point of T. If we remove the point, we get four half-lines. Equivalently, any lift of the map obtained by removing the point, has the symmetries, S, SR, TS, TSR, where R is given in (1.2.2.2.2). The fixed lines of Fix(S), Fix(SR) correspond to the two half-lines of S, and those of Fix(TS), Fix(TSR) to the half-lines of TS. In symmetry coordinates (z, θ) for any of the above four symmetries, the generating function for T^q, for any q, has the symmetry

$$\tau(\theta, \theta') = \tau(-\theta', -\theta). \tag{1.2.3.4.1}$$

So

$$\frac{d}{d\theta}\tau(\theta, \theta)\Big|_{\theta=0} = 0 \tag{1.2.3.4.2}$$

and T has a periodic point of type (p, q) for each p/q in the range of the twist, on each symmetry line.

A remarkable observation (Greene, private communication, Shenker and Kadanoff, 1982) is that for most of the examples studied, on one of the half-lines there is a point of non-negative residue for each coprime pair (p, q). Following Shenker and Kadanoff (1982) I call it the *dominant* half-line. Suppose, wlog, that it is $\text{Fix}(S)$. Then furthermore, on $\text{Fix}(SR)$, one finds a point of non-negative residue for q even, and a point of non-positive residue for q odd. On $\text{Fix}(TS)$ one finds a point of non-positive residue for p odd, and a point of non-negative residue for p even, while on $\text{Fix}(TSR)$ there is a point of non-positive residue for p even and a point of non-negative residue for p odd. This is illustrated in figure 1.2.3.4.1. These other half-lines I call *subdominant*. In easy cases, for example small and large k in the standard map (Greene, 1979a), one can verify that "a point of non-positive/non-negative residue" in the above can be replaced by "a minimizing/minimaximizing point" (§1.2.2.2). I will assume that this is true in all cases, though the behaviour has not yet been explained. Thus, knowledge of the half-lines helps greatly in locating the Birkhoff orbits. Table 1.2.3.4.1 gives the choice of lines to use. Note that:

$$\text{Fix}(SR^j) = R^{-j/2}\text{Fix}(S) \tag{1.2.3.4.3}$$

where

$$R^{\frac{1}{2}}(\theta, z) = (\theta - \tfrac{1}{2}, z). \tag{1.2.3.4.4}$$

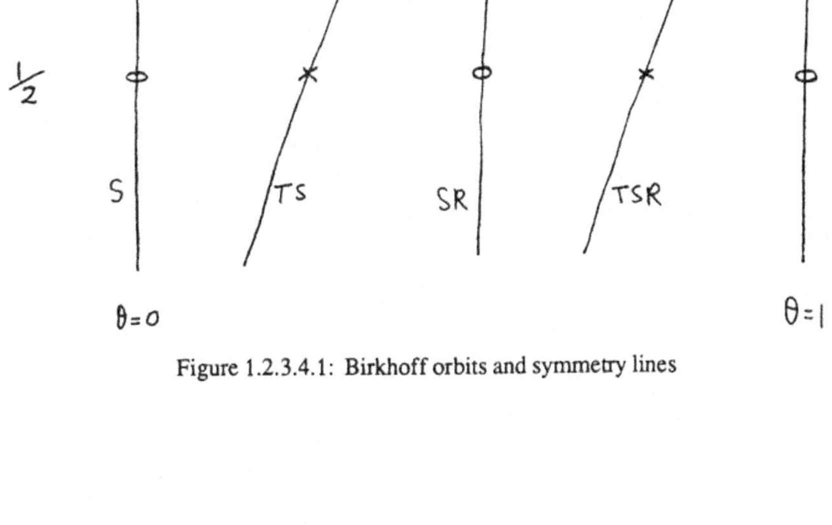

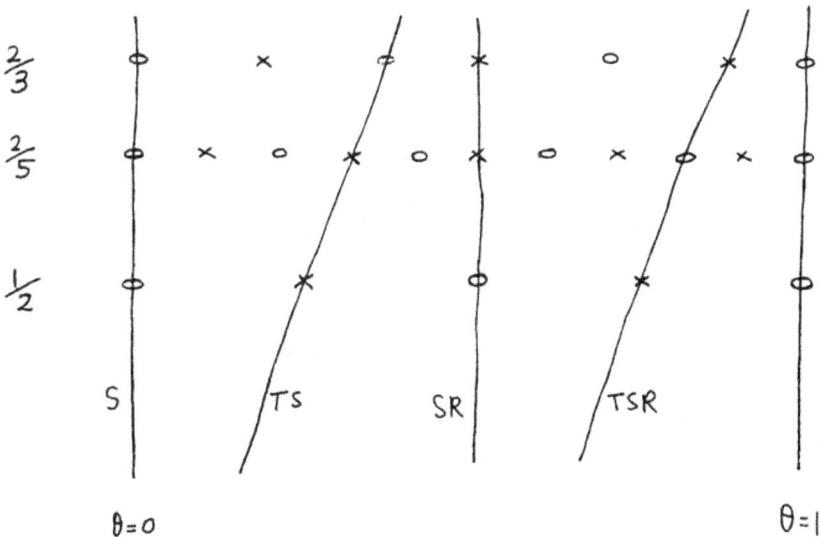

Figure 1.2.3.4.1: Birkhoff orbits and symmetry lines

Table 1.2.3.4.1: Symmetry lines for Birkhoff periodic orbits of type (p, q), p, q coprime

stability	p, q	initial line	final line	number of iterations
minimax	p even	S	TSR^p	$(q+1)/2$
		TS	SR^p	$(q-1)/2$
	q even	S	SR^p	$q/2$
		SR	SR^{p+1}	$q/2$
	both odd	S	TSR^p	$(q+1)/2$
		TSR	SR^{p+1}	$(q-1)/2$
min	p even	SR	TSR^{p+1}	$(q+1)/2$
		TSR	SR^{p+1}	$(q-1)/2$
	q even	TSR	TSR^{p+1}	$q/2$
		TS	TSR^p	$q/2$
	both odd	TS	SR^p	$(q-1)/2$
		SR	TSR^{p+1}	$(q+1)/2$

Beware that there can be other periodic orbits than the Birkhoff orbits. Figure 1.2.3.4.2 shows y_{154} against x_0, for the quadratic map in DeVogelaere form, for initial condition on the dominant half-line. So each zero crossing gives a point of period 308 on the dominant line. Apart from the pair on the far right, which I did not check, they are all of type (117,308)! This means that one has to be careful, to get the Birkhoff orbits. I did not have any systematic approach to this problem, but just threw out points which gave inconsistent results. Note that generic bifurcation from a fixed point (§1.2.4.4) produces only Birkhoff orbits, so the extra ones arise later, by tangent bifurcation because (117,308) are coprime. The graph in figure 1.2.3.4.2 looks non-smooth, but this is only because I took large steps in x.

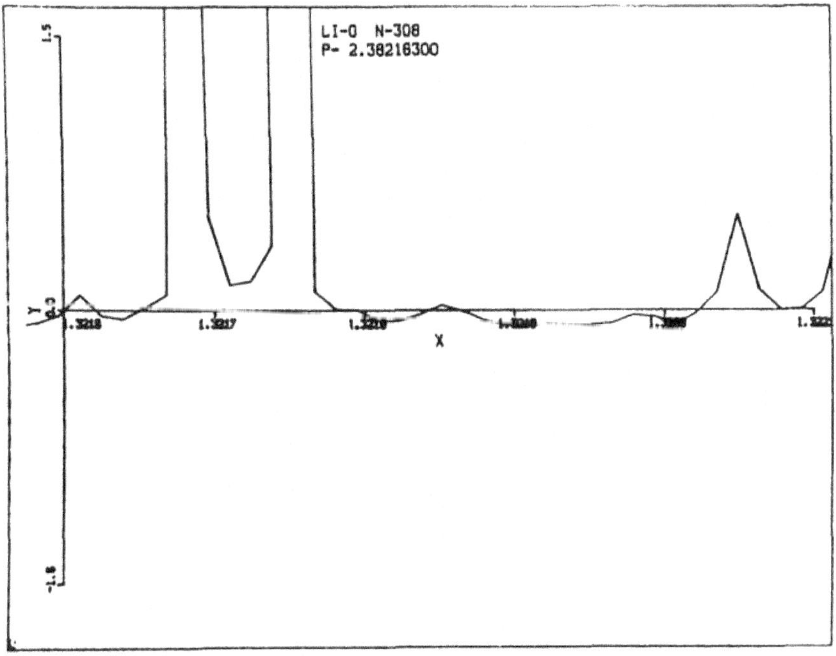

Figure 1.2.3.4.2: A whole bunch of periodic points of type (117,308)

Note that it is not actually necessary to work in polar coordinates to find the Birkhoff orbits. All one needs is the half-lines, and a method for counting revolutions. The latter is easy to devise, if, for example, you know that no point rotates more than half a revolution in one step. Then you can count crossings of one of the symmetry lines to give the number of half-revolutions.

§1.2.3.5 Counting periodic orbits in the quadratic map. I frequently use the quadratic map:

$$Q: x' = f(x) - y \qquad y' = x - f(x') \tag{1.2.3.5.1}$$

where

$$f(x) = a - x^2$$

as a typical one parameter family of reversible area preserving maps, which is about the simplest possible. It is useful to know how many periodic orbits of various types it has.

Bézout's theorem (e.g. Moser, 1960, Simo, 1979) shows that there are precisely 2^n fixed points of Q^n in the complex plane (counted according to multiplicity). Note from §1.4.3 that they are all real and hyperbolic for $a > \frac{5}{4} + \frac{\sqrt{5}}{2}$. To find out the number of periodic points of period n one has to subtract off the points with period a proper factor of n. This is given in the first three columns of table 1.2.3.5.1, for n up to 12.

One might also want to know the numbers of symmetric periodic orbits. The quadratic map Q has the symmetry:

$$S: x' = x, y' = -y. \tag{1.2.3.5.2}$$

Let us begin with odd periods. A symmetric periodic orbit of odd period has precisely one fixed point on any symmetry line. A point (x_0, y_0) on Fix(QS) is fixed under Q^{2n+1} iff its nth iterate (x_n, y_n) lies on Fix(S). Thus I require:

$$\begin{aligned} y_0 &= x_0 - f(x_0) \\ y_n &= 0. \end{aligned} \tag{1.2.3.5.3}$$

Using

$$\begin{aligned} x_1 &= f(x_0) - y_0 \\ x_k &= 2f(x_{k-1}) - x_{k-2} \\ y_n &= x_{n-1} - f(x_n) \end{aligned} \tag{1.2.3.5.4}$$

one sees that y_n is a polynomial of degree 2^{n+1} in x_0. So there are 2^{n+1} fixed points of Q^{2n+1} on Fix(QS) (in $\mathbb{C}$). Again one has to subtract off points with period a proper factor of $2n+1$. Hence the entries in table 1.2.3.5.1 for n odd.

In the same way one can count symmetric periodic orbits of even period. They either have precisely two points on Fix(S) and none on Fix(QS) or vice versa. One finds 2^{n+1} fixed points of Q^{2n} on Fix(S) and 2^n on Fix(QS) (in $\mathbb{C}$). Hence the entries in table 1.2.3.5.1 for n even. Note that 6 is the smallest period for which there are non-symmetric periodic orbits.

Index arguments can be useful if you want to know about the stability of the periodic orbits. The total index of fixed points for a map of a surface isotopic to the identity has to be equal to its Euler characteristic. For the plane it is 2. So the total index of fixed points of Q^n is 2, for all n. Again, one would like to subtract off the effect of periodic points of period a factor of n. This is slightly more messy:

$$\text{Index}(f^n) = \sum_{m|n} m(e_m + i_m - h_m) - 2\sum_{2m|n} mi_m \qquad (1.2.3.5.5)$$

when f has e_m, i_m, h_m elliptic, inversion hyperbolic, and regular hyperbolic periodic orbits of period m ($m|n$ means m divides n, i.e. m is a factor of n, which can include n itself). This follows from the fact that the index of a fixed point is the sign of its residue (§1.2.1.5). For example, if all 2^n fixed points of Q^n are real and hyperbolic, one obtains the entries i_n, h_n of inversion and regular hyperbolic periodic orbits of table 1.2.3.5.1.[10]

Table 1.2.3.5.1: Numbers F_n of fixed points of Q^n, c_n of n-cycles of Q, SF_n of symmetric fixed points of Q^n on Fix(S) for n odd, and on Fix(S), Fix(QS) for n even, SC_n of symmetric n-cycles, divided as for SF_n, and i_n, h_n of inversion and regular hyperbolic periodic orbits when they are all real and hyperbolic.

n	F_n	C_n	SF_n		SC_n		i_n	h_n
1	2	2	2		2		1	1
2	4	1	4	2	1	0	1	0
3	8	2	4		2		1	1
4	16	3	8	4	2	1	2	1
5	32	6	8		6		3	3
6	64	9	16	8	5	2	5	4
7	128	18	16		14		9	9
8	256	30	32	16	12	6	16	14
9	512	56	32		28		28	28
10	1024	99	64	32	27	12	51	48
11	2048	186	64		62		93	93
12	4096	335	128	64	54	27	170	165

10 The numbers of bifurcations at period n can also be counted (MacKay and Shardlow, 1992).

§1.2.4 **Bifurcation of periodic orbits.** In this section I discuss bifurcation of periodic orbits, meaning the branching of periodic orbits as a parameter is varied. First it is necessary to find under what conditions a periodic orbit persists under perturbation. I analyse bifurcations in the simplest systems: 1-D maps, and then in area preserving maps. Reversibility leads to a few modifications, as does possession of a commutator.

§1.2.4.1 Persistence and bifurcation. One is often interested in following periodic orbits of a system as parameters are varied, and in particular, finding where they are created or destroyed, or collide. Such occurrences are called *bifurcations*. A preliminary point to establish is that periodic points persist under perturbation most of the time, so bifurcations are special events.

If a fixed point x_0 of a map f has no multipliers equal to $+1$, the implicit function theorem guarantees that it has a neighbourhood such that all small enough perturbations of f will also have a unique fixed point in that neighbourhood, depending smoothly on the perturbation, and with multipliers depending smoothly on the perturbation. Crudely speaking, it is given by:

$$x_0 + \delta x = f(x_0 + \delta x) + \delta f(x_0 + \delta x) \qquad (1.2.4.1.1)$$

i.e.

$$\delta x \sim Df_{x_0} \cdot \delta x + \delta f_{x_0} \qquad (1.2.4.1.2)$$

so

$$\delta x \sim (1 - Df_{x_0})^{-1} \, \delta f_{x_0}. \qquad (1.2.4.1.3)$$

We already saw (§1.2.1.2) that fixed points with no multipliers equal to $+1$ are isolated. So now we see that they also persist under perturbation. Similarly, for a periodic point of period n, say, by considering it as a fixed point of f^n, one deduces that provided none of its multipliers are $+1$, the periodic point persists under perturbation, and has a neighbourhood in which there are no other periodic points of the same period or factors of it. Note that there could be other periodic points arbitrarily close, but not of the same period, nor a factor of it.

In conclusion, the only time that a periodic orbit could be created or destroyed or collide with another one of the same period or a submultiple is when it has a multiplier $+1$. Similarly, by considering it as a fixed point of f^k, a fixed point x (wlog) of a map f can have a periodic orbit of period k branching from or colliding with it only when one of its multipliers is a kth root of unity. This is because, as a fixed point of f^k, x is isolated unless Df_x^k has an eigenvalue +1, or equivalently unless DF_x has an eigenvalue λ such that $\lambda^k = 1$. Thus one should analyse in more detail (i.e. nonlinearly) the cases where *a* fixed point has a multiplier +1 or a higher root of unity.

I will discuss *generic* bifurcations in one parameter families. I will use the term "generic" to mean that they are the only forms that one will see if one excludes a certain codimension 1 space of "degenerate" systems. This is not the usual sense in which it is used in mathematics, where a property is said to be *generic* in a class of systems if it holds on a countable intersection of open dense sets in that class. This weaker notion is useful because theorems about generic bifurcations, for example, make a statement about all bifurcations of all periodic orbits of the systems.

Note that index (§1.2.1.5) can be useful in bifurcation analysis. The index of a closed curve passing through no fixed points depends continuously on parameters, so being integer-valued it is constant. Thus the sum of the indices of the fixed points in a given region is constant as parameters are varied, provided none of them cross the boundary of the region. This can give instant information about the possible bifurcations. For example, when the residue of a fixed point of a map T decreases through -1, its index does not change, so the number of fixed points in any neighbourhood remains the same. But its index as a fixed point of T^2 changes from +1 to -1. So there must be some other fixed points of T^2 around. They must be of period two, by the previous comment. Thus at least two periodic points of period two are either created (with index +1) or destroyed (with index -1).

§1.2.4.2 One Dimensional Maps. As a simple example of bifurcation analysis I will consider maps on a one dimensional manifold. This will be relevant to §2.2.2. In this case Df_x is a real scalar, so the only possible roots of unity are ± 1. So there are only these two cases to analyse.

(i) Multiplier +1. For a fixed point x_0 of f with multiplier +1, $f'(x_0) = +1$. So shifting the fixed point to the origin, we get:

$$f(x) = x + a_2 x^2 + \ldots \qquad (1.2.4.2.1)$$

Consider a general perturbation

$$\delta f(x) = \varepsilon_0 + \ldots \qquad (1.2.4.2.2)$$

Then look for fixed points of the perturbed map:

$$x = \tilde{f}(x) = f(x) + \delta f(x) = x + a_2 x^2 + \ldots + \varepsilon_0 + \ldots \qquad (1.2.4.2.3)$$

One finds the maximal balance:

$$a_2x^2 + \varepsilon_0 = 0 \tag{1.2.4.2.4}$$

provided

$$a_2 \neq 0,\ \varepsilon_0 > \varepsilon_1^2, \text{etc.} \tag{1.2.4.2.5}$$

Thus we get two fixed points,

$$x \sim \pm\sqrt{-\varepsilon_0/a_2} \tag{1.2.4.2.6}$$

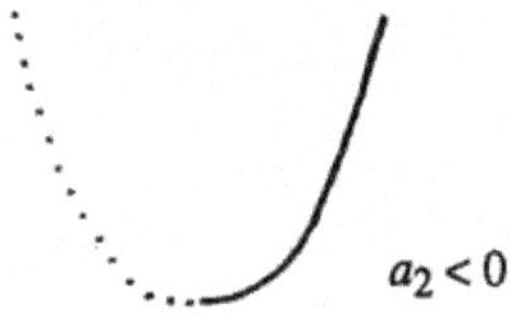

when ε_0 has the opposite sign from a_2, and none for the same sign. Their stability is given by

$$D\tilde{f}_x = 1 + 2a_2x + \ldots \sim 1 \pm 2a_2\sqrt{-\varepsilon_0/a_2}. \tag{1.2.4.2.7}$$

This gives a *saddle-node* or *tangent bifurcation*. The above sketch gives the position x (horizontal) of the fixed points as a function of the parameter ε_0 (vertical). In this and future sketches, solid line indicates linear stability, dotted line regular hyperbolic, and dashed will represent inversion hyperbolic (making the natural extension from area preserving to 1-D maps). Note that the minimal number of terms was kept above. In general one cannot tell how many terms to keep until a maximal balance is found.

If $a_2 = 0$, one has to keep more terms.

Fixed points: $$a_3x^3 + \varepsilon_0 \sim 0 \text{ i.e. } x \sim \left(-\frac{\varepsilon_0}{a_3}\right)^{1/3}. \tag{1.2.4.2.8}$$

Stability: $$D\tilde{f}_x = 1 + 3a_3x^2 + \ldots \sim 1 + 3a_3\left(-\frac{\varepsilon_0}{a_3}\right)^{2/3} \tag{1.2.4.2.9}$$

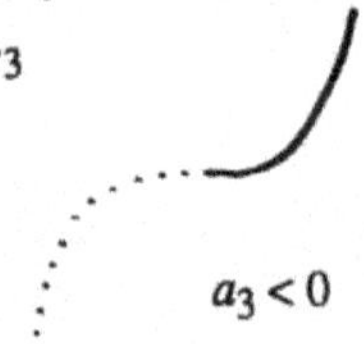

In cases with symmetry, not only is $a_2 = 0$, but also $\varepsilon_0 = 0$. Thus one finds:

Fixed points: $$a_3x^3 + \varepsilon_1 x \sim 0 \text{ i.e. } x = 0 \text{ and } x \sim \pm\left(-\frac{\varepsilon_1}{a_3}\right)^{1/2}. \tag{1.2.4.2.10}$$

Stability: $$D\tilde{f}_x = 1 + 3a_3x^2 + \varepsilon_1 = 1 + \varepsilon_1 \text{ and } 1 - 2\varepsilon_1 \text{ respectively} \quad (1.2.4.2.11)$$

This is a *pitchfork bifurcation.*

$a_3 < 0$

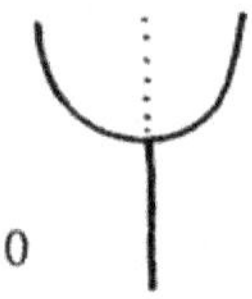

$a_3 > 0$

(ii) Multiplier -1. Again I put the fixed point of f at the origin.

$$f(x) = -x + a_2x^2 + a_3x^3 \ldots \qquad (1.2.4.2.12)$$

$$\delta f(x) = \varepsilon_0 + \varepsilon_1 x + \ldots \qquad (1.2.4.2.13)$$

Fixed point: $$2x = \varepsilon_0 \text{ i.e. } x = \varepsilon_0/2 \qquad (1.2.4.2.14)$$

Stability: $$D\tilde{f}_x = -1 + 2a_2x + \varepsilon_1 + \ldots \sim -1 + a_2\varepsilon_0 + \varepsilon_1 \qquad (1.2.4.2.15)$$

Period 2: $$\tilde{f}^{\,2}(x) = -(-x + a_2x^2 + a_3x^3 + \ldots + \varepsilon_0 + \varepsilon_1 x + \ldots)$$

$$+ a_2(x^2 - 2a_2x^3 - 2\varepsilon_0 x + \ldots) - a_3x^3 + \varepsilon_0 + \varepsilon_1(-x) + \ldots \qquad (1.2.4.2.16)$$

$$= x - 2(a_3 + a_2^2)x^3 - 2(\varepsilon_1 + a_2\varepsilon_0)x + \ldots$$

Hence period 2 points: $$x \sim \pm\left(-\frac{\varepsilon_1 + a_2\varepsilon_0}{a_3 + a_2^2}\right)^{1/2}. \qquad (1.2.4.2.17)$$

Stability: $$D\tilde{f}^{\,2} = 1 - 6(a_3 + a_2^2)x^2 - 2(\varepsilon_1 + a_2\varepsilon_0) + \ldots$$

$$\sim 1 + 4(\varepsilon_1 + a_2\varepsilon_0). \qquad (1.2.4.2.18)$$

This is the *period doubling* or *flip* bifurcation. It produces a stable 2-cycle or unstable 2-cycle according as:

$$a_3 + a_2^2 \gtrless 0. \qquad (1.2.4.2.19)$$

These two cases are known as the *direct* and *inverse* period doubling bifurcations, respectively. They are illustrated below.

direct

inverse

§1.2.4.3 Normal Forms. In the analysis of period doubling, since one is guaranteed persistence of a fixed point with no multipliers +1, I could have shifted it to the origin not only in the critical case f but also in its perturbation $\tilde{f}$. Thus I could choose coordinates to make $\varepsilon_0 = 0$. Note that this would not be possible in general for the tangent bifurcation, as one is not guaranteed persistence there.

Furthermore, given

$$x' = g(x) = a_1 x + a_2 x^2 + a_3 x^3 + \dots, \text{ with } a_1 \neq 0, +1 \qquad (1.2.4.3.1)$$

the coordinate change:

$$y = x + \alpha x^2, \ \alpha = \frac{a_2}{a_1(1 - a_1)} \qquad (1.2.4.3.2)$$

transforms it to

$$y' = h(y) = a_1 y + \left(a_3 + \frac{2a_2^2}{1 - a_1} \right) y^3 + \dots, \qquad (1.2.4.3.3)$$

eliminating the quadratic term. So for any one dimensional map with a fixed point with multipliers not equal to 0 or +1, there is a coordinate change to the *normal form*:

$$x' = a_1 x + a_3 x^3 + \dots \qquad (1.2.4.3.4)$$

The use of normal forms often greatly simplifies the analysis of bifurcations. For example I will redo the period doubling analysis in one dimensional maps using the above normal form.

$$f(x) = -x + a_3 x^3 + \dots$$
$$\delta f(x) = \varepsilon_1 x + \dots \qquad (1.2.4.3.5)$$

Fixed point: $$x = 0 \qquad (1.2.4.3.6)$$

Stability: $$D\tilde{f}_x = -1 + \varepsilon_1 \quad (1.2.4.3.7)$$

Period 2: $$\tilde{f}^{\,2}(x) = -(-x + a_3x^3 + \varepsilon_1 x) + a_3(-x^3) + \varepsilon_1(-x)$$

$$= x - 2a_3x^3 - 2\varepsilon_1 x \quad (1.2.4.3.8)$$

Period 2 points: $$x = \pm\left(-\frac{\varepsilon_1}{a_3}\right)^{1/2} \quad (1.2.4.3.9)$$

Stability: $$D\tilde{f}^{\,2}_x = 1 - 6a_3x^2 - 2\varepsilon_1 = 1 + 4\varepsilon_1. \quad (1.2.4.3.10)$$

Normal forms will be important in the analysis of bifurcations in area preserving maps.

§1.2.4.4 Area preserving maps. For periodic orbits of an area preserving map T, all roots of unity are attainable. Remember that the multipliers are given by the trace t of the derivative at any point x of the orbit (1.2.1.4.2):

$$t = \mathrm{Tr}\, DT^n_x \quad (1.2.4.4.1)$$

Before analysing the bifurcations, I describe a useful trick (Meyer, 1970) to save work in evaluating t when the multipliers are close to ± 1. The problem is that often the first order effects on the diagonal terms of DT cancel in evaluating t, so it would appear to be necessary to calculate them to higher order. To avoid this, one can use the off-diagonal terms, as follows. Writing

$$DT^n_x = \begin{vmatrix} A & B \\ C & D \end{vmatrix} \quad (1.2.4.4.2)$$

area preservation implies that $AD - BC = 1$. Thus

$$t^2 = (A + D)^2 = 4AD + (A - D)^2 = 4 + 4BC + (A - D)^2. \quad (1.2.4.4.3)$$

So when t is close to ± 2, it is given to leading order by

$$t = \pm 2\left(1 + \frac{BC}{2} + \frac{(A - D)^2}{8} + \ldots\right). \quad (1.2.4.4.4)$$

The bifurcations will be illustrated by sketches in the text, and also in figure 1.2.4.4.3.

(i) Multipliers +1. Put the linear terms of T into Jordan normal form. The typical case is non-diagonal. So take

$$\begin{aligned} x' &= x + y + a_1 x^2 + \dots + \varepsilon_0 + \dots \\ y' &= \quad\; y + a_2 x^2 + \dots + \eta_0 + \dots \end{aligned} \qquad (1.2.4.4.5)$$

Although area preservation imposes restrictions on the quadratic terms as a whole, it is easy to check that a_1 and a_2 are unrestricted, so typically non-zero.

The maximal balance is $$x^2 \sim y \sim \varepsilon_0 \sim \eta_0 \qquad (1.2.4.4.6)$$

Fixed points: $$a_2 x^2 + \eta_0 = 0 \ \text{ so } \ x = \pm\left(-\frac{\eta_0}{a_2}\right)^{1/2} \qquad (1.2.4.4.7)$$

$$y = -\varepsilon_0 - a_1 x^2 = -\varepsilon_0 + \frac{a_1}{a_2}\eta_0 \qquad (1.2.4.4.8)$$

Stability: $$DT = \begin{vmatrix} 1 + 2a_1 x & 1 \\ 2a_2 x & 1 \end{vmatrix}. \qquad (1.2.4.4.9)$$

I cannot evaluate t directly from this, as I have left out a lot of the quadratic terms. However, the above expression contains enough to use the trick (1.2.4.4.4). Hence,

$$t = +2\,(1 + a_2 x + \dots) = 2\,(1 \pm \sqrt{-\eta_0 a_2}). \qquad (1.2.4.4.10)$$

This is a *tangent bifurcation*.

$a_2 < 0$

(ii) Multipliers -1. For this analysis I follow Meyer (1970). Put the linear terms in Jordan normal form, and wlog (by persistence, §1.2.4.1) keep the fixed point at 0. I assume non-diagonal Jordan normal form.

$$\begin{aligned} x' &= -x + y + ax^2 + \dots \\ y' &= -y + bx^2 + cxy + dx^3 + \dots + \varepsilon x + \dots \end{aligned} \qquad (1.2.4.4.11)$$

Fixed point: $(0, 0)$

Stability:
$$DT = \begin{vmatrix} -1 + 2ax & +1 \\ 2bx + cy + 3dx^2 + \varepsilon & -1 + cx \end{vmatrix} = \begin{vmatrix} -1 & 1 \\ \varepsilon & -1 \end{vmatrix}. \quad (1.2.4.4.12)$$

Using the trick again,

$$t = -2\,(1 + \varepsilon/2). \quad (1.2.4.4.13)$$

Now I look at the second iterate. I expect a maximal balance $x^2 \sim y \sim \varepsilon$.

$$x'' = -(x + y + ax^2) + (-y + bx^2) + a(x^2) + 0(\varepsilon^{3/2})$$

$$= x - 2y + bx^2 + 0(\varepsilon^{3/2}) \quad (1.2.4.4.14)$$

$$y'' = -(-y + bx^2 + cxy + dx^3 + \varepsilon x) + b(x^2 - 2xy - 2ax^3)$$

$$+ c\,(-x)\,(-y + bx^2) + d(-x^3) + \varepsilon(-x) + 0(\varepsilon^2)$$

$$= y - (2d + 2ab + bc)\,x^3 - 2bxy - 2\varepsilon x + 0(\varepsilon^2) \quad (1.2.4.4.15)$$

Period 2 points:
$$2y = bx^2 + 0(\varepsilon^{3/2})$$

$$(2d + 2ab + bc + b^2)\,x^2 = -2\varepsilon + 0(\varepsilon^{3/2}) \quad (1.2.4.4.16)$$

Stability:
$$DT^2 = \begin{vmatrix} 1 + 2bx & -2 \\ -3(2d + 2ab + bc)x^2 - 2by - 2\varepsilon & 1 - 2bx \end{vmatrix} \quad (1.2.4.4.17)$$

$$t = 2\,(1 + 3x^2(2d + 2ab + bc) + 2by + 2\varepsilon + 2b^2x^2) = 2\,(1 - 4\varepsilon). \quad (1.2.4.4.18)$$

This is a *period doubling bifurcation*. It is *direct* or *inverse* according as:

$$2d + 2ab + bc + b^2 \lessgtr 0 \quad (1.2.4.4.19)$$

A typical direct period doubling bifurcation will be shown in figure 3.1.2.1.

Takens (Meyer, 1975) showed that in fact the quadratic terms can all be removed by coordinate change at multipliers -1, which somewhat simplifies the analysis. Diagonal Jordan normal form is a special case which gives different bifurcations. It occurs, for example, in maps with a square root (see also §1.2.4.7).

(iii) Higher order bifurcations. Now I consider bifurcations from a point whose multipliers are $e^{\pm 2\pi i m/n}$, with m, n coprime and $n > 3$. Following Meyer (1970), when $n \geq 3$ there is a coordinate change locally to put any area preserving map with a fixed point whose multipliers λ satisfy:

$$\lambda^i \neq 1, \quad 0 < i < n \tag{1.2.4.4.20}$$

into the following normal form:

$$\theta' = \theta + \alpha + \sum_{j=1}^{[(n-2)/2]} \beta_j \rho^j + \gamma \cos n\theta \, \rho^{(n-2)/2} + o(\rho^{(n-2)/2}) \tag{1.2.4.4.21}$$

$$\rho' = \rho + 2\gamma \sin n\theta \, \rho^{n/2} + o(\rho^{n/2}).$$

Here, (ρ, θ) are "canonical polars", cf. (1.1.3.1.3), i.e.

$$x = (2\rho)^{1/2} \cos \theta, \quad y = (2\rho)^{1/2} \sin \theta \tag{1.2.4.4.22}$$

and I adopt the convention that $\rho, \rho^{\frac{1}{2}} > 0$. The multipliers of the fixed point at 0 are $e^{\pm i\alpha}$. The coefficients β_j are independent of the order n of the normal form (subject to (1.2.4.4.20)), but γ does depend on n. The normal form can be derived along the lines of Birkhoff (1927, §III.8). See for example Siegel and Moser (1971, §23), Chenciner (1981) and Arnold (1980).

To analyse the case of multipliers close to a prime n^{th} root of unity, I take the normal form to order n, with

$$\alpha = 2\pi m/n + \varepsilon \tag{1.2.4.4.23}$$

and look at T^n. Although the coefficients β_j, γ will also be perturbed, the important change is in α. I will write β_1 as β. Note that T^n has diagonal normal form, so we will not get the case (i) above.

$n = 3$

$$\theta^{(3)} = \theta + 3\varepsilon + 3\gamma \cos 3\theta \, \rho^{\frac{1}{2}} + \ldots$$

$$\rho^{(3)} = \rho + 6\gamma \sin 3\theta \, \rho^{3/2} + \ldots \tag{1.2.4.4.24}$$

Thus there are period 3 points given by:

$$\sin 3\theta = 0 \text{ and } \rho^{\frac{1}{2}} = -\frac{\varepsilon}{\gamma}\cos 3\theta \qquad (1.2.4.4.25)$$

$$\text{i.e. } \rho = \left(\frac{\varepsilon}{\gamma}\right)^2 \text{ and } \theta = 0,\ \pi/3, 2\pi/3 \text{ for } \varepsilon,\ \gamma \text{ of opposite signs} \qquad (1.2.4.4.26)$$

$$\text{and } \theta = \pi/6,\ \pi/2, 5\pi/6 \text{ for } \varepsilon,\ \gamma \text{ of the same sign}$$

Stability: Evaluating the derivative in canonical polars,

$$DT^3 = \begin{vmatrix} 1 - 9\gamma \sin 3\theta\, \rho^{\frac{1}{2}} & 3/2\ \gamma \cos 3\theta\, \rho^{-\frac{1}{2}} \\ 18\, \gamma \cos 3\theta\, \rho^{3/2} & 1 + 9\gamma \sin 3\theta\, \rho^{\frac{1}{2}} \end{vmatrix} \qquad (1.2.4.4.27)$$

$$t = +2\,(1 + 27/2\ \gamma^2 \cos^2 3\theta\, \rho) = 2\,(1 + 27/2\, \varepsilon^2). \qquad (1.2.4.4.28)$$

So we get a bifurcation diagram like:

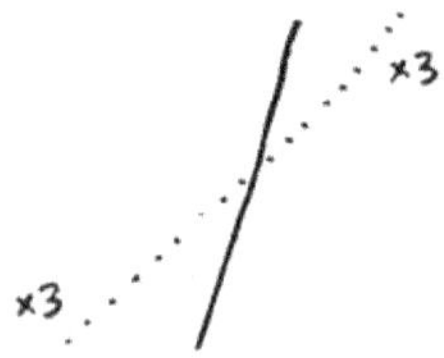

Note that often one sees the 3-cycle undergo tangent bifurcation a little further away, but this is outside the range of a local analysis.

$\underline{n = 4}$

$$\theta^{(4)} = \theta + 4\varepsilon + 4\gamma \cos 4\theta\, \rho + 4\beta\rho + \ldots$$
$$\rho^{(4)} = \rho + 8\gamma \sin 4\theta\, \rho^2. \qquad (1.2.4.4.29)$$

Period 4 points:
$$\sin 4\theta = 0 \text{ and } \rho = \frac{-\varepsilon}{\beta + \gamma \cos 4\theta}. \qquad (1.2.4.4.30)$$

Thus, for $|\beta| > |\gamma|$, we get two periodic orbits for ε and β of opposite sign, and none for the same sign:

$$\rho = \left| \frac{\varepsilon}{\beta \pm \gamma} \right|, \ \theta = k\pi/2, (k+\tfrac{1}{2})\pi/2 \ \text{according as } \pm \qquad (1.2.4.4.31)$$

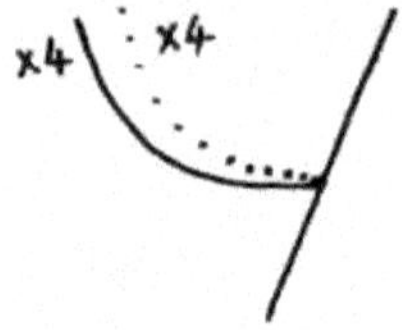

When $|\beta| < |\gamma|$ we get one 4–cycle when ε, γ have opposite signs:

$$\rho = \left| \frac{\varepsilon}{\gamma + \beta} \right|, \ \theta = k\pi/2 \qquad (1.2.4.4.32)$$

and one when they have the same sign:

$$\rho = \left| \frac{\varepsilon}{\gamma - \beta} \right|, \ \theta = (k+\tfrac{1}{2})\,\pi/2 \qquad (1.2.4.4.33)$$

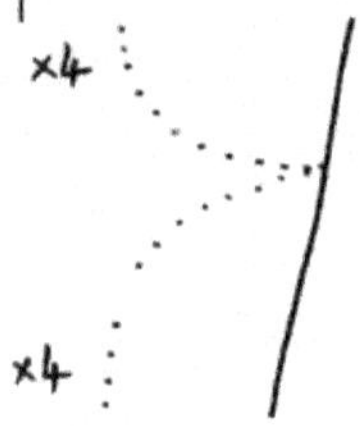

Stability:

$$t = 2\,(1 + 64\,\gamma \cos 4\theta\, \rho^2\, (\gamma \cos 4\theta + \beta)) \qquad (1.2.4.4.34)$$

$$= 2\,(1 + 64\,\frac{\gamma \cos 4\theta}{\gamma \cos 4\theta + \beta}\,\varepsilon^2).$$

Hence the above bifurcation diagrams. Again one often sees a tangent bifurcation on one branch in the second case. This gives pretty pictures like figures 1.2.4.4.1, 1.2.4.4.2.

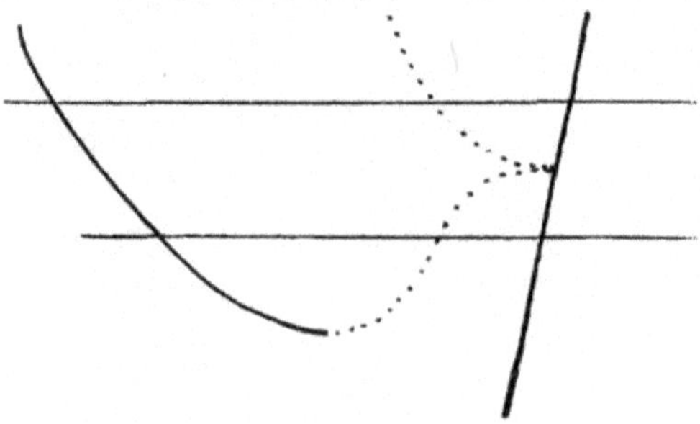

figure 1.2.4.4.2

figure 1.2.4.4.1

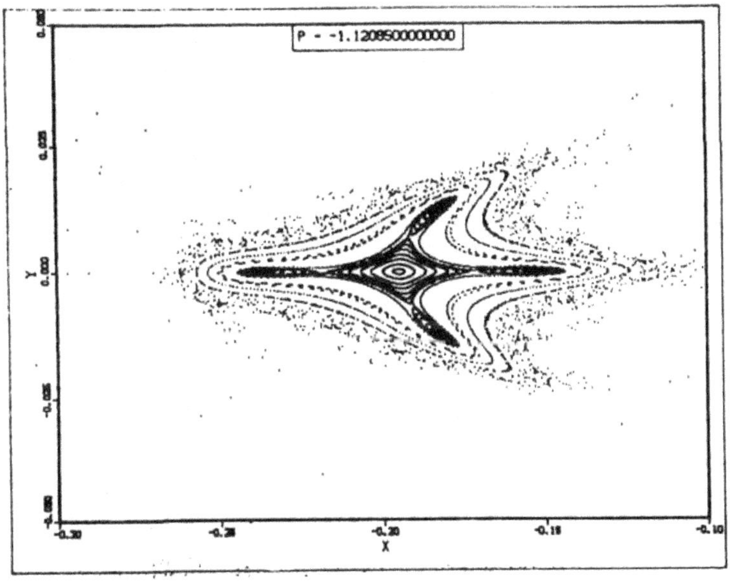

Figure 1.2.4.4.1: Just before a 1/4-bifurcation

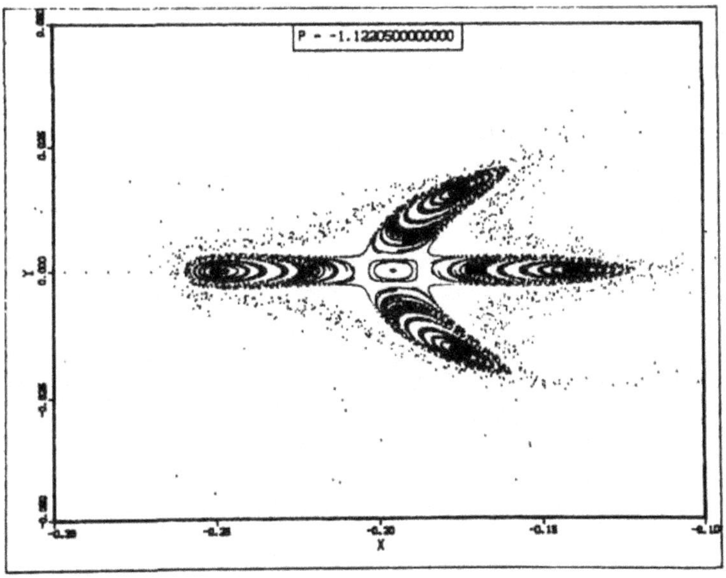

Figure 1.2.4.4.2: Just after a 1/4-bifurcation

$\underline{n > 5}$

$$\theta^{(n)} = \theta + n\varepsilon + n\beta\rho$$

$$\rho^{(n)} = \rho + 2n\gamma \sin n\theta\, \rho^{n/2} \qquad (1.2.4.4.35)$$

Period n points: $$\sin n\theta = 0 \text{ and } \rho = -\frac{\varepsilon}{\beta} \qquad (1.2.4.4.36)$$

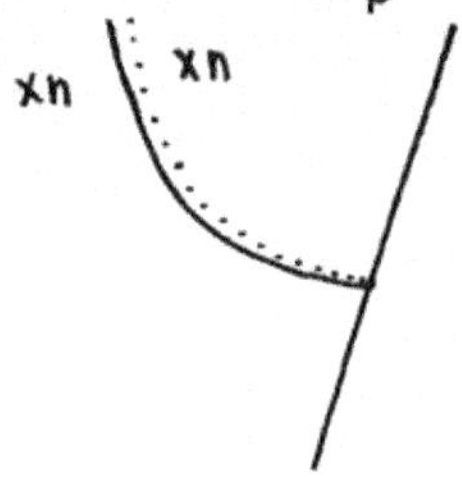

Stability:

$$t = 2\,(1 + n^3\,\gamma\beta \cos n\theta\, \rho^{n/2}) = 2\left(1 + n^3\,\gamma\beta \cos n\theta \left(\frac{-\,\varepsilon}{\beta}\right)^{n/2}\right). \qquad (1.2.4.4.37)$$

These are all the generic bifurcations in area preserving maps. Some of them are sketched (from cases in the quadratic map) in figure 1.2.4.4.3. One can label the periodic orbits which arise by bifurcation other than tangent bifurcation, by their ancestry, i.e. the sequence of "branching ratios" k/m by which they were produced. Thus, given a periodic orbit π, its daughters by k/m-bifurcation are labelled as $\pi : k/m$. A plus or minus may be added to indicate the sign of the residue (equivalently the index) of the orbit.

In special cases one might see different bifurcations from the above. This happens in reversible maps, to be discussed in §1.2.4.6, and in maps with simple commutators (see §1.2.4.7). First I discuss other approaches to bifurcations in conservative systems.

§1.2.4.5 Other approaches in conservative systems. Meyer has two ways of analysing the bifurcation at multipliers +1. In the first (Meyer, 1970) he represents area preserving maps by Poincaré's generating function, and in the second (Meyer, 1975) he associates a Hamiltonian flow with the map. In both cases he shows that elliptic fixed points correspond to extremes, and hyperbolic fixed points to saddles. Hence the fixed points of an area preserving map can be analysed by looking at the critical points of functions of two variables.

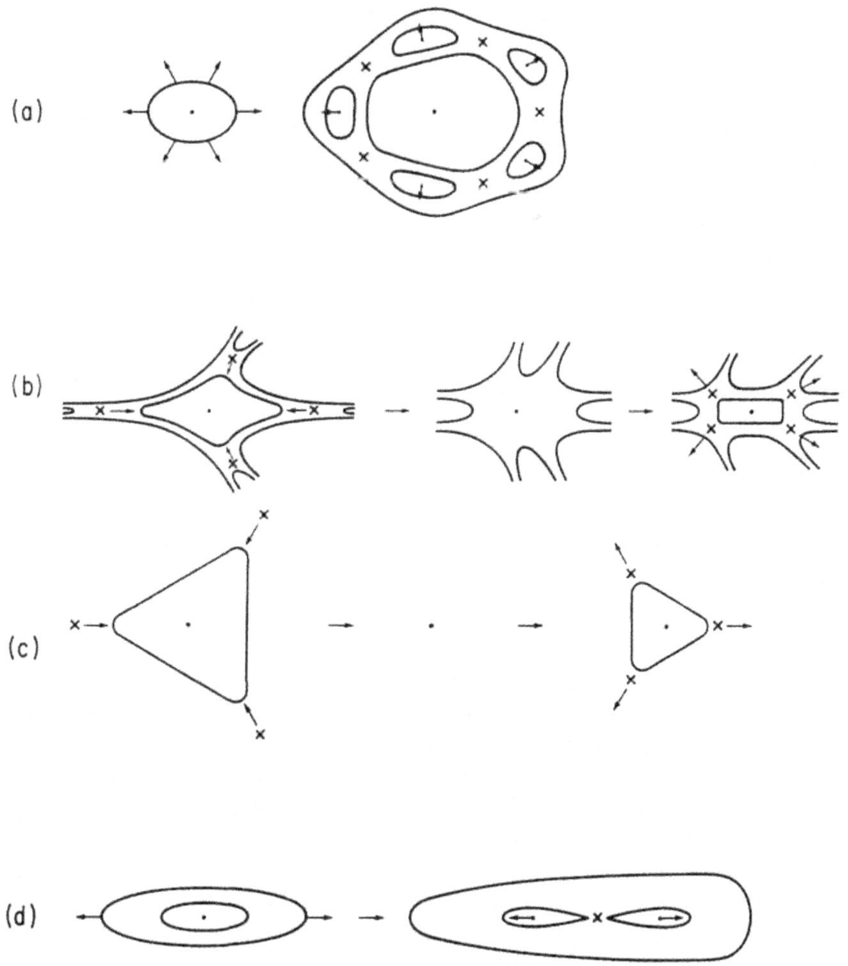

Figure 1.2.4.4.3: Typical bifurcations of an elliptic fixed point, showing the main periodic points and some representative invariant curves associated with bifurcations of orders a) 1/5 (or 2/5); b) 1/4; c) 1/3; d) 1/2.

Arnold (1978, App. 7D and F)[11] considers periodically time dependent Hamiltonian systems, and uses a truncation of Moser's normal form (Moser, 1958, lemma 4, p. 109) for the Hamiltonian in a neighbourhood of a periodic orbit with frequencies β:

$$H(\zeta, \bar{\zeta}, t) = \Sigma\, \gamma_{k\ell} \exp\left(-\, i s_{k\ell} t\right) \zeta^k\, \bar{\zeta}^\ell \qquad (1.2.4.5.1)$$

where the sum extends over k, ℓ such that

$$s_{kl} \equiv \beta.(k-\ell) \in \mathbb{Z}. \qquad (1.2.4.5.2)$$

Moving into a rotating frame permits one to remove one more time dependent term. Neglecting the higher order terms, H becomes time independent and hence simpler. For example, a system with one degree of freedom becomes integrable, and the solution can be illustrated by drawing the level curves of the Hamiltonian.

§1.2.4.6 Reversible maps. Rimmer (1978, 1979) analysed bifurcations of symmetric periodic orbits in reversible area preserving maps. The major difference comes in the case of multipliers $+1$, when there are three typical cases (all branches shown are fixed points):

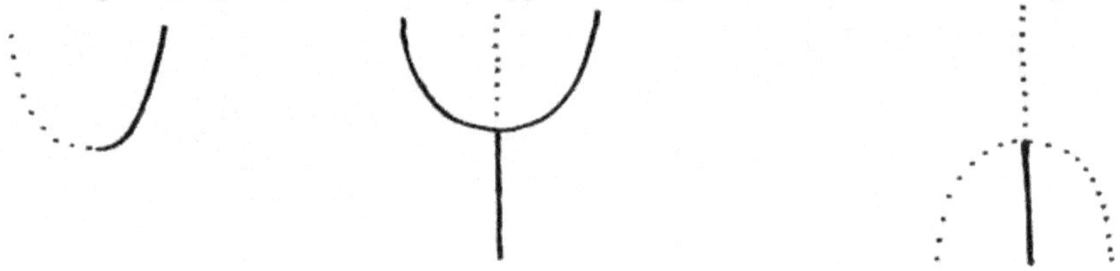

tangent direct Rimmer inverse Rimmer

All periodic orbits produced by generic bifurcations of a symmetric periodic orbit are also symmetric except the daughters of the Rimmer bifurcations.

One can see a Rimmer bifurcation in the quadratic map, for example. Counting periodic orbits (§1.2.3.5) shows that there are two 6-cycles belonging to QS. One of them is born by 1/6-bifurcation from the fixed point and has negative index, so does not bifurcate. The other is born at $p = -\frac{1}{2}$ (in the parametrisation of (3.1.2.1)) by period doubling of a 3-cycle, and is elliptic until $p = -0.5104$. Its residue increases to a maximum, and then returns through 1 (at $p = -0.9959$) and through 0 (at $p = -1$) to negative values. At residue 1, it gives birth to a hyperbolic 12-cycle (inverse period doubling), and at residue 0 to a pair of unsymmetric elliptic 6-cycles (direct Rimmer bifurcation).

As reversibility alone implies that the product of the multipliers of any symmetric periodic orbit is 1 (apart from a trivial case), one might ask what happens if one drops area preservation. One finds that there is no change in the bifurcation at multipliers ± 1 (see §3.1.1 for period doubling), but I have not yet verified the normal form for higher roots of unity.[12]

[11] This approach was also developed by Takens (1974).

[12] This has been done by Roberts and Quispel (1992).

§1.2.4.7 Maps with simple commutators/Anomalies of the standard map. Various authors (Benettin et al., 1980a, Schmidt and Bialek, 1982, Greene, private communication, Karney, private communication, Bak and Jensen, 1981) have reported anomalous bifurcations of periodic orbits in generalised standard maps. A common type of non-generic bifurcation is what I call the *double k/m-bifurcation*. It is simplest described as the bifurcation which occurs in the square of a map undergoing normal $k/2m$-bifurcation. Thus for $m \geq 5$, for example, the periodic orbit gives birth to twice as many periodic orbits as generically. The Rimmer bifurcation can be considered as a double $1/1$-bifurcation, because it looks like a period doubling bifurcation, but the orbits it produces have the same period as the parent.

Observations for standard map can be summarised as follows:

(i) The fixed point at 0 and the 2-cycle at $(0, \pi)$, (π, π) have normal k/m bifurcations at multipliers $e^{\pm 2\pi i \nu}$, when $\nu = k/m, k, m$ coprime, m even, but double k/m-bifurcations when m is odd.

(ii) Their daughters of period an even multiple of the parent behave as follows: the multipliers, born at $+1$, travel round the unit circle as a complex conjugate pair, pass through each other at -1, with change of sign of the Krein signature, continue round to $+1$, and then split along the positive real line. At this transition a pair of stable periodic orbits of the same period as their parent is born (Rimmer bifurcation). Furthermore, writing the multipliers as $e^{\pm 2\pi i \nu}$ when on the unit circle, with ν determined continuously by:

$$2 \cos 2\pi\nu = \text{Trace } DT$$

$$\text{sign}(\sin 2\pi\nu) = \sigma \quad (\cos 2\pi\nu \neq \pm 1) \tag{1.2.4.7.1}$$

$$\nu = 0 \text{ at birth}$$

one finds normal k/m-bifurcations as ν passes through $k/m, k, m$ coprime, k even, but double k/m-bifurcations when k is odd. The behaviours at multipliers ± 1 are in fact also included in this statement as the cases $\nu = 1/2$ and $1/1$, respectively.

Double k/m-bifurconre not stable to perturbation. For $m > 2$ they can be split into a normal k/m-bifurcation and a tangent bifurcation as in figure 1.2.4.7.1a,b. For $m = 2,1$ they can be split as in figure 1.2.4.7.1c,d. If one restricts one's attention to symmetric periodic orbits of reversible maps, this is still true, with the exception of the case $m = 1$, since double $1/1$-bifurcations are stable to reversible perturbations.

I will explain these oddities in terms of a commutator for the standard map. It is well-known that the standard map can be considered as acting not on a plane, but on a torus, since it is invariant under the coordinate changes:

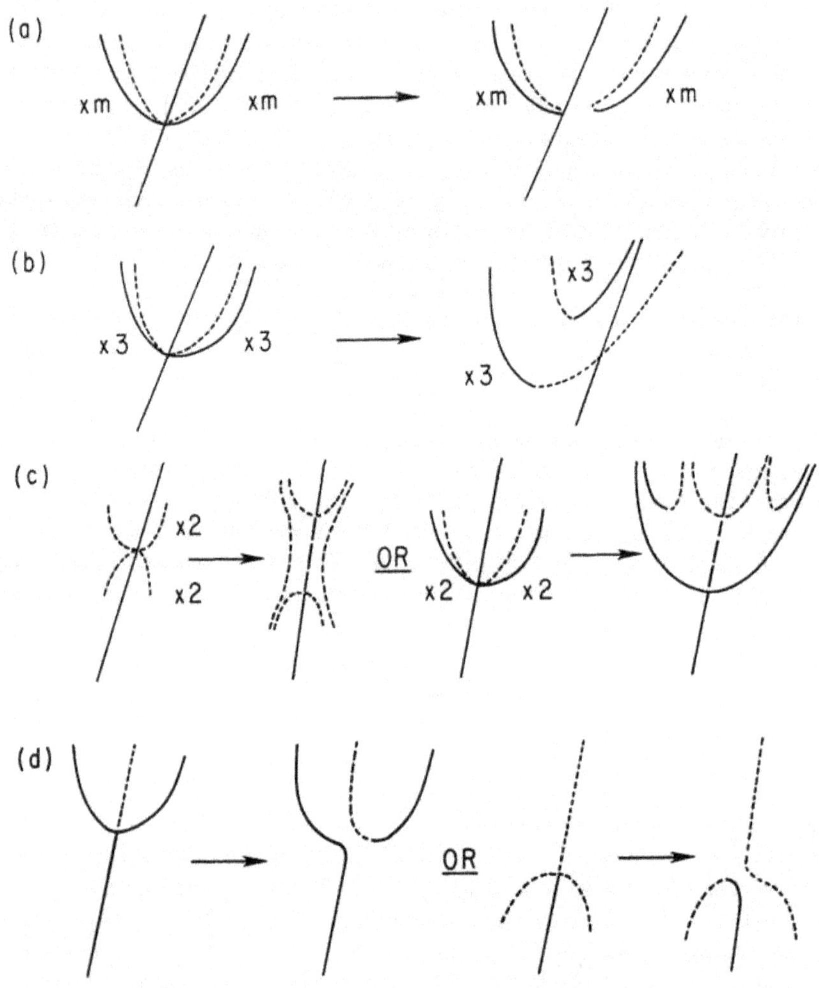

Figure 1.2.4.7.1: Splitting of double m-bifurcations for a) $m > 4$, b) $m = 3$, c) $m = 2$, d) $m = 1$ ($m = 4$ can resemble either a) or b)), into:
a), b) a normal m-bifurcation and a tangent bifurcation
c) two normal 2-bifurcations and a tangent bifurcation
d) a tangent bifurcation and a point of constant index
Continuous line indicates elliptic, dotted line regular hyperbolic, and dashed line inversion hyperbolic.

$$C_{mn}: y' = y + 2m\pi \qquad m, n \in \mathbb{Z}$$
$$x' = x + 2n\pi. \tag{1.2.4.7.2}$$

Equivalently, it commutes with C_{mn}:

$$C_{mn} T = T C_{mn}. \tag{1.2.4.7.3}$$

The coordinate changes C_{mn} are examples of *commutators*. It has another commutator, however, possessed by all generalised standard maps with f, g odd:

$$x' = x + f(y)$$
$$y' = y + g(x') \tag{1.2.4.7.4}$$

viz.

$$Z: x' = -x, \ y' = -y. \tag{1.2.4.7.5}$$

This is responsible for many nongeneric bifurcations.

I say C is a *commutator* of T if C is invertible and

$$C^{-1} T C = T. \tag{1.2.4.7.6}$$

A commutator generates a group $\{C^n: n \in \mathbb{Z}\}$ of commutators. One could have a group of commutators with more than one generator e.g. $\{C_{mn}: m, n \in \mathbb{Z}\}$ for the standard map. Given a group G on a set M, one can define the quotient M/G as the set of equivalence classes x_G of M, under the equivalence relation:

$$x \sim y \text{ if } \exists C \in G \text{ s.t. } x = Cy. \tag{1.2.4.7.7}$$

Thus M/G is the set of orbits of G on M, or equivalently M/G is M with points identified if they are related by an element of G. For example,

$$\mathbb{R}^2/\{C_{mn}\} \sim \mathbb{T}^2, \text{ a 2-torus.} \tag{1.2.4.7.8}$$

If a map T on M has a group G of commutators, then it induces a map T_G on M/G, by

$$T_G x_G = (Tx)_G. \tag{1.2.4.7.9}$$

So one can consider T as acting on M/G.

If M is a manifold, then M/G can often be a manifold. The only problems come at points x_G such that x is a fixed point of some element of G. Then M/G may not be differentiable there (it could have a "corner"), or depending on your point of view, T_G may be non-differentiable there. For example, T is a commutator for T, and T_T (letting T stand for the group it generates) is the identity, but in general M_T will be very complicated.

Thus one could consider the generalised standard maps with f, g odd as maps on $\mathbb{R}^2/Z$. This is equivalent to $\mathbb{R}^2$ except for a corner at 0. In the case of the standard map one can consider the group G of commutators generated by Z, C_{01}, C_{10}. One finds that

$$\mathbb{R}^2/G \sim S^2, \text{ a 2-sphere, but with 4 corners.} \tag{1.2.4.7.10}$$

This can be seen by identifying points of a torus which are related by Z, as in figure 1.2.4.7.2. Note that $R_1 = R_2, S_1 = S_2$ under this identification, so the symmetries reduce to a single family. Thus except when dealing with the corner points, I propose that one should think of the orbits of the standard map as living on this sphere.

The commutator Z is an involution. Suppose an orbit $\{T^n x: n \in \mathbb{Z})$ is Z-symmetric, i.e.

$$\{ZT^n x: n \in \mathbb{Z}\} = \{T^n x: n \in \mathbb{Z}\}. \tag{1.2.4.7.11}$$

Then

$$Zx = T^m x, \text{ for some } m. \tag{1.2.4.7.12}$$

Thus

$$T^{2m} x = x \text{ i.e. } x \text{ is peridoic.} \tag{1.2.4.7.13}$$

Suppose it has period p. Then without loss of generality, $0 \le m < p$. But $2m$ is a multiple of p, so there are two cases:

(a) Strongly symmetric orbit: $m = 0$ and the orbit of x is composed of fixed points of Z.

(b) Weakly symmetric orbit: $m = p/2$, p even, and $ZT^k x = T^{k+p/2} x, \ \forall k$.

Both cases have anomalous bifurcations. I discuss the strongly symmetric orbits first.

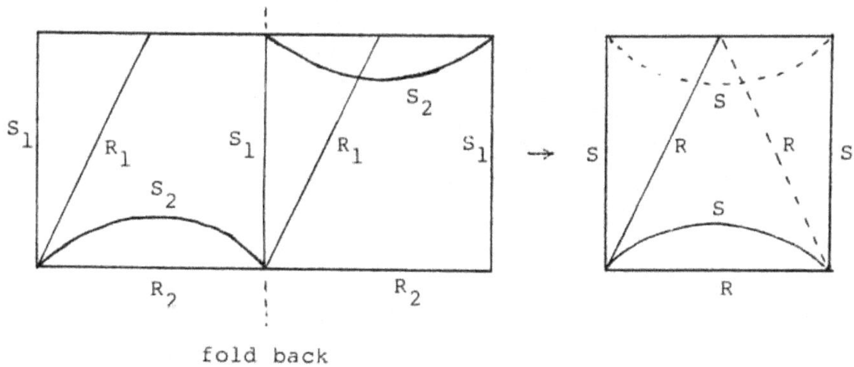

Figure 1.2.4.7.2: Identification of points on the torus related by Z. The top half of the unit square has been removed. The symmetry lines of the standard map are shown to help in visualisation (cf. figure 1.1.4.3.1). Dotted lines are on the back of the four-cornered sphere.

(i) Strongly symmetric orbits. The problem reduces to finding generic bifurcations from a fixed point 0 of area preserving maps with commutator Z. I suspect that generically at multipliers $e^{\pm 2\pi i\nu}$, $\nu = k/m$, k, m coprime, there is a normal k/m-bifurcation for m even, giving weakly symmetric daughters, and a double k/m-bifurcation for m odd, giving pairs of unsymmetric duaghters. I have not, however, worked through the analysis. Restricting to reversible maps where the symmetry commutes with Z as well, could affect the generic bifurcations for $m = 1$ and 2. The standard map has four fixed points under Z, giving two fixed points of T_Z and a 2-cycle. So this would explain their anomalous bifurcations.

(ii) Weakly symmetric orbits. We see from the previous section that generically all daughters of strongly symmetric periodic orbits with period an even multiple of their parent's are weakly symmetric. To analyse their bifurcations I consider the corresponding orbits of T_Z.

Every periodic orbit of T gives a periodic orbit of T_Z. Its period is the same except in the case of weakly symmetric orbits when it is halved. Conversely, if $x_Z, \ldots T_Z^{n-1} x_Z$ is a periodic orbit of T_Z, then

$$T^n x_Z = x_Z \Rightarrow \text{either i) } T^n = x \neq Zx$$

$$\text{or ii) } T^n x = Zx \neq x$$

$$\text{or iii) } T^n x = x = Zx \qquad (1.2.4.7.14)$$

In case i) the orbit lifts to two periodic orbits of the same period.
In case ii) it lifts to one weakly symmetric orbit of twice the period.
In case iii) it lifts to one strongly symmetric orbit of the same period.

If x is a periodic point of T of period n, then it is a fixed point of T^n, but not of any lower power of T. But if x belongs to a weakly symmetric orbit (so n is even), then x is also a fixed point of $ZT^{n/2}$, a square root for T^n. So generic behaviour of x as a fixed point of $ZT^{n/2}$ would lead to exactly the anomalous behaviour described earlier.

Next I make a few comments about connections between double reversibility (§1.1.4.3) and commutators, since the standard map has both. Double reversibility always gives a commutator, as R, S symmetries of T implies that RS is a commutator. For the standard map, using R_1 and R_2, R_1R_2 happens to be an involution, but this need not happen in general. Conversely, if a reversible map T, with symmetry S, has a commutaotr C, then SC is also a symmetry if it is an involution, so T is doubly reversible (the symmetries are independent provided $T^n \neq C$ for all integers n).

If T is the square of a reversible map Q, then T is doubly reversible (provided $Q^{2n-1} \neq 1, \forall n \in Z$), since:

$$Q = RS,\ R^2 = S^2 = 1 \Rightarrow T = R.SRS = RSR.S,$$

$$\text{and } R = T^nS \text{ iff } T^n = Q. \qquad (1.2.4.7.15)$$

The existence of a square root map is an important thing to notice for a map (see also Vivaldi and Ford, 1981).

Note that commutators other than involutions can be important. For example, the Hénon-Heiles system (1964) has a commutator of period 3, and one could do similar analysis to the above for its periodic orbits.[13]

I conclude by mentioning two other anomalies of the standard map, and their possible resolution by using the commutator Z. The first is to do with the breakup of invariant circles. As I will discuss in §1.3 and §4.4.1, the approximating periodic orbits to a critical noble circle, have residues converging to 0.250088... . This was not observed, however, for the case of circles surrounding the fixed point of the standard map. Instead the residues converged to a 3-cycle. This discrepancy can be removed by considering the appropriate orbits of T_Z.

The second anomaly is one found by Grebogi and Kaufman (1981). They find that the ratio of a correlation decay rate to a Liapunov exponent is roughly independent of λ for both the standard and Rannou maps, provided it is large enough, but is half as big for the former as for the latter. I would guess that this has something to do with the fact that $\mathbb{T}^2/Z$ has half the area of $\mathbb{T}^2$.

[13] For example, one can explain the anomalous period-4 cycle in the residues of periodic orbits appraoching a noble torus, observed by Greene (1979b), along the lines of MacKay (1984).

§1.3 INVARIANT CIRCLES

In this section, I discuss the second type of orbit commonly found in area preserving maps, namely, those which lie on invariant circles. In particular I discuss the KAM theorem which guarantees persistence of invariant circles for small enough perturbations in area preserving twist maps. Then I discuss some connections between invariant circles and the Birkhoff orbits, which have practical value in determining where there are invariant circles.

§1.3.1 **Invariant circles.** Invariant circles are essential for nonlinear stability or confinement in area preserving maps. Integrable maps have a foliation by invariant circles, but this is exceptional. The properties of an invariant circle depend crucially on its rotation number. Generically, there are no rational circles. But sufficiently irrational circles persist for small enough perturbation in periodic area preserving twist maps. I conclude with two variational principles for quasiperiodic orbits. If there is not a circle of the given rotation number, these principles still give a quasiperiodic orbit. Its closure, however, is a Cantor set.

§1.3.1.1 Confinement. In pictures such as figures (i) and 1.3.2.1.1 are visible orbits which appear to fill up a closed curve, topologically equivalent to a circle. Such *invariant circles* are very significant, because they separate the surface (modulo questions of the degree of connectedness of the surface). If the orbit of one point stays inside, then so must the orbits of all points inside. Higher dimensional conservative systems have analogous *invariant tori*, but they do not separate the space, so they are not quite as significant. As mentioned in §1.1.3.1, many systems involving invariant circles can be expressed as periodic twist maps. Thus, for the whole of §1.3, I will restrict attention to the class A of end preserving, area preserving twist maps of a cylinder with zero Calabi invariant.

If a map of class A has an invariant circle encircling the cylinder, then the points below it form an invariant set. Conversely, Birkhoff (1932) (see also Mather, 1984) showed that an encircling invariant circle is necessary for confinement below some level of the orbits of all points below some lower level. More precisely,

Theorem (Birkhoff): For an area preserving map of a cylinder satisfying a *tilt* condition, the boundary of any open invariant set homeomorphic to $\mathbb{R} \times S^1$ and containing all points below some level, is the graph $\{(\theta, \zeta(\theta)) : \theta \in S^1\}$ of a Lipschitz function $\zeta(\theta)$.

So the boundary is an invariant circle. I will not state the tilt condition here, but it is implied by the twist condition.[14]

---

[14] To apply the theorem to a confined set U, take $\bigcup_{-\infty}^{\infty} f^n(U)$ and fill in the holes by taking the complement of the component of the complement containing the top end. This is homeomorphic to $\mathbb{R} \times S^1$.

In all that follows I will restrict attention to the class of invariant circles which encircle the cylinder. There can be others, as in island chains, but these can be treated by finding polar coordinates locally for which they are encircling. Note that zero Calabi invariant is necessary for existence of any circle of this class. An encircling invariant circle divides the cylinder into two invariant components. Applying Birkhoff's theorem to one of them shows that every encircling invariant circle is the graph of a Lipschitz function. In particular, it is single-valued.

§1.3.1.2 Integrable systems. There are systems which are foliated by invariant circles. A map of a surface is said to be *integrable* if it possesses a differentiable invariant function which is not constant on any open set. For example, an axisymmetric magnetic field has a *flux function* which is invariant, so its return map is integrable.

A typical extremum of the invariant of an integrable map is surrounded by many circles on which the invariant is constant. They are invariant under some power of the map, and so is the region between any two. Liouville (Arnold, 1978) showed (actually for continuous time!) that if the derivative of the invariant is non-singular on a compact connected invariant set, then there exist coordinates (θ, z), called *angle-action variables*, in which the map takes the standard integrable form:

$$(\theta', z') = (\theta + \omega(z), z) \tag{1.3.1.2.1}$$

So the set is foliated by the invariant circles $z =$ constant.

The goal of classical perturbation theory was to find coordinate changes to bring perturbations of such systems into the same form. This is impossible in general, however, because integrable maps are very special. To show this I need to introduce the concept of rotation number.

§1.3.1.3 Rotation number. If a map of a surface is a homeomorphism (i.e. continuous with continuous inverse), then so is its restriction to an invariant circle. Poincaré (Nitecki, 1971) showed that for an orientation preserving homeomorphism of g of a circle (or really, for a lift of g to a periodic homeomorphism of the line), the limit:

$$\omega = \lim_{q \to \infty} \frac{g^q(\theta_0)}{q} \tag{1.3.1.3.1}$$

exists for all initial points θ_0, and is the same for all θ_0. It is called the *rotation number* of the circle. Rotation number will be seen to be a crucial quantity for invariant circles.

Uniform rotation on a circle:

$$\theta' = \theta + \omega \tag{1.3.1.3.2}$$

clearly has rotation number ω, but one would like to know to what extent rotation number ω implies equivalence to rotation. Maps f, g are said to be *topologically/differentiably conjugate* if there exists a homeomorphism/diffeomorphism h such that:

$$hf = gh \tag{1.3.1.3.3}$$

i.e. they are the same apart from the coordinate change h (a diffeomorphism is a differentiable map with differentiable inverse).

For rational rotation number, conjugacy to rotation is not the typical case. For example, for rotation number p/q, in lowest terms, there could be a pair of periodic orbits of period q, one attracting and the other repelling. But uniform rational rotation has every point periodic, and none of them attracting or repelling. Incidentally, for rational rotation number p/q, there is always at least one periodic orbit of period q, and all periodic orbits have period q.

In the irrational case, the first answer came from Denjoy. He showed (Nitecki, 1971) that if g is an orientation preserving C^1-diffeomorphism of a circle with irrational rotation number, and g' has bounded variation, then g is topologically conjugate to rotation with that rotation number.[15] Thus the map is *transitive*, i.e. it has a dense orbit. In fact, every orbit is dense. Herman (1976) has shown much stronger conjugacy results under stronger differentiability conditions, for a set of rotation numbers of full measure (see §4.7.1). If the differentiability conditions are dropped, the result of Denjoy's theorem does not necessarily hold. There is another possibility, namely, the limit set could be a Cantor set on the circle. In this case g is said to be *intransitive*. A *Cantor set* is a non-empty compact set, with no isolated points, but which is totally disconnected. The standard example is Cantor's middle third set, but all Cantor sets are homeomorphic to each other. For further discussion of conjugacy of maps on a circle to rotation, see §4.7.1.

One can also define rotation number for some other orbits of a periodic map T than those on invariant circles. I say (θ, z) (or its orbit) has rotation number:

$$\lim_{q \to \infty} \frac{\pi_1 T^q(\theta, z)}{q} \tag{1.3.1.3.4}$$

if the limit exists, where π_1 is the projection onto the first coordinate:

$$\pi_1(\theta, z) = \theta. \tag{1.3.1.3.5}$$

Note that the limit need not exist. In §1.4.4, for example, I find an orbit of the standard map, for which the above quantity comes arbitrarily close to all points in $[0, 1]$. Note also

15 For a renormalisation proof, see MacKay (1988).

that this definition of rotation number is compatible with the previous definition for invariant circles, by Birkhoff's theorem above. Finally, note that rotation number depends on the lift chosen, but the only effect of changing lifts is to add an integer, the same for all orbits.

§1.3.1.4 Instability of rational circles. The invariant circles of the standard integrable map (1.3.1.2.1) have continuously varying rotation number $\omega(z)$. Furthermore, the motion on the circles is smoothly conjugate to rotation, even for rational rotation number. Thus the rational circles have every point periodic, and of residue 0. But generically all periodic points have non-zero residue and are isolated from points of the same period. So integrable maps are special.

In fact, generically there are no rational circles of any sort. This is because the only way to get a rational circle with no periodic points of residue zero, is by saddle connections, as in the sketch below.[16] But generically, there are no saddle connections (Robinson, 1970). They can be broken by arbitrarily small perturbations giving transverse intersection of the stable and unstable manifolds, a state which is stable to perturbation.

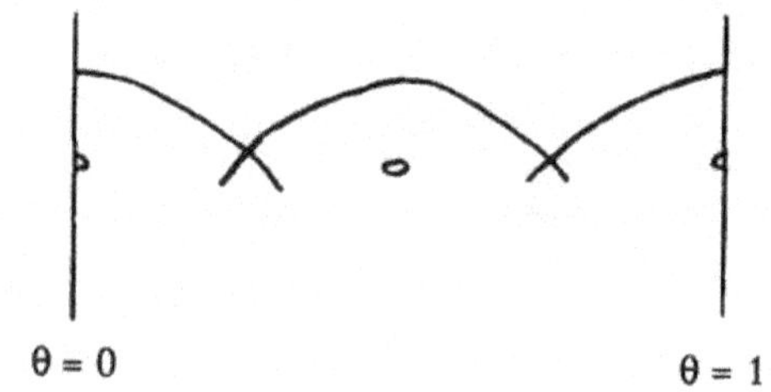

§1.3.1.5 Persistence of sufficiently irrational circles. Although integrable systems are very special, conditions close to integrable are very common. For example, by using the Birkhoff normal form of (1.2.4.4.21), one sees that a map is arbitrarily close to integrable, sufficiently near to any typical elliptic point. A remarkable theory, due to Kolmogorov (1954), Arnold (1963) and Moser (1962), shows that systems close enough to an integrable one possess an arbitrarily large fraction of the invariant circles of the integrable system. The particular result most relevant to this thesis is Moser's twist theorem (Moser, 1973).

First, I introduce some terminology. ω is called a *Diophantine number* (Niven, 1963), if:

$$\exists C>0,\ \tau \text{ such that } \left|\omega - \frac{p}{q}\right| > \frac{C}{q^{\tau}}\ \forall p, q \in \mathbb{Z},\ q>0. \quad (1.3.1.5.1)$$

One might think, at first, that there cannot be any Diophantine numbers, since given C, τ,

[16] Herman (1983) gives a more complete argument.

(1.3.1.5.1) excludes a neighbourhood of each rational. But in fact, for $\tau > 2$, the measure of the excluded set goes to zero as $C \to 0$, leaving most of the irrationals (see §4.7.8).

Theorem (Moser): Given $C > 0$, τ, and an integrable twist map (i.e. $\omega'(z) \neq 0$ in (1.3.1.2.1)), then all small enough C^r perturbations ($r > 2\tau-1$) in class A possess invariant circles close to the original ones for all rotation numbers satisfyfing the Diophantine condition (1.3.1.5.1).

For $\tau > 2$, this gives a set of invariant circles whose complement has arbitrarily small measure as $C \to 0$. Note that Moser (1973) also proved the same theorem for reversible twist maps, not necessarily area preserving. Proofs of results like this often also show that these invariant circles are smooth. This means more than just the circle being the graph of a smooth function. An invariant circle is called *smooth* if the motion on it is sufficiently differentiably conjugate to rotation. The number of derivatives depends on τ for a Diophantine rotation number, and the application. The circles of Moser's theorem, he showed to be at least C^1, but one can do a lot better. For example, Herman (1981, 1982) gets invariant circles with $\tau = 2$ for C^r, $r > 3$, which are C^{r-1}-conjugate to rotation. In the analytic case, the circles can be shown to be analytic (Arnold & Avez, 1968, App. 34).

Thus, for example, typical elliptic periodic orbits are stable (Meyer, 1971). This is because for an elliptic point with multipliers not a third or fourth root of unity there is a local coordinate change to Birkhoff normal form (1.2.4.4.21), and the coefficient β_1 exists. It is typically non-zero, so the Birkhoff normal form can be seen to be a perturbation of an integrable twist, the perturbation being arbitrarily small close enough to the fixed point. Thus Moser's theorem gives invariant circles arbitrarily close.

Given a smooth invariant circle of Diophantine rotation number ω in a C^r twist map, there is a coordinate change locally to the normal form:

$$\begin{aligned} \theta' &= \theta + \omega + z \\ z' &= z + z^n f(\theta, z) \end{aligned} \qquad (1.3.1.5.2)$$

for any $n \leq r$ (Mather, private communication). Thus close enough to any smooth Diophantine circle, a map looks close to integrable. This implies (modulo differentiability conditions) that:

Smooth Diophantine circles persist, for small enough perturbation in class A.

It implies many other results too, to which I will refer as needed.

On the other hand, there are maps of class A with no (encircling) invariant circles. For example, the standard map:

$$z' = z - \frac{k}{2\pi}\sin 2\pi\theta$$
$$\theta' = \theta + z' \tag{1.3.1.5.3}$$

has no invariant circles for $|k| > 2\pi$, because then it has an accelerator mode (Chirikov, 1979), that is, an orbit such that:

$$z_{n+1} = z_n + 1, \ \theta_{n+1} = \theta_n \text{ (mod 1)}. \tag{1.3.1.5.4}$$

Existence of an invariant circle would imply that no orbits can go from $-\infty$ to $+\infty$. In fact, Mather (1984) has proved non-existence of invariant circles for $|k| \geq 4/3$, by applying Birkhoff's theorem of §1.3.1.1. His result applies to the class of second difference equations:

$$\Delta^2 \theta_n \equiv \theta_{n+1} - 2\theta_n + \theta_{n-1} = h(\theta_n) \tag{1.3.1.5.5}$$

and states that (1.3.1.5.5) has no invariant circles unless:

$$m > -2 \tag{1.3.1.5.6}$$

and

$$\frac{2}{2+m} + \frac{2+m}{2} \leq M \qquad (1.3.1.5.7$$

where

$$m = \min h'$$
$$M = \max h'. \tag{1.3.1.5.8}$$

The size of the perturbations allowed by Moser's twist theorem depends only on C and τ in the Diophantine condition, and the local twist. It is largest for τ small and C large. By a scale change, the twist can be normalised to 1. For example, Rüssmann (1981) proves persistence of a Diophantine circle for perturbations of size less than somethinglike:

$$A_1 A_2^{\tau} \left(\frac{C}{\Gamma(\tau)} \right)^2 \tag{1.3.1.5.9}$$

for some constants $A_1, A_2 < 1$ ($\Gamma(\tau)$ is the gamma function).

In any interval, the number (or numbers) for which τ can be taken smallest and C largest is always a *noble* number (terminology due to Percival (1982)). These are the numbers whose *continued fraction expansion*:

$$\omega = m_0 + \cfrac{1}{m_1 + \cfrac{1}{m_2 + \dots}} = [\, m_0, m_1, m_2, \dots], \; m_i \in \mathbb{Z}, m_i \geq 1 \text{ for } i \geq 1 \tag{1.3.1.5.10}$$

has $m_i = 1$ for all large enough i. They satisfy a Diophantine condition with $\tau = 2$, the smallest possible. The noblest of them all is the *golden ratio*:

$$\gamma = [\,(1,)^{\infty}] = \frac{\sqrt{5}+1}{2} = 1.6180339\dots \tag{1.3.1.5.11}$$

which has the largest possible value for C (when $\tau = 2$) of γ^{-2} (Prasad, 1948). This leads one to suspect that noble circles may typically be the most robust, in the sense that the last circle to break up in any region, as a parameter varies, will be a noble. I will support this idea in §4.5.2.

The proofs in KAM theory generally give unrealistically low estimates of the perturbation sizes sufficient for persistence of invariant circles. For example, Rüssmann (1981), who obtains the strongest results of which I am aware, requires (as one of his conditions) that:

$$\sup \| D^4 \varphi \| < 0.19 \times 10^{-10} \tag{1.3.1.5.12}$$

For persistence of a golden circle, for maps in the normal form (1.3.1.5.13) to be discussed below. Thus he gets a golden circle in the standard map for $|k| < (2\pi)^{-3} \times 0.19 \times 10^{-10}$. But the standard map appears to have a golden circle for $|k| \leq 0.9716\dots$ (Greene, 1979a).[17] For a review of some approximate methods for determining the boundary, see Lichtenberg (1979).

In Chapter 4, I develop a new approach to KAM theory which, I believe, gives the boundary in class A of the set of maps with an invariant circle of given rotation number. I have worked this out only for the particular case of nobles, but other quadratic irrationals could be dealt with in the same way, and there are hopes that a modified treatment would apply to the rest of the Diophantines.[18]

---

[17] KAM estimates have been greatly improved by Herman (1986), Chierchia and Celletti (1988) and de la Llave & Rana (1990), for example.

[18] Indeed, this has been achieved by Haydn (1990).

Finally, note a useful (though not necessarily area-preserving) normal form for twist maps:

$$\begin{aligned} \theta' &= \theta + R \\ R' &= R + \varphi(\theta, R). \end{aligned} \tag{1.3.1.5.13}$$

Given the twist map:

$$\begin{aligned} \theta' &= \theta + t(r) + \varphi_1(\theta, r) \\ r' &= r + \varphi_2(\theta, r) \end{aligned} \tag{1.3.1.5.14}$$

the coordinate change:

$$R = t(r) + \varphi_1(\theta, r) \tag{1.3.1.5.15}$$

is invertible provided:

$$\left| \frac{\partial \varphi_1}{\partial r} \right| < | t'(r) | \tag{1.3.1.5.16}$$

and transforms it into the above normal form.

§1.3.1.6 Variational principles for quasiperiodic orbits. In this section, I discuss a couple of generalisations of the variational principle for periodic orbits, described in §1.2.2, to quasiperiodic orbits. Analogues of the Poincaré-Birkhoff theorem prove existence of quasiperiodic orbits, with all irrational rotation numbers in the range of the twist. In the case that such a quasiperiodic orbit does not lie on an invariant circle, its closure is a Cantor set, called a *cantorus* by Percival.

Percival (1979, 1980) introduced a variational principle for invariant circles, which has been greatly developed by Mather (1982b, 1986 and others). A transitive invariant circle of a map with action $\tau(\theta, \theta')$, with rotation number ω, can be written in the form:

$$\{(\varphi(t), -\tau_1(\varphi(t), \varphi(t+\omega)) : t \in S^1\} \tag{1.3.1.6.1}$$

for some continuous order preserving $\varphi(t)$ which conjugates the map to rotation. For an intransitive circle, the minimal set on the circle can also be written in this form, but with $\varphi(t)$ only order preserving. Mather showed that $\varphi(t)$ globally minimizes the *Lagrangian*:

$$\int_0^t dt\, \tau(\varphi(t), \varphi(t+\omega)) \tag{1.3.1.6.2}$$

over weakly order preserving $\varphi(t)$. Percival has used this to locate invariant circles numerically.

A related variational principle was introduced by Aubry et al. (1982 & 1983). They showed that for any finite segment $\{\theta_m,..., \theta_n\}$ of any orbit on an invariant circle,

$$\sum_{i=m}^{n-1} \tau(\theta_i, \theta_{i+1}) \tag{1.3.1.6.3}$$

is globally minimal over all sequences $\{\theta_m,..., \theta_n\}$ with the same end points θ_m, θ_n. This can be used to prove non-existence of invariant circles. For example, it provides a simple proof that for a billiard ball on a convex table whose boundary has a point of zero curvature, given $\varepsilon > 0$, there is an orbit which comes within angles ε, $\pi-\varepsilon$ of the tangent to the boundary (Mather, private communication; see also Mather (1982d), for a proof using Birkhoff's theorem only). Newman and Percival (1982) have also used this result numerically to find regions in which a map has no invariant circles. For the standard map, they find that there can be no invariant circles for $|k| > 1.04$. With the necessary estimates, this could probably be turned into a proof.[19]

Mather (1982) showed that the Lagrangian always has a minimum, unique for ω irrational. If the minimizing $\varphi(t)$ is continuous, it gives a transitive invariant circle of rotation number ω. Otherwise, it gives an invariant Cantor set on which the map is conjugate, in the sense of order-preservation, to rotation at rate ω. Similarly, Aubry et al. (1982) found an orbit for any rotation number, whose finite segments are minimizing. Its closure is an invariant circle or Cantor set. I will show some pictures in §1.3.2.3.

They also prove existence of orbits homoclinic to the Cantor set, in the discontinuous case. These are minimaximizing in the following sense. Given a Cantor set, write $\underline{x}^0$, $\underline{x}^1$ for the orbits of the endpoints of one of its gaps. For sequences $\underline{x}$ satisfying:

$$x_i^0 \leq x_i \leq x_i^1 \tag{1.3.1.6.4}$$

the sum:

$$G(\underline{x}) = \sum_{-\infty}^{\infty} (\tau(x_i, x_{i+1}) - \tau(x_i^0, x_{i+1}^0)) \tag{1.3.1.6.5}$$

is convergent and non-negative. Also $G(\underline{x}^0) = G(\underline{x}^1) = 0$. Consider the sets:

---

19 I carried this out in MacKay and Percival (1985).

$$X_a = \{\underline{x}: G(\underline{x}) \le a\} \tag{1.3.1.6.6}$$

and let

$$a_{min} = \inf \{a: \underline{x}^0 \text{ and } \underline{x}^1 \text{ lie in the same connected component of } X_a\}. \tag{1.3.1.6.7}$$

The important thing is that the infimum is attained. Then any $\underline{x}$ for which $G(\underline{x}) = a_{min}$ can be shown to be an orbit homoclinic to the endpoints.

In a special case, the homoclinic orbits can fill in the gaps of the Cantor set to give an intransitive invariant circle. Define:

$$\Delta W_\omega = \max \{a_{min}\} \tag{1.3.1.6.8}$$

taking the maximum over gaps in the Cantor set. The maximum is attained. Then Mather showed that the Cantor set lies on an invariant circle iff $\Delta W_\omega = 0$. Equivalently, it does not lie on an invariant circle iff there is some gap for which a_{min} is positive. We can include the case where the minimizing $\varphi(t)$ is continuous, by making the natural definition:

$$\Delta W_\omega = 0. \tag{1.3.1.6.9}$$

Thus, there is an invariant circle of irrational rotation number ω iff $\Delta W_\omega = 0$.

Aubry derived a similar criterion. For a gap, define the *Peierls energy barrier* to be:

$$\max_{\xi} \min_{\{\underline{x}: x_0 = \xi\}} G(\underline{x}). \tag{1.3.1.6.10}$$

The minimaximizing sequence need not be an orbit, but the Cantor set does not lie on an invariant circle iff the Peierls energy barrier is positive somewhere.

§1.3.2 **Connections between invariant circles and periodic orbits.** In this section I discuss two intimate connections between invariant circles of a twist map, and the nearby Birkhoff periodic orbits. The first is a conjecture of Greene, connecting the existence of an invariant circle with irrational rotation number with the stability of a particular sequence of Birkhoff periodic orbits whose rotation numbers are the convergents of the irrational. The second is a new theorem of Mather, giving a necessary and sufficient condition for existence of an invariant circle of irrational rotation number in terms of the differences in actions of nearby Birkhoff orbits.

§1.3.2.1 Greene's conjecture. Greene (1979a, 1979b, 1980) suggested a connection between existence of invariant circles and the stability of the nearby Birkhoff periodic orbits. This has also been followed up extensively by Schmidt (1980) and Bialek (Schmidt and Bialek, 1982). Their "fractal diagram" will be discussed in §4.5.1.

Essentially, the Birkhoff orbits and their associated island chains are the reason that a neighbourhood of every rational has to be excluded in the KAM theorem. They correspond to the resonant denominators of perturbation theory.

Heuristically, the island chains of types (p_1, q_1), (p_2, q_2) interact nonlinearly to produce an island chain of type $(p_1 + p_2, q_1 + q_2)$. They produce other island chains with higher order sums and differences, but this one will be the most important. One can measure the strength of an island chain by the residue of either of its periodic orbits (which are typically roughly equal in modulus). Chirikov popularised measurement of the "island width", but its order of magnitude can be related to the residue via the twist (Greene, 1980, Lichtenberg, 1979). The residue is preferable because it is coordinate independent, and we shall see that it obeys some very exact relations.

The strength of the "sum" of two island chains is roughly the product of their strengths. Thus for weak island chains the daughter island chain is weaker, but for strong ones it is stronger. The daughter will interact with each of its parents. Let's restrict to one parent in particular. Thus one gets a sequence of island chains each of which is the "sum" of the previous two, and they get successively weaker or stronger depending on the initial strengths, except in a critical case dividing the two behaviours. If they get weaker, then the map gets closer to integrable as the scale is reduced, so one expects to find an invariant circle in the limit. On the other hand, if they get stronger, they would squeeze gaps in any invariant circle, so there can't be a circle as their limit (cf. island overlap criterion of Chirikov (1979)). The p_n, q_n for the island chains form two Fibonacci sequences, i.e.:

$$p_{n+1} = p_n + p_{n-1} \tag{1.3.2.1.1}$$

and similarly for q_n. Thus:

$$\frac{p_n}{q_n} \to \frac{p_1 + \gamma p_2}{q_1 + \gamma q_2} \tag{1.3.2.1.2}$$

where γ is golden ratio (1.3.1.5.11). This is a noble, provided:

$$|p_2 q_1 - p_1 q_2| = 1. \tag{1.3.2.1.3}$$

With this motivation, I proceed to numerical results. Given ω irrational, there is a preferred sequence of rationals p_n/q_n converging to it, called the *convergents* of ω. They are the successive truncations of its continued fraction expansion (1.3.1.5.10):

$$\frac{p_n}{q_n} = m_0 + \cfrac{1}{m_1 + \cfrac{1}{\dots + \cfrac{1}{m_n}}} \equiv [m_0, m_1, \dots, m_n] \qquad (1.3.2.1.4)$$

The convergents have the property that they are the closest rationals to ω compared to rationals with the same or smaller denominator. But note that for numbers other than nobles, there are other rationals besides the convergents which have this property too (e.g. Kappraff and Marzec, 1982). In §1.3.2.4 I give algorithms for computing convergents.

Let us restrict attention to the Birkhoff periodic orbits of type (p_n, q_n). Calling their residues $R_n^{\pm}$, one finds numerically one of three cases:

(i) <u>subcritical</u>: $R_n^{\pm} \to 0$, and it looks as if the island chains converge to a smooth invariant circle of rotation number ω.

(ii) <u>critical</u>: $R_n^{\pm}$ are eventually bounded away from 0 and $\pm\infty$ (in the case of nobles they converge to limits given in (4.4.1.8)), and it looks as if the island chains converge to a non-smooth invariant circle of rotation number ω.

(iii) <u>supercritical</u>: $R_n^{\pm} \to \pm\infty$, and it looks as if there is no circle of rotation number ω.

Greene's conjecture is that under suitable conditions, one could replace "and it looks as if" in the above by "which implies that". A partial result in this direction follows from Moser's twist theorem, and the normal form that Mather told me about (§1.3.1.5), namely, if there is a smooth Diophantine circle, then $R_n^{\pm} \to 0$ (in fact, faster than exponentially) (Mather, private communication), so in the critical and supercritical cases there is no smooth circle (for Diophantine rotation number). The converse, however, is not known.[20]

The critical case is shown in figure 1.3.2.1.1 for $\omega = \gamma^{-2} = [0, 2, (1,)^{\infty}]$ for the quadratic map in the form (1.2.2.2.4). The convergents are 2/5, 3/8, 5/13, ..., and these island chains can be seen to be converging to the outermost invariant circle. It is non-smooth in the sense that it has thin spots (smoothness refers to the conjugacy, not the graph). An easy way to find a critical case in a one parameter family is to find parameter values μ_n for which $R_n^{+} = 1$, say. Then the μ_n typically converge to a limit, corresponding to a critical case. Then to plot the critical circle, one still has to find an initial condition for it. For figure 1.3.1.2.1, I located an initial condition by taking the limit of the periodic points of type (p_n, q_n) lying on the dominant half-line (§1.2.3.4). The symmetry lines are the lines passing through the fixed point, and the dominant half-line is the one heading into the lower right corner. The initial condition is given in table 1.3.2.3.1.

[20] The current status of Greene's conjecture is summarised in MacKay (1992b).

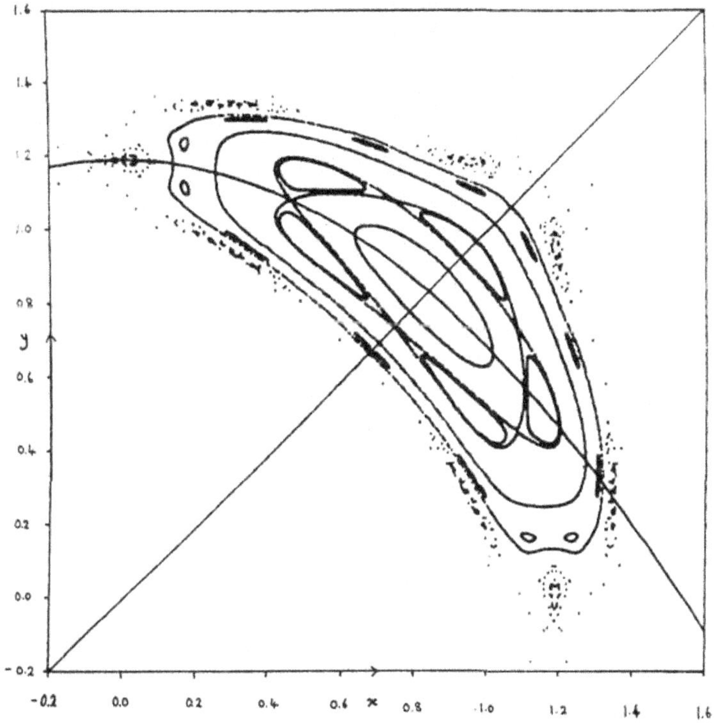

Figure 1.3.2.1.1: Birkhoff orbits in the quadratic map, converging to a critical noble circle (the outermost one). Also shown are two symmetry lines.

§1.3.2.2 Mather's theorem. A necessary and sufficient condition for existence of an invariant circle with irrational rotation number has recently been proved by Mather (1986), based on another property of nearby island chains (it can probably be extended to rationals too). For ω irrational, ΔW_ω was defined in (1.3.1.6.8). For rationals, define $\Delta W_{p/q}$ to be the difference in actions between the two Birkhoff orbits of type (p, q):

$$\Delta W = W_{\text{minimax}} - W_{\text{min}} \tag{1.3.2.2.1}$$

Theorem (Mather): ΔW_ω is continuous in ω at irrationals. Thus, given a sequence of rationals $p_n/q_n \to \omega$, irrational, there exists an invariant circle of rotation number ω iff $\Delta W_{p_n/q_n} \to 0$.

If there is no invariant circle, then $\Delta W_{p_n/q_n}$ converges to the positive value of ΔW_ω for the Cantor set (§1.3.1.6).

§1.3.2.3 Comments. In the subcritical and critical cases of §1.3.2.1, I find numerically that $\Delta W_{p_n/q_n} \to 0$ (faster than exponentially in the subcritical case, as Mather's normal form implies for a smooth circle), so there is an invariant circle, and in the supercritical case $\Delta W_{p_n/q_n}$ tends to a positive limit, so there is no circle, but there is a Cantor set. This is shown in figure 1.3.2.3.1 for $\omega = \gamma^{-2}$, and four parameter values in the quadratic map, one subcritical, one critical, and two supercritical. To locate an initial condition for the noble orbit in the supercritical cases, I found the limit of the periodic points on one of the subdominant lines, in fact the other half of the dominant half-line. One can't use the limit point on the dominant half-line, as the noble cantorus has a gap there. The parameter values and initial conditions are given in table 1.3.2.3.1. Note that in order to separate the orbits in the figure, I included a parameter dependent scale change and translation. Another point to note is that the supercritical orbits are unstable (see also §4.5.1), so one can not iterate them for long before falling off.

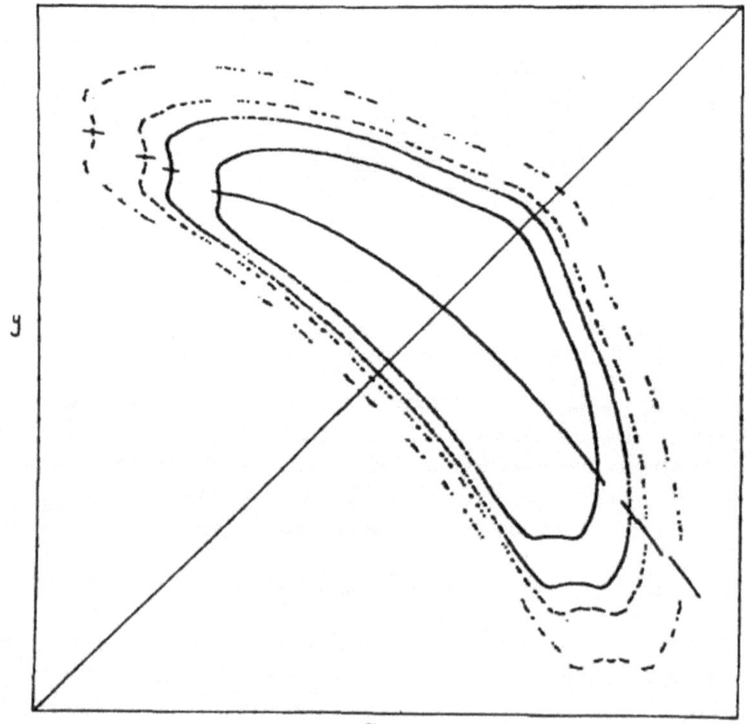

Figure 1.3.2.3.1: An orbit of rotation number γ^{-2} for four parameter values in the quadratic map, one subcritical, one critical, and two supercritical

Table 1.3.2.3.1: Parameter values and initial conditions for an orbit of rotation number γ^{-2} in the quadratic map. The initial conditions marked "dom" are located on the dominant half-line, and those marked "sub" on the other half of that line. For figure 1.3.2.3.1, the initial conditions were iterated N times in each direction of time.

parameter value	case	initial condition	line	N
2.37216325159	sub	1.307669756528	dom	700
2.38216325159	crit	1.31603217783688	dom	700
2.39072266209	super	0.1358731	sub	700
2.40484962405184	super	0.127463	sub	200

Existence of a smooth Diophantine circle is stable to perturbation, by Moser's twist theorem. But also non-existence of an invariant circle of given irrational rotation number is stable to perturbation. This is because Mather also showed that ΔW_{ω} depends continuously on C^1 perturbations of the map at irrationals. Thus, if it is non-zero for one map, then it is non-zero for all small enough perturbations. It also follows from Birkhoff's theorem of §1.3.1.1 (Mather, private communication). So one expects, and I find numerically, that the subcritical and supercritical cases are open sets in class A.

Finding periodic orbits and evaluating their residues and actions is a relatively straightforward procedure, especially if one takes advantage of symmetries the system may possess, using the results of §1.2.3. These approaches to finding where there are invariant circles have the advantage that they generalise directly to continuous time systems. There is no need to choose a surface of section or evaluate a return map, as periodic orbits can be found relatively easily in continuous time, and their residues and actions by integrating the derivative of the flow, and the Lagrangian, respectively, around the orbit.

§1.3.2.4 Calculation of convergents. In this subsection, I give algorithms for calculating convergents. The most obvious way is to calculate from the top, using:

$$[a_0, \ldots, a_n] = \frac{h_n}{k_n} \qquad (1.3.2.4.1)$$

where

$$\begin{aligned} h_{-2} &= 0,\ h_{-1} = 1,\ h_i = a_i h_{i-1} + h_{i-2} \\ k_{-2} &= 1,\ k_{-1} = 0,\ k_i = a_i k_{i-1} + k_{i-2}. \end{aligned} \qquad (1.3.2.4.2)$$

One can also calculate them from the bottom:

$$[b_1, \ldots, b_m] = \frac{g_0}{m_0} \tag{1.3.2.4.3}$$

where

$$g_n = 1, g_{i-1} = b_i g_i + m_i$$
$$m_n = 0, m_{i-1} = g_i. \tag{1.3.2.4.4}$$

The two can be combined, using (Niven, 1963):

$$[\underline{a}, x] = \frac{h_{n-1} + x\, h_n}{k_{n-1} + x\, k_n} \tag{1.3.2.4.5}$$

to give:

$$[\underline{a}, \underline{b}] = \frac{m_0\, h_{n-1} + g_0\, h_n}{m_0\, k_{n-1} + g_0\, k_n}. \tag{1.3.2.4.6}$$

This is useful if one wishes to add terms in the middle. We can use:

$$[x, \underline{b}] = \frac{x\, g_0 + m_0}{g_0} \tag{1.3.2.4.7}$$

to deduce that:

$$[\underline{a}, \underline{c}, \underline{b}] = \frac{Q\, g_0\, h_{n-1} + (P g_0 + Q\, m_0)\, h_n}{Q\, g_0\, k_{n-1} + (P g_0 + Q\, m_0)\, k_n} \tag{1.3.2.4.8}$$

if

$$[\underline{c}] = \frac{P}{Q}. \tag{1.3.2.4.9}$$

It is particularly useful if one wants to evaluate $[\underline{a}, (\underline{c})^s, \underline{b}]$, $s = 0,1,2\ldots,$ because after evaluating the cases $s = 0$ and 1 as above, one can use the relation:

$$[\underline{a}, (\underline{c})^{s+1}, \underline{b}] = P[\underline{a}, (\underline{c})^s, \underline{b}] \oplus Q^2[\underline{a}, (\underline{c})^{s-1}, \underline{b}] \tag{1.3.2.4.10}$$

where

$$\frac{a}{b} \oplus \frac{c}{d} = \frac{a+c}{b+d} \tag{1.3.2.4.11}$$

to generate the rest successively.

§1.4 STOCHASTIC BEHAVIOUR

Finally, I discuss the third type of orbit in area preserving maps, those which are stochastic in some sense. I give a brief outline of a hierarchy of notions of stochasticity in measure preserving systems, and then discuss horseshoes, giving two examples where they can be seen explicitly. Lastly I discuss Birkhoff's zones of instability.

§1.4.1 **Notions of stochasticity**. A very striking and common feature of area preserving maps is orbits which appear to cover densely some region of positive measure. Such orbits can often be seen, for example, around island chains in the regions between invariant circles. Loosely speaking, these orbits are called *stochastic*, and the regions they fill *stochastic regions*. Generally, initially nearby orbits in a stochastic region diverge exponentially.

There is a beautiful hierarchy of notions of stochasticity in measure preserving systems (e.g. Arnold and Avez, 1968), to which I will give a brief introduction in this section. Unfortunately, it is unclear how often any of these notions apply, as they are not easy to verify. There is, however, one very common form of stochastic behaviour, which is fairly easily checked. This is the behaviour in a horseshoe, to be described in §1.4.2. Horseshoes have measure zero, so they are unimportant in a measure theoretic sense, but they affect the behaviour in a neighbourhood, in something like the same way as periodic orbits do. I will come back to a discussion of the regions between invariant circles in §1.4.5.

I will restrict attention to measure preserving systems, although some of the notions are more general. Each of the notions to be discussed will be defined for the whole system, but they apply equally well to invariant subsets.

(i) Topological transitivity. A system is said to be *topologically transitive* if it has a dense orbit. As mentioned above, it would be interesting enough to find an orbit whose closure has positive measure, giving an invariant subset on which the map is topologically transitive.

(ii) Ergodicity. For a measure preserving map T, Birkhoff (Halmos, 1956) proved the following theorem:

Theorem (Birkhoff ergodic theorem): Given any measurable function $f(x)$, the time average along orbits:

$$\bar{f}(x) = \lim_{n \to \infty} \frac{1}{n} \sum_{i=0}^{n} f(T^i x) \tag{1.4.1.1}$$

exists for almost all initial conditions, and:

$$\int \bar{f}(x) = \int f(x). \tag{1.4.1.2}$$

A measure preserving system is said to be *ergodic* if there is no decomposition into invariant subsets of positive measure. Then the above theorem implies that every invariant function must be constant almost everywhere. In particular, the time average $\bar{f}$ is an invariant function, and so constant almost everywhere for an ergodic system, and therefore equal to the spatial average of f.

A measure preserving system can be decomposed into ergodic components, but they need not have positive measure (for example, the decomposition for the standard integrable map (1.3.1.2.1) is given by the circles z = constant). Note that ergodicity implies that almost every orbit is dense, but topological transitivity does not imply ergodicity. There are examples with a decomposition into an invariant Cantor set of positive measure and its complement.

(iii) Bernoulli. There are intermediate notions between ergodicity and Bernoulli, such as mixing, and positivity of the metric entropy, but I will not dwell on them. Suffice it to say that positive entropy corresponds to systems for which predictive power given knowledge, however accurate short of perfect, of the initial conditions, is lost exponentially in time, or uniformly in time if measured in information theory terms. For details, see Arnold and Avez (1968).

A *Bernoulli shift* is a measure preserving map acting on a space of doubly infinite sequences $\{I_j: j \in \mathbb{Z}\}$ of two or more (or infinitely many) symbols $I = 0,1,2,\ldots$. The measure is that defined by assigning weights $p_1, p_2, \ldots$ to the symbols, with $\Sigma p_I = 1$. So the measure of the set of sequences with $I_0 = I$ is p_I, and the measure of more complicated sets is computed by treating the I_j as independent. The map is defined by:

$$I_j' = I_{j+1} \tag{1.4.1.3}$$

which is just the left-shift on the sequences. A measure preserving system is said to be *Bernoulli* if there is a measure preserving transformation taking it to a Bernoulli shift.

Bernoulli systems are as random as a sequence of coin tosses, if you can make any sense of the word "random", because with two symbols the Bernoulli shift corresponds to looking at the outcome of a doubly infinite sequence of coin tosses after successive tosses. With 37 symbols, it is equivalent to a roulette wheel (except I have been told that in Las Vegas, roulette wheels have two zeroes!).

Bernoulli systems are ergodic, mixing and have positive entropy. Katok (1979) has constructed examples of Bernoulli maps on all two dimensional manifolds, but apart from the Anosov systems on the torus, to be discussed shortly, all known examples are very special. For example, Katok's example on the annulus has zero twist to all orders. Bernoulli systems could be quite common, however, because Pesin (1977) has shown a remarkable decomposition result for measure preserving diffeomorphisms for the subset on which all Lyapunov exponents are non-zero. This could be very interesting, provided there are systems for which the subset can be shown to have positive measure. First I will define the Lyapunov exponent.

For a measure preserving map T, the limit:

$$\lim_{n \to \infty} (DT^n_x (DT^n_x)^t)^{1/2n} \tag{1.4.1.4}$$

exists almost everywhere (Oseledec, 1968). It is called the *Lyapunov matrix* at x. Note that the reason for multiplying DT^n by its transpose is so that the root of the matrix can be taken, which can be defined only for symmetric matrices. The logarithms of its eigenvalues are called the *Lyapunov exponents*. For example, the Lyapunov exponents of a hyperbolic periodic orbit of period N are $1/N$ times the logarithms of the absolute values of its multipliers. Possession of a positive Lyapunov exponent implies that most nearby orbits diverge away exponentially fast. Note that being invariant functions, they are constant almost everywhere for an ergodic system. Pesin showed that the set on which no Lyapunov exponent is zero can be decomposed into a countable number of ergodic components of positive measure. Furthermore, on each ergodic component T is essentially Bernoulli, that is, the component can be decomposed into a number N of pieces which are permuted cyclically by T, and T^N is Bernoulli on each of them.

(iv) Anosov. Lastly, I come to the Anosov systems. A periodic orbit is said to be *hyperbolic* if it has no multipliers on the unit circle. This agrees with the definition for area preserving maps given in §1.2.1. Then the tangent space splits as the direct sum of the expanding and contracting spaces of the derivative round the orbit. The notion of hyperbolicity can be extended to other invariant sets than periodic orbits. Loosely speaking, an invariant set is said to be *hyperbolic* if it has invariant expanding and contracting directions at each point. More precisely, the tangent space at each point should split as:

$$E_x = E^s_x \oplus E^u_x \tag{1.4.1.5}$$

with

$$T\,E^s_x = E^s_{Tx}\,,\quad T\,E^u_x = E^u_{Tx} \tag{1.4.1.6}$$

and there should exist $\lambda < 1$ and C, independent of x, such that:

$$\| DT^n_x v \| < C\,\lambda^n \,\| v \| \ \text{ for } v \in E^s_x,\ n \geq 0$$

(1.4.1.7)

$$\| DT^{-n}_x v \| \leq C\ \lambda^n \,\| v \| \text{ for } v \in E^u_x,\ n \geq 0.$$

Hyperbolicity is fairly easy to test. For example, a criterion of Newhouse will be used in §§1.4.3, 1.4.4, involving construction of a suitable conefield. Note that hyperbolicity of an invariant set implies that all its periodic orbits are hyperbolic.

A system for which the whole manifold is hyperbolic is called *Anosov*. A simple example is the map:

$$\underline{x}' = L.\underline{x} \tag{1.4.1.8}$$

where L is any hyperbolic matrix of integers with Jacobian 1, and points $\underline{x}$ with coordinates differing by an integer vector are identified. Anosov systems have very strong properties. They are structurally stable, i.e. any C^1 perturbation of an Anosov system is topologically conjugate to that system. Restricting to measure preserving Anosov systems, they are Bernoulli and the periodic points are dense. But they can live on only certain manifolds. In two dimensions they can live only on the torus.

§1.4.2 **Horseshoes**. A very common form of stochastic behaviour in area preserving maps is due to horseshoes. These are invariant sets under some power of the map on which it is equivalent to a shift on symbol sequences. They are not Bernoulli systems, however, as they have measure zero.

Horseshoes arise in a very easy way (e.g. Newhouse, 1980). Given a hyperbolic fixed point x_0 of T, the stable and unstable directions E^s, E^u of DT_{x_0} can be continued nonlinearly into the *stable* and *unstable manifolds* of x_0. Actually they are not usually manifolds, as they can pile up on themselves (a manifold is a topological space every point of which has a neighbourhood homeomorphic to $\mathbb{R}^n$ for some n, so piling up is prohibited). They are constructed by first defining local stable and unstable manifolds (which are manifolds). This is done by choosing a local coordinate system (x, y) such that $y = 0$ is tangent to E^s and $x = 0$ to E^u. Then there is a unique function $w(x)$, such that:

$$W^s_{\text{loc}} \equiv \{(x, w(x)):\ x \ \text{small}\} \tag{1.4.2.1}$$

passes through the fixed point, is tangent there to E^s and maps into itself under T (e.g. Lanford, 1983). A local unstable manifold is constructed in the same way, by considering T^{-1}. The local manifolds can be extended to global ones by:

$$W^s = \bigcup_{n \leq 0} T^n W^s_{\text{loc}}$$
$$(1.4.2.2)$$
$$W^u = \bigcup_{n \geq 0} T^n W^u_{\text{loc}}.$$

They are the sets of points whose orbits converge to x_0 in forwards and backwards time, respectively.

W^s (and likewise W^u) can not cross itself, but they could intersect each other. An intersection of W^s and W^u for some fixed point is called a *homoclinic point*. Its orbit is asymptotic to the fixed point in both directions of time. As a special case, W^s and W^u could join up exactly on one side of the fixed point, giving a *homoclinic connection*, but generically this does not happen. All the images of a homoclinic point are homoclinic points too, but in fact there are lots more, because when W^u comes close to itself it gets stretched along itself, causing long loops which intersect W^s yet again. This is sketched in figure 1.4.2.1.

Now suppose a fixed point f has a transverse homoclinic point h (i.e. W^s and W^u intersect transversally at h). Consider a closed strip Q containing f, h and the intermediate part of W^s (figure 1.4.2.2). Under iteration, Q gets squished in along W^s and stretched out along W^u, so that after a few iterations $T^n(Q)$ crosses Q at h. Let us write U for T^n and label the components of $U(Q) \cap Q$ containing f and h by Q_0 and Q_1, respectively, and write $N = Q_0 \cup Q_1$. Then we see that $U(N) \cap N$ consists of four vertical strips, two in each of its components. Similarly, $U^{-1}(N) \cap N$ consists of four horizontal strips. One can proceed indefinitely. The set:

$$H = \bigcap_{-\infty}^{+\infty} U^i(N) \tag{1.4.2.3}$$

is an invariant set. If it is hyperbolic (§1.4.1), as is guaranteed for Q thin enough, it is topologically equivalent to the product of two Cantor sets, and is called a *horseshoe*.

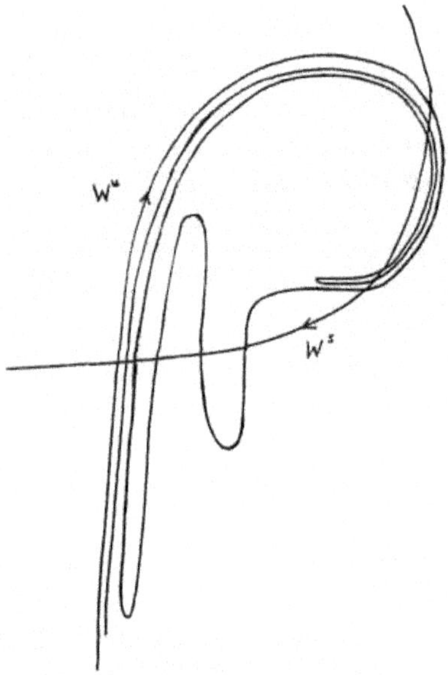

Figure 1.4.2.1: Homoclinic oscillations

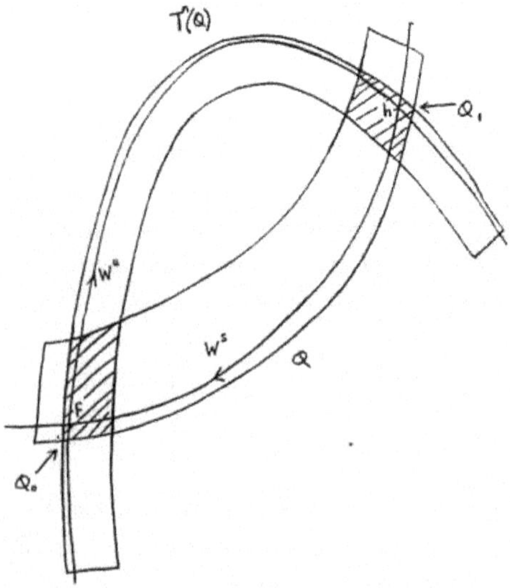

Figure 1.4.2.2: Construction of a horseshoe

The points of the horseshoe can be labelled by symbol sequences. Given a point $x \in H$, define its symbol sequence $\{a_i : i \in \mathbb{Z}\}$ by $a_i = 0$ or 1 according as $U^i x \in Q_0$ or Q_1. In fact, every doubly infinite sequence $\{a_i : i \in \mathbb{Z}\}$ of 0s and 1s, occurs precisely once. To see this, consider the set:

$$\bigcap_{i=-m}^{+n} U^i(Q_{a_i}) \tag{1.4.2.4}$$

which is a little rectangle. Hyperbolicity implies that its diameter goes to zero as $n, m \to \infty$. Thus the sequence defines a unique point of H. Two points of H are close iff their sequences are close, in the sense that they agree on a large interval around $i = 0$. Furthermore, the action of U on H is precisely the left-shift on the symbol sequences.

Consequently, we see that:

(i) the periodic points of a horseshoe are dense in it (take periodic sequences), and are all hyperbolic.
(ii) the points homoclinic to the fixed point of the horseshoe are dense in it (take sequences beginning and ending with all 0s). Similarly points homoclinic to each periodic orbit in the horseshoe are dense.
(iii) there are uncountably many aperiodic orbits (take aperiodic sequences).

Horseshoes are very common. For example, a typical elliptic periodic orbit has horseshoes arbitrarily close (Zehnder, 1973). They can often be constructed explicitly (see §1.4.3).

Given two different periodic orbits, one can consider the intersections of their stable and unstable manifolds, called *heteroclinic points*. If they join up exactly they form a *saddle connection*, but this is unlikely again. Typically intersections are transverse. If there is a *chain* of intersections, i.e. a chain of periodic orbits such that the unstable manifold of one intersects the stable manifold of the next, then one gets very similar behaviour to the horseshoe. Incidentally, horseshoes contain lots of heteroclinic points, corresponding to sequences which are eventually periodic in both directions, but with different repeat patterns, and every pair of periodic orbits forms a chain. I will give an example of a chain of two fixed points in §1.4.4.

One can construct similar things conjugate to the shift on more than two symbols (even infinitely many), and to subshifts, i.e. shifts on sets of sequences where only certain transitions are allowed.

If H is not hyperbolic, it is called a *pseudo-horseshoe*. Although it has a point corresponding to each symbol sequence, there could be more than one point. So it could contain stable periodic orbits, for example. The formation of a horseshoe as a parameter is varied may involve passing through a pseudo-horseshoe stage (e.g. Devaney and Nitecki, 1979). It also involves many tangent bifurcations (Newhouse, 1979, 1980). I will touch on this again in §3.3.3.

§1.4.3 **A horseshoe in the quadratic map.** Newhouse (1980) set an exercise which was to show that for large enough parameter value in the quadratic map, the only bounded orbits are those of a horseshoe. I give the construction here as it is a good example of something useful. Newhouse was kind enough to give me substantial hints. Devaney and Nitecki (1979) independently did the same construction.

Consider the quadratic map in the form:

$$f: x' = y$$
$$y' = c - y^2 - bx. \tag{1.4.3.1}$$

I allow the dissipative cases with $b \neq 1$. For this analysis I will restrict to $b > 0$, but $b < 0$ can be treated in the same way giving the same answer with b replaced by $|b|$.

(i) All bounded orbits lie in the square:

$$Q: |x|, |y| \leq \beta \equiv \frac{1+b}{2} + \left(\frac{(1+b)^2}{4} + c\right)^{1/2}. \tag{1.4.3.2}$$

To show this, write the map in second difference form:

$$y_2 + by_0 = c - y_1^2. \tag{1.4.3.3}$$

For $\lambda > 1$, find β_λ such that $|y_1| > \beta_\lambda$ implies at least one of y_2, y_0 satisfies $|y| > \lambda |y_1|$. Then, by induction, $|y_n| \to \infty$ either forwards or backwards in time. To find β_λ, the minimum of $\max(|y_2|, |y_0|)$ subject to (1.4.3.3) is attained when:

$$y_2 = y_0 = \frac{c - y_1^2}{1+b}. \tag{1.4.3.4}$$

This exceeds $\lambda |y_1|$ if:

$$|y_1| > \beta_\lambda \equiv \frac{1+b}{2}\lambda + \left(\frac{(1+b)^2}{4} + c\right)^{1/2}. \tag{1.4.3.5}$$

For any $|y_1| > \beta - \beta_1$, $\exists \lambda > 1$ s.t. $|y_1| \geq \beta_\lambda$. Hence result (1.4.3.2).

(ii) Outside the strip $|y| \leq \frac{\lambda + b}{2}$, $\lambda > 1$, Df preserves the sector $|\delta y| \geq |\delta x|$ and expands vectors in it by at least λ, using the norm:

$$\|(\delta x, \delta y)\| = \max(|\delta x|, |\delta y|). \tag{1.4.3.6}$$

Similarly, outside the strip $|x| \leq \frac{b\lambda + 1}{2}$, $\lambda > 1$, Df^{-1} preserves the sector $|\delta x| > |\delta y|$, and expands vectors in it by at least λ. Thus a theorem of Newhouse (1980) implies hyperbolicity of any invariant set avoiding these strips.

(iii) The picture so far is given in figure 1.4.3.1. Q is the square $ABCD$. $A^{\pm}$ $B^{\pm}$ $C^{\pm}$ $D^{\pm}$ are its forward and backward images. To make the correspondence with figure 1.4.2.2, the hyperbolic fixed point f is A, and Q_0, Q_1 are the vertical strips $Q \cap f(Q)$. You can imagine where the stable and unstable manifolds of A go. The result will follow if we find parameter values such that $Q \cap f(Q) \cap f^{-1}(Q)$ avoids the strips, for some $\lambda > 1$. For the horizontal strip this is satisfied if:

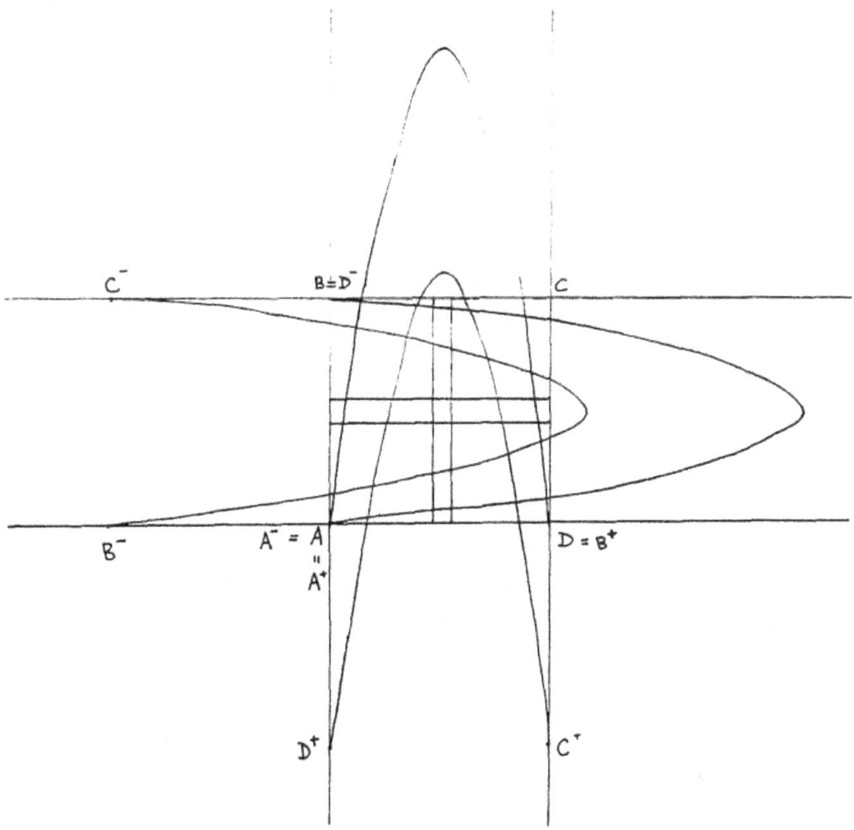

Figure 1.4.3.1: Construction of a horseshoe in the quadratic map

$$c - (1+b)\beta > \left(\frac{\lambda + b}{2}\right)^2 \tag{1.4.3.7}$$

and for the vertical strip, if:

$$c - (1+b)\beta > \left(\frac{b\lambda + 1}{2}\right)^2. \tag{1.4.3.8}$$

Thus, for:

$$c - (1+b)\beta > \left(\frac{b + 1}{2}\right)^2 \tag{1.4.3.9}$$

there exists $\lambda > 1$ satisfying (1.4.3.7) and (1.4.3.8). The relation (1.4.3.9) reduces to:

$$c > \frac{(1+b)^2}{4}(5 + 2\sqrt{5}). \tag{1.4.3.10}$$

Thus for these parameter values, all bounded orbits belong to a horseshoe on which f is equivalent to the shift on $2^{\mathbf{Z}}$. So in particular, there are no invariant circles, and all periodic orbits are hyperbolic. For $b < 0$, one finds the same result with b replaced by $|b|$. Finally, note that $|b| > 1$ can be converted to $|b| < 1$ by considering the inverse, so b in (1.4.3.10) could be strengthened to $\min(|b|, |b|^{-1})$.

§1.4.4 **Orbits without rotation number in the standard map.** I used Newhouse's idea to construct a horseshoe in the standard map for $k > 2\pi$, connected with heteroclinic orbits between two fixed points with different rotation numbers. This leads to existence of orbits without rotation number. The idea is to find a region Q whose image around the cylinder at least once, intersecting Q in two strips. This will give an invariant set whose points can be labelled by sequences $\{a_i : i \in Z\}$ of 0s and 1s, on which the map acts as a shift. Existence of the limit of θ_n/n is equivalent to that of:

$$\lim_{n \to \infty} \frac{1}{n} \sum_0^n a_i \tag{1.4.4.1}$$

and the limits are the same when they exist. Thus the point whose symbol sequence starting at $i = 0$ is 0, 1, 1, 0, 0, 0, 0, 1, 1, 1, 1, 1, 1, 1, 1,... has no rotation number.

Similarly there is an orbit for which θ_n/n comes arbitrarily close to all points of [0, 1].

I write the standard map in coordinates such that it has reflection symmetry about $\theta = 0$:

$$T : y' = y - \frac{k}{2}(\sin\theta + \sin\theta') = \theta' - \theta - \frac{k}{2}\sin\theta'$$

$$\theta' = \theta + y - \frac{k}{2}\sin\theta \tag{1.4.4.2}$$

Then I define $Q = ABCD$ as the region:

$$|y - x| \leq |\frac{k}{2}\sin\theta| + \pi, \quad \frac{\pi}{2} < \theta < \frac{3\pi}{2} \tag{1.4.4.3}$$

which is illustrated in figure 1.4.4.1 for $k > 2\pi$. The images of the curves bounding Q are:

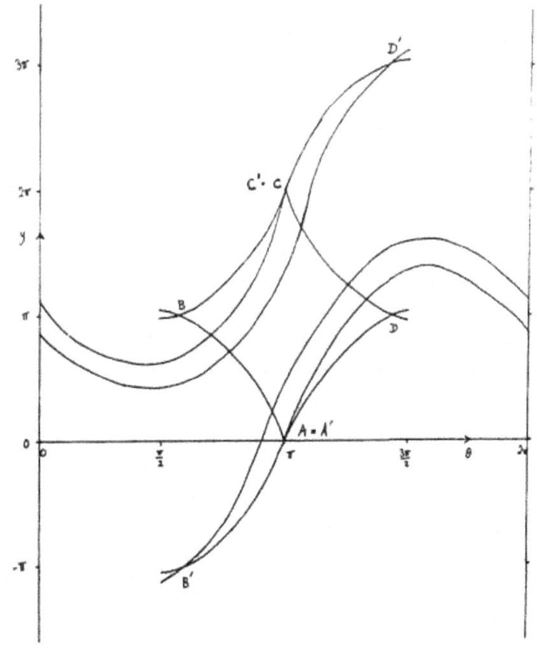

Figure 1.4.4.1: Construction of a horseshoe in the standard map

$$y = \pi - \frac{k}{2}\sin\theta \pm \pi \tag{1.4.4.4}$$

and

$$\begin{aligned} \theta &= \theta_0 - k\sin\theta_0 \\ y &= \theta - \theta_0 - \frac{k}{2}\sin\theta + \pi \pm \pi \end{aligned} \tag{1.4.4.5}$$

where the curves (1.4.4.5) are parametrised by $\pi \le \theta_0 \le \frac{3\pi}{2}$ and $\frac{\pi}{2} \le \theta_0 \le \pi$, respectively. It can be checked that $f(Q)$ wraps once round the cylinder and crosses Q twice as shown. $f^{-1}(Q)$ is the reflection of $f(Q)$ about $\theta = \pi$, as Q is symmetric. Thus the successive images intersect Q in narrower and narrower strips. The width of the strips goes to zero, as $\delta\theta.\delta y \gtrless 0$ form good forward and backward cones for Newhouse's theorem, in $k\cos\theta \le \varepsilon < 0$, using the norm

$$\|(\delta\theta, \delta y)\| = |\delta\theta| + |\delta y|. \tag{1.4.4.6}$$

Thus $\bigcap_{-\infty}^{+\infty} T^i(Q)$ is a hyperbolic set and its points x can be coded by symbol sequences $\{a_i\}$ with $a_i = 0$ or 1 according as $T^i x$ is in the top or bottom component of $T(Q) \cap Q$. The map T acts as a shift on the symbol sequences.

§1.4.5 **Birkhoff zones**. In this section I return to the question of what happens in the regions between invariant circles for a twist map. This is still largely unanswered. Although I will have nothing new to add, it deserves some discussion.

Following Birkhoff (1932), an invariant annulus having no decomposition into invariant annuli is called a *zone of instability*. This name is a little presumptuous, however, as apart from Katok's example with zero twist, no-one has an example or even existence proof of a zone which is unstable in any sense, nor even of a zone containing a subset of positive measure which is unstable. I list what is known:

(i) Given $\varepsilon > 0$, there is an orbit coming within ε of each boundary.[21] This is because otherwise there would be an invariant set which kept away from one boundary by at least ε. So Birkhoff's theorem (§1.3.1.1) would give an invariant circle in the annulus, which would divide it.

[21] Mather (1990) has strengthened this to show that there is an orbit which converges in forward time to one boundary and in backwards time to the other, plus many variations on the theme.

(ii) There are at least two periodic orbits of each rational rotation number between the rotation numbers of the boundaries (Poincaré-Birkhoff §1.2.2.2). So there are island chains. Of course there need not be any elliptic points, but I generalise island chains to include that case.

(iii) There is a Cantor set for each irrational rotation number between those of the boundaries (§1.3.1.6). The measure of those which are unstable, however, in the sense that they have positive Lyapunov exponent (§1.4.1), is zero (Katok, 1982, Aubry et al., 1982).

(iv) Generically, at least one of the Birkhoff orbits of each rotation number is hyperbolic. There cannot be any saddle connections as they would divide the annulus. In any case, generically the hyperbolic orbits have transverse homoclinic points (Zehnder, 1973). So there are lots of horseshoes (§1.4.2). Similarly there are likely to be a lot of heteroclinic points connecting orbits with different rotation numbers as in §1.4.4. But horseshoes have measure zero, so there is a lot of space left!

Chapter 2
Introduction to renormalisation

Renormalisation is the study of asymptotic self-similarity. It has important applications in several areas of physics. More recently, its relevance to dynamical systems has been discovered. In particular I outline its relevance to area preserving maps, the theme of this thesis. Then I discuss various techniques for renormalisation in dynamical systems.

§2.1 RENORMALISATION IN PHYSICS

Renormalisation has emerged as an important new tool in physics. It was invented in quantum field theory, and has also been extensively applied to the study of criticaal phenomena in statistical mechanics. I briefly discuss the possible role of renormalisation in turbulence theory too. This section is included to give the historical background to renormalisation, but it is only a rapid sketch.

§2.1.1 **Introduction**. Renormalisation was introduced in physics to cope with systems having many interacting degrees of freedom. A *renormalisation operator* on a space of systems is an operator whose action on a system is to remove its small scale behaviour and to rescale the remaining variables to preserve some normalisation. The new system has exactly the same large scale behaviour, apart from the scale change, but the effective number of interacting degrees of freedom is less. Iteration of the renormalisation[1] will bring the system into a form with few interacting degrees of freedom which can probably be treated exactly.

The only case where the number of interacting degrees of freedom is not reduced is the critical case when the correlation length is infinite. In fact, this is the case for which renormalisation really comes into its own. This is because the rescaling is done in such a way as to make the new system look as close as possible to the old one. Typically it is chosen to preserve some normalisation on the system, hence the term "renormalisation". The hope is that in the critical case, the system may actually converge under iteration of the renormalisation to some limiting behaviour, the simplest case being a fixed point, i.e. a system which is invariant under renormalisation. The significance of a fixed point is that all systems attracted to it under successive renormalisation will have the same large scale behaviour, so the behaviour is *universal*. Since the renormalisation relates different scales, the universal behaviour is *self-similar*. The fixed point may be impossible to find exactly, but could be found by approximate methods.

The idea of renormalisation was invented in 1953 in quantum field theory, in an attempt to understand the electron (Stueckelberg and Petermann, 1953). In 1966, Kadanoff (1966)

---

[1] The iterates of a renormalisation operator form a semigroup, often called the *renormalisation group*.

introduced it into statistical mechanics, where it has been very significant in the understanding of critical phenomena.[2] Its use has also been suggested in the study of turbulence (Nelkin, 1974). The most recent advance is that its relevance to dynamical systems was discovered, by Feigenbaum (1978) and Coullet and Tresser (1978). Applications of renormalisation to this field have been multiplying rapidly, and it is to such problems that this thesis is devoted. It is noteworthy that its applications to dynamical systems are much cleaner mathematically, than to the other three areas mentioned above.

For historical interest, I will now attempt a brief review of the applications to the first three areas. As my understanding of each of these fields is very limited, you should not trust too deeply any of the things I will say! Useful references to renormalisation in field theory and statistical mechanics are Wilson and Kogut (1974) and Amit (1978). For applications to turbulence theory, see Sulem et al. (1979).

§2.1.2 **Critical phenomena**. I will begin with the application of renormalisation to statistical mechanics, because I understand it best.[3] The study of second order phase transitions is difficult because the correlation length goes to infinity at the transition. Thus one has to treat a system with infinitely many degrees of freedom within a correlation length. We are really interested only in macroscopic properties, so it would be nice if we could somehow replace the microscopic degrees of freedom by a smaller number of effective degrees of freedom, with interactions adjusted to give the same macroscopic result. This can be done. For a system with variables σ and Hamiltonian $H(\sigma)$, the macroscopic physics is given by the partition function:

$$Z = \int e^{-H(\sigma)} \tag{2.1.2.1}$$

where I have normalised the Hamiltonian to the temperature of the heat bath. The partition function is a function of the temperature and external parameters of the Hamiltonian. If one performs the integration over short scale lengths, one may be able to write the new integrand as:

$$e^{-H'(\sigma')} \tag{2.1.2.2}$$

where σ' are some new variables representing the remaining scale lengths, and H' is a new Hamiltonian. For example, for a spin system on a lattice, one could sum over alternate spins. There is an arbitrariness of scale in choosing the new variables. This can be resolved by choosing some normalisation for the Hamiltonians, and then scaling the new variables to preserve that normalisation. For example, in a lattice spin system, the spins can be rescaled to make the nearest neighbour interaction have a coefficient of unity.

2 It was developed by Wilson, for which he received the Nobel prize in 1982.

3 A very nice reference is the lecture notes of Fisher (1983). Another is Anderson (1984).

The transformation from H to H' is a renormalisation R. Suppose it has a fixed point H^*. Then the system with Hamiltonian H^* looks the same as itself on longer scales, so it has either infinite or zero correlation length. In the case of zero correlation length, it is easily solved (ideal gas). So all systems attracted to the fixed point under R behave like an ideal gas, with parameters related by the renormalisation to the original parameters.

The case of infinite correlation length is the case of a second order phase transition. So all critical systems attracted to this fixed point will have the same large scale behaviour, and it is self-similar behaviour. Typically, such a fixed point has only a small number of unstable directions under R. Let us consider the case of one unstable direction. Then the fixed point has a stable manifold W^s of codimension 1 (figure 2.1.2.1). Points on W^s converge to the fixed point. Points on either side diverge away. Thus a one parameter family crossing W^s, parametrised let's say by temperature T, will have a critical temperature T_c, given by the crossing point, dividing the large scale behaviour of the family into subcritical, critical and supercritical. Furthermore, for T close to T_c, the renormalisation brings the one parameter family close to the unstable manifold W^u, but just stretched along it. Thus one gets universal behaviour near the critical temperature too. This is usually expressed by critical exponents.

In a Fourier representation, the renormalisation can be done by integrating over the shortest wavelengths. In fact it is best to do a "partial integration", so as not to introduce a sharp cutoff. The range of wave numbers over which one integrates can also be made infinitesimal, giving a *renormalisation flow*, instead of a renormalisation operator, but the same ideas apply.

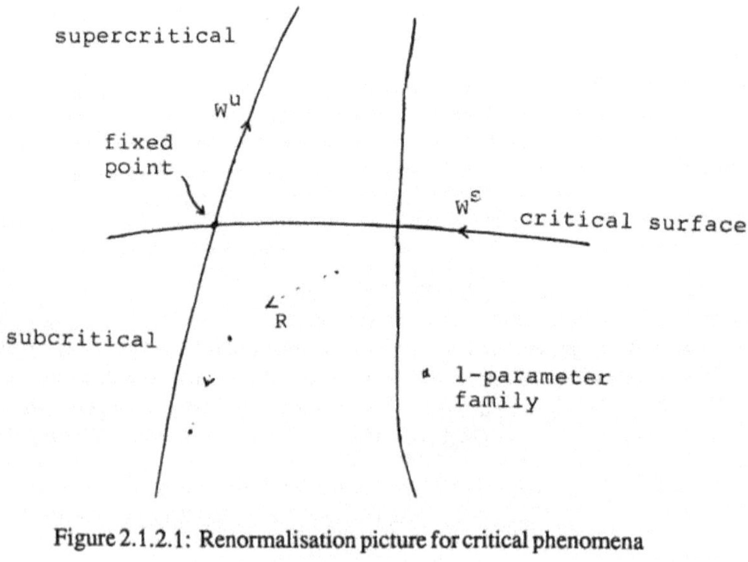

Figure 2.1.2.1: Renormalisation picture for critical phenomena

§2.1.3 **Field theory**. Field theory describes particles as eigenstates of the Hamiltonian for a field. It is important to understand the interaction of particles, i.e. the evolution of an initial state consisting of two well separated particles. One would like to represent the interaction as solely that of the initial particles, but in fact the interaction produces more particles. The new particles interact and create more. The resulting cloud screens the original particles, adjusting their effective coupling (cf. Debye shielding in plasmas). The screening depends on how close the particles come to each other (equivalently the momentum transfer), so the effective coupling depends on the scale at which one is looking. Calculating the screening in perturbation theory requires knowledge of the effective coupling at all smaller scales, so that approach looks a mess. A much tidier approach would be to start at the small scales and work outwards. This is done by renormalisation. I will give a general outline of renormalisation in field theory.

The evolution of the field φ is given by a Lagrangian $L(\varphi, \partial\varphi)$. The answers to scattering and other problems can be expressed in terms of the *generating functional*:

$$Z = \int [\, d\varphi \,] \exp\left(\frac{2\pi\, i}{h}\, S\, [\varphi] \right) \tag{2.1.3.1}$$

where the integral is taken over all paths from an initial to a final state, and

$$S\,[\varphi] = \int d^4x\, L(\varphi, \partial\varphi) \tag{2.1.3.2}$$

is the action for a field configuration. Z is a function of parameters in the Lagrangian, such as external fields.

The idea of renormalisation comes in the evaluation of this path ingegral. Suppose that it is possible to integrate over all wiggles smaller than some scale (cutoff). So Z is expressed as a path integral over longer scales only. The integrand can be written in terms of an effective Lagrangian depending on these longer scales only. As there is some freedom in choosing scales, one usually normalises the fields in some way, for example, so that the coefficient of the kinetic term in the Lagrangian is $\frac{1}{2}$. Now we are all set to renormalise. Integrating over scales just below the cutoff, and rescaling to preserve the normalisation, produces a new effective Lagrangian. So the answer for the original system is the same as that for the renormalised system, apart from a change of scale. The variation of the effective Lagrangian with cutoff corresponds to the screening effects of perturbation theory. Study of the flow under the renormalisation will typically reduce the problem to that for a fixed point. Also one can follow it in either direction. More complicated limiting behaviours than fixed points are possible, e.g. limit cycles and even aperiodic motion.

§2.1.4 **Turbulence**. Fluid turbulence involves the interaction of motions over a wide range of scales. It is observed to have universal self-similar properties in the "inertial range" of length scales, where energy production and dissipation are negligible. This suggests a renormalisation approach. Viscosity imposes a cutoff at a maximum wavenumber. The forcing of the highest wavenumbers can be written in terms of lower wavenumbers, and thus their back-reaction calculated. So the short wavelengths can be removed from the description of the system. With an appropriate rescaling, this gives a renormalisation. One expects convergence to a fixed point on going to longer scales, but no-one seems to have found it yet. It would give the large scale equations of motion for turbulent flows, a problem of considerable significance.[4]

§2.2 RENORMALISATION IN DYNAMICAL SYSTEMS

The idea of renormalisation in dynamical systems is to look at a system on longer timescales and smaller spatial scales. I describe the remarkable discovery of universally self-similar period doubling sequences in $1\text{-}D$ maps, and its explanation in renormalisation terms. The significance of these results extend to higher dimensional dissipative systems. Finally, I give an outline of the relevance of renormalisation to area preserving maps, as an introduction to Chapters 3 and 4.

§2.2.1 **General idea.** In 1978 a discovery was made which is revolutionising the subject of dynamical systems, namely, a class of dynamical systems was found, exhibiting universal self-similar behaviour. It was realised that this could be explained in renormalisation terms. Since then, universal self-similarity has been found and explained by renormalisation, in several more classes of systems. This thesis represents the state of the art in the subject.[5] I begin by discussing general reasons for the relevance of renormalisation to dynamical systems.

A *dynamical system* is a group (or semi-group) of transformations on a topological space, usually a manifold. Typically the group has one parameter called time. It may have integer values, giving a *map* and its iterates, or be a real number, parametrising a *flow*. One is often interested in the asymptotic long-time behaviour of a dynamical system, for example, stability of the orbits, or equilibration of a distribution. Let us for simplicity consider a discrete time system, i.e. a map f. Then the short timescale behaviour can be removed by considering a higher iterate f^n of the map. It has the same longtime behaviour as the original map, except with time rescaled by a factor n. The behaviour of f^n is likely to be simpler than that of f. For example, if the orbits of f are asymptotically periodic, then those of f^n are too, but with shorter transients, and generally shorter period.

The interest focusses, however, on cases when the system has non-trivial behaviour on all timescales. Then f^n will be just as complicated as f. But it may be that an appropriate coordinate change to smaller spatial scales, will make it look almost the same as f. The

[4] See Yakhot and Orszag (1986) and Avellaneda & Majda (1991) for some developments.

[5] For important subsequent developments see in particular Sullivan (1991) and references therein, and Rand (1987, 1988, 1992).

operation of composing n times and rescaling is a renormalisation. In such a case, it may be that in fact f converges under the renormalisation to a fixed point of the renormalisation. Then f would have asymptotically self-similar behaviour on longer timescales and smaller space scales. As it would be the same for all maps converging to the fixed point, the behaviour is said to be universal.

If the fixed point has a number m of unstable directions under the renormalisation, then any m-parameter family crossing the stable manifold will have asymptotically universal behaviour near enough the critical parameter value, in the same way as discussed in §2.1.2. Incidentially, a family with a larger number k of parameters, will have a critical $(k-m)$-dimensional set of parameters attracted to the fixed point, and in the transverse parameter directions the family will look like the universal one.

§2.2.2 **Period doubling in 1-*D* maps**. Now I give the example first discovered of a class of dynamical systems exhibiting universal self-similar behaviour, namely, single-humped maps of a line to itself. I first read about iteration of 1-*D* maps in 1974 (May, 1974). I thought they were really cute, but surely no-one could expect to be paid to play around with them! Their significance, however, as a prototype and as a class to which many other systems reduce, is now becoming clear. For reviews, see May (1976), Collet and Eckmann (1980), and Eckmann (1983).

The simplest behaviour of an orbit of such a map is convergence to a fixed point. This is illustrated graphically in figure 2.2.2.1. The stability of periodic orbits was discussed in §1.2.1. As a parameter varies, the fixed point may lose stability via multiplier -1. For a special class of maps, those with negative *Schwarzian derivative* Sf:

$$Sf = \frac{f^{(3)}}{f'} - \frac{3}{2}\left(\frac{f''}{f'}\right)^2 \tag{2.2.2.1}$$

there must be a direct period doubling (compare §1.2.4.2), producing a stable orbit of period 2. The new orbit may itself lose stability. Since negative Schwarzian derivative is inherited by every power of a map with negative Schwarzian derivative (e.g. Collet and Eckmann, 1980), the new orbit also has a direct period doubling. Typically, one can find as many successive period doublings as one wants.

It was already well-known (Metropolis et al., 1973) that one parameter families of one dimensional maps can have infinite period doubling sequences. But the remarkable numerical discovery of May and Oster (1976), Grossman and Thomae (1977), Feigenbaum (1978), and Coullet and Tresser (1978), was that the period doubling sequences are asymptotically self-similar. Furthermore, the latter two papers showed that they appear to have identical form for a large open set of one parameter families. Thus the limiting form is called universal. For example, the parameter values μ_n at which the nth doubling occurs, converge asymptotically geometrically, with a universal ratio δ:

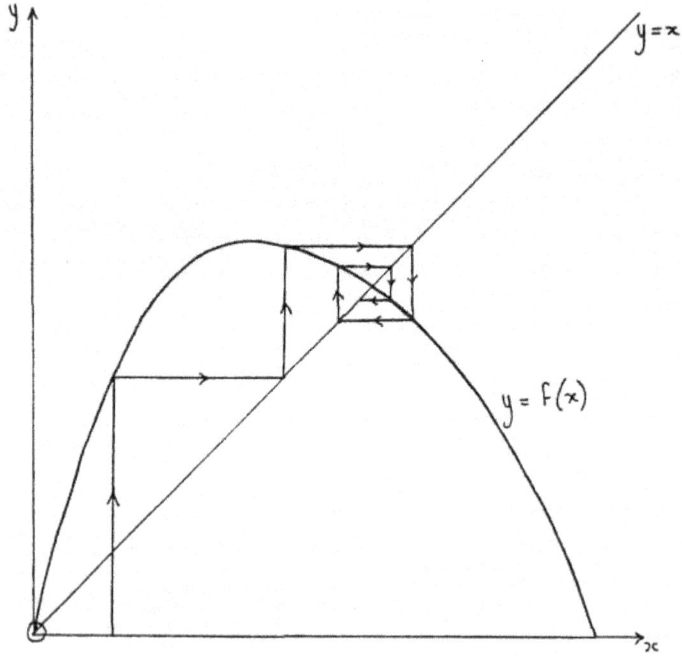

Figure 2.2.2.1: Graphical iteration of a 1-*D* map

$$\frac{\mu_{n+1} - \mu_n}{\mu_{n+2} - \mu_{n+1}} \to \delta = 4.669\ldots . \qquad (2.2.2.2)$$

There is a similar scaling in position. Without loss of generality, let us suppose that the maximum of f is at 0, for all μ. Then at μ_∞, the positions x_n of the point of period 2^n closest to the maximum $x = 0$, converge asymptotically geometrically to the maximum with universal ratio α:

$$\frac{x_n}{x_{n+1}} \to \alpha = -2.50290788. \qquad (2.2.2.3)$$

In fact, $f^{2^{n+1}}$ looks the same as f^{2^n}, on a smaller scale. Specifically, there appears to be a universal limit for the following sequence of maps, at μ_∞:

$$\alpha^n f^{2^n}\left(\frac{x}{\alpha^n}\right) \to f^*(x). \tag{2.2.2.4}$$

Feigenbaum (1978, 1979) (with help from Cvitanovic) and Coullet and Tresser (1978) realised that this might be understood in renormalisation terms. They suggested considering the renormalisation N, an operator that given one map f produces a new map:

$$N(f) = a f(f(\frac{x}{a})). \tag{2.2.2.5}$$

Here, a is a scale change chosen to normalise f^2 in some sense. For example,[6]

$$a = \frac{1}{f(f(0))} \tag{2.2.2.6}$$

forces

$$N(f)(0) = 1. \tag{2.2.2.7}$$

Alternatively, restricting to the space of maps for which $f(0) = 1$, one can take:

$$a = \frac{1}{f(1)}. \tag{2.2.2.8}$$

Then at μ_∞, we expect $N^n f \to f^*$, a fixed point of N, and $a \to \alpha$, the value of (2.2.2.6) for the fixed point f^*.

Now suppose that f^* has only one unstable direction under N (see figure 2.2.2.2). In other words, the derivative DN of the operator N (2.2.2.5) has only one simple eigenvalue δ not contained inside the unit circle. Let W^u be its (one-dimensional) unstable manifold, the family of maps whose preimages under N converge to the fixed point. Furthermore, let W^s be its (codimension 1) stable manifold. If W^u crosses once and transversally (by which I include non-zero speed) a surface Σ_0, defined by a coordinate independent property (e.g. existence of a fixed point with multiplier -1), then it also crosses the surfaces $\Sigma_n = N^{-n}\Sigma_0$, on which f^{2^n} has the same property. These surfaces pile up on W^s, asymptotically at rate δ.

[6] Another choice would be $a = -f^{2''}(0)$ which enforces normalisation $N(f)''(0) = -1$. Coullet & Tresser in fact used an affine coordinate change rather than a linear one, corresponding to moving the origin to the rightmost maximum of f^2 and rescaling about it.

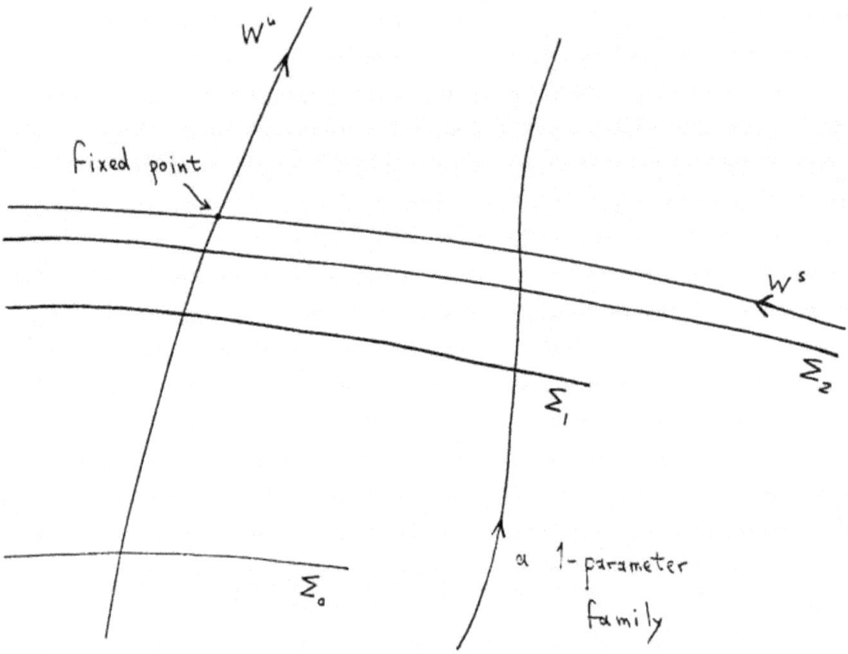

Figure 2.2.2.2: Renormalisation picture for period doubling

Suppose f_μ is a one parameter family crossing W^s transversally, at parameter μ_∞, say. Then for all large enough n, f_μ crosses Σ_n transversally at μ_n, with:

$$\mu_n \sim \mu_\infty + \bar{\mu}\, \delta^{-n} \tag{2.2.2.9}$$

where $\bar{\mu}$ is a constant. So this would explain the observed self-similarity of period doubling sequences.

The way I prefer to look at it is that under the renormalisation, a one parameter family gets stretched by a factor approaching δ, where it crosses W^s, and moved closer to the fixed point by a factor given by the largest eigenvalue inside the unit circle (figure 2.2.2.3). Thus, after a few iterations it gets stretched along W^u, which can be regarded as the *universal one parameter family* when parametrised according to the relation:

$$f^*_{\mu.\delta} = a f^*_\mu (f^*_\mu (\frac{x}{a})) \tag{2.2.2.10}$$

with a as in (2.2.2.6), and the initial conditions:

$$f_0^* = f^*, \quad \left.\frac{df^*}{d\mu}\right|_{\mu_\infty} = v \tag{2.2.2.11}$$

where v is any eigenvector of DN with eigenvalue δ. The choice of v sets the scale in parameter.

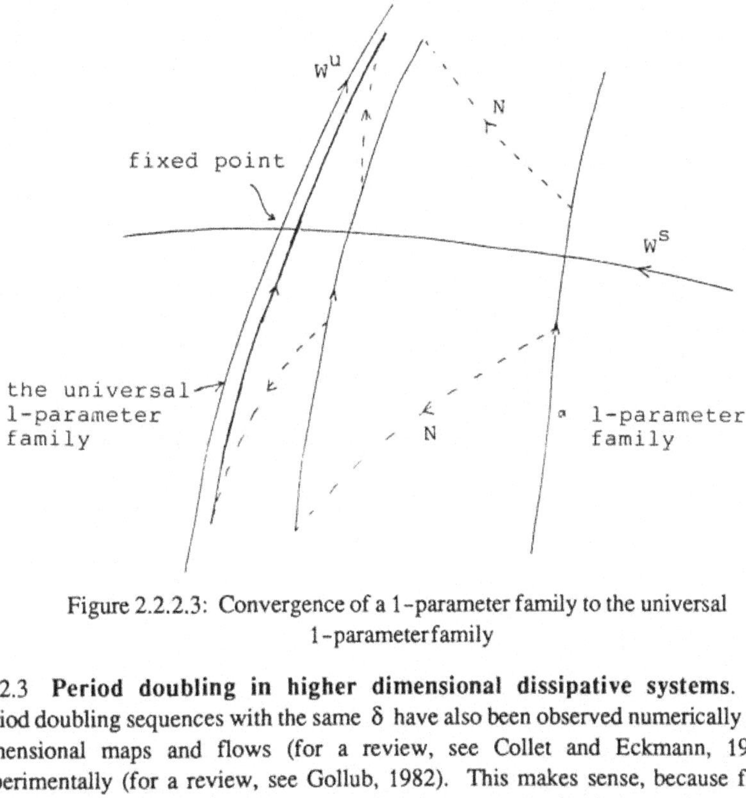

Figure 2.2.2.3: Convergence of a 1-parameter family to the universal 1-parameter family

§2.2.3 **Period doubling in higher dimensional dissipative systems.** Infinite period doubling sequences with the same δ have also been observed numerically in higher dimensional maps and flows (for a review, see Collet and Eckmann, 1980) and experimentally (for a review, see Gollub, 1982). This makes sense, because firstly we already saw how flows can often be reduced to maps by finding a surface of section (§1.1.2.1). Secondly, if a map is strongly contracting in all but one direction, then all initial conditions end up on essentially a 1-D subset. On this subset, the map is essentially 1-D, giving the same behaviour as for real 1-D maps.

I built a little electronic demonstration of period doubling, which was loosely modelled on Rössler's equations (1979). Electronic systems appeal to me because they are continuous time systems, unlike systems solved on a digital computer (even if they are supposed to be flows). They are unlikely to have any particular symmetries which could give rise to pathological results. Furthermore, they are infinite dimensional, unlike a computer implemented system which does not even have a continuous state space, since it is designed to operate in a finite state space (or give a hardware error if it does not!). Of course, electronic components are also designed to behave in a low dimensional way, but only to within 5% or so. Also, electronic systems will reveal only those features which are stable to noise. Lastly, I like to listen to dynamical systems, so I think it is rather fun to design them for audio frequencies. I also like to watch them on an oscilloscope, though you can do this on a computer terminal too. I do not call it analogue computing, because I do not have in mind some exact equations which they are supposed to be solving. They are physical systems in their own right.[7]

The circuit of my model is given in figure 2.2.3.1. I do not give parameter values because mine were not particularly well chosen. You would do better to design your own system. I varied the parameter R. Below some value, every initial condition converges to a stable equilibrium. At a certain value it has a Hopf bifurcation (e.g. Chenciner, 1981), producing an attracting limit cycle. This limit cycle can then be observed to period double as the parameter is increased, with a corresponding drop of an octave in the pitch. I was able to distinguish several period doublings, up to period 16. Figure 2.2.3.2 shows the apparatus exhibiting a stable 4-cycle, obtained after two period doublings. The results are given in table 2.2.3.1, and are consistent with asymptotically geometric scaling at rate δ. I could also identify several of the universal stable windows beyond period doubling, which I did not have time to discuss here, and even some of the period doublings within them.

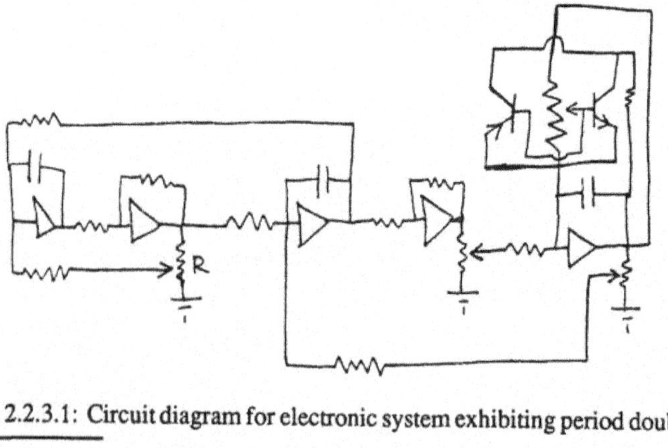

Figure 2.2.3.1: Circuit diagram for electronic system exhibiting period doubling

[7] This circuit and another one to demonstrate breakup of invariant tori in non-conservative systems (see figure 4.7.6.1) were demonstrated at a Plasma Physics Lab. Open Day in 1981 and at Dynamics Days, La Jolla (Jan. 1982). The use of electronic circuits to demonstrate and discover new phenomena in dynamics has since mushroomed.

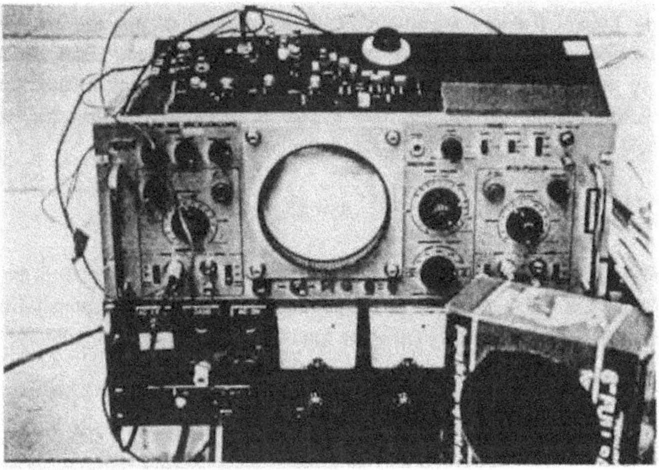

Figure 2.2.3.2: Period doubling, live!

Table 2.2.3.1: Parameter values for successive period doubling bifurcations in an electronic system, showing asymptotically geometric scaling

bifurcation	# of revolutions of new orbit	parameter value	difference	ratio
Hopf	1	113.0		
period doubling	2	579.0	466.0	
period doubling	4	699.5	120.5	3.87
period doubling	8	724.0	24.5	4.92
period doubling	16	729.5	5.5	4.5

§2.2.4 **Renormalisation in area preserving maps**. The subject of this thesis is renormalisation in area preserving maps. The existence of structure on all scales in area preserving maps has been appreciated for a long time. For example, Poincaré (1899), commenting on the intersections of the stable and unstable manifold of a hyperbolic point (see §1.4.2), said that:

> "The intersections form a kind of lattice, web or network with infinitely tight loops; neither of the two curves must ever intersect itself, but it must bend in such a complex fashion that it intersects all the loops of the network infinitely many times. One is struck by the complexity of this figure, which I won't even attempt to draw."

In figure 2.2.4.1, I give a schematic of the progression of possible behaviours in area preserving maps, as non-linearity is increased, in some sense. Each situation can be described by renormalisation! Beginning at the top with tangent bifurcation (§1.2.4.4), Zisook and Shenker (1982) showed how it can be considered in renormalisation terms. Close enough, the map looks like the time-1 map of the flow:

$$\dot{x} = y, \ \dot{y} = x^2 \tag{2.2.4.1}$$

which is self-similar. Reducing the scales in x, y by factors of α^{-2}, α^{-3}, gives the same flow but a factor of α^{-1} slower. So for example it gives a fixed point of the doubling operator ($\alpha = 2$). It has essentially only one unstable direction with eigenvalue 16.

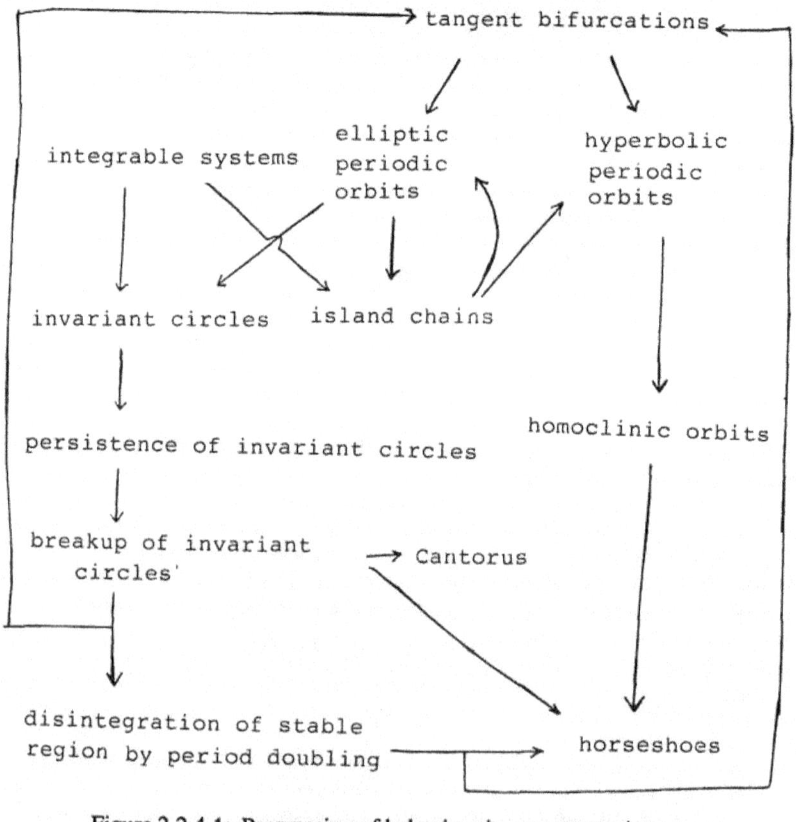

Figure 2.2.4.1: Progression of behaviour in area preserving maps, as nonlinearity is increased

The product of a tangent bifurcation is a pair of periodic orbits, one hyperbolic and the other elliptic. Typically the hyperbolic one has transverse homoclinic points and hence horseshoes (§1.4.2). Horseshoes are the product of two Cantor sets, and hence self-similar. It does not have differentiable self-similarity, however, as the residues of the periodic orbits go to infinity asymptotically geometrically with their period, so I have not worked out a renormalisation description. Perhaps homoclinic tangencies have a renormalisation description.

Close enough to the elliptic periodic orbit, the map is close to integrable (§1.3.1.5). Maps near integrable will be described by the simple fixed point of §4.3. They have invariant circles. For the case of the neighbourhood of an elliptic orbit, these arise by bifurcation from that orbit. Island chains have elliptic and hyperbolic periodic orbits, hence we get return loops in figure 2.2.4.1.[8]

The breakup of invariant circles is given by a critical fixed point, to be described in §4.4. Breakup of an invariant circle leads to horseshoes between the island chains it used to separate, and also a Cantorus (§1.3.1.6), which is self-similar. I think it may also give universal accelerator modes (cf. (1.3.1.5.4)), recycling us back up to tangent bifurcations. Finally, the entire stable region associated with the original orbit disintegrates by period doubling (§1.2.4.4). This is described by a fixed point in Chapter 3. There are universal islets of stability, sending us up to the top again. The end result is a horseshoe.

So renormalisation seems to be unifying concept in the understanding of area preserving maps. In this thesis, I will concentrate on three of the above steps. The disintegration of islands by period doubling will be discussed in Chapter 3. Chapter 4 begins with a renormalisation approach to persistence of smooth invariant circles, and this leads on to discussion of breakup of invariant circles.

§2.3 RENORMALISATION TECHNIQUES

This section contains a collection of techniques which are useful in the analysis of any renormalisation on dynamical systems. The simplest limiting behaviour under iteration of a renormalisation is convergence to a fixed point. So I concentrate on the question of finding fixed points, and proving their existence.

§2.3.1 **Simple example.**

I begin this section with a really simple example of renormalisation. The problem is to show that the graph $y = f(x)$ of any analytic function f, with $f''(0) \neq 0$, looks locally like a parabola. Consider the operator:

$$N : \tilde{f}(x) = 4f\left(\frac{x}{2}\right). \tag{2.3.1.1}$$

This expands by factors of 2 in x and 4 in y. It has an analytic fixed point, in fact a whole line of them:

---

8 The island-round-island hierarchy leads to formation of "adding machines", analogous to the period-doubling accumulation, which have asymptotically universal small-scale structure, understandable by renormalisation. This is touched on in §3.4.1, and for more see Meiss (1986).

$$f(x) = Cx^2. \tag{2.3.1.2}$$

By choosing the scale we can take $C = 1$, provided $f''(0) \neq 0$.

Next examine the stability of the fixed point under N. If:

$$f(x) = x^2 + \delta f(x) \tag{2.3.1.3}$$

then

$$\widetilde{\delta f}(x) = 4\delta f\left(\frac{x}{2}\right). \tag{2.3.1.4}$$

If we expand the functions in power series about 0:

$$\delta f(x) = \Sigma\, a_n\, x^n \tag{2.3.1.5}$$

then (2.3.1.4) translates into:

$$\tilde{\underline{a}} = \begin{vmatrix} 4 & 0 & & & & \\ 0 & 2 & 0 & & & \\ 0 & 0 & 1 & 0 & & \\ 0 & 0 & 0 & \frac{1}{2} & 0 & \\ & & & & \cdot & \\ & & & & & \cdot \end{vmatrix} \cdot \underline{a}. \tag{2.3.1.6}$$

So this fixed point has three non-attracting directions, viz. $\delta f(x) = 1,\ x,\ x^2$, with eigenvalues 4, 2, 1 respectively. This makes sense, because $\delta f(x) = 1$ implies that $f(0) \neq 0$, so we are rescaling about a point which is not on the curve. $\delta f(x) = x$ corresponds to $f'(0) \neq 0$, so the rescaling is not aligned with the proper directions. Lastly, $\delta f(x) = x^2$ corresponds to a scale change, which is a neutral direction.

Given f with $f''(0) \neq 0$, we could shift it in y to get $f(0) = 0$, shear in y to get $f'(0) = 0$, and fix the scale to get $f''(0) = 2$. Then f would have no component in the above directions, and we get convergence to the fixed point. Hence every f with $f''(0) \neq 0$ looks locally like $f(x) = x^2$.

This example gives the essence of all the renormalisation calculations to be described in this thesis. Note that if one works in the larger space of functions which are only continuous, say, then this renormalisation becomes much less trivial, giving "fractals" (Mandelbrot, 1977). For example,

$$f(x) = x^2\, u(\log_2 x) \tag{2.3.1.7}$$

is a fixed point of (2.3.1.1) for any function u of period 1. For the renormalisations I will consider, very complicated behaviour can arise even in spaces of analytic functions.

§2.3.2 **Fixed point analysis**. The simplest limiting behaviour under iteration of a renormalisation is convergence to a fixed point. There are plenty other possibilities, such as divergence to infinity (e.g. §4.5.3), and convergence to a periodic orbit (e.g. §3.4.1) or more complicated limit sets. But I will be concerned mainly with fixed points.

The significance of a fixed point depends strongly on its stability under the renormalisation N, in particular the number of unstable directions. If N is C^1, the linear stability of a fixed point is given by the (Frechet or functional) derivative DN. For infinite dimensional (complete normed) spaces, this is defined just as in finite dimensional spaces. N is said to be differentiable at f_0 if there exists a linear operator DN such that:

$$\| Nf - Nf_0 - DN(f - f_0) \| = o(\|f - f_0\|) \qquad (2.3.2.1)$$

and DN is called its derivative. Note that differentiability of N is not automatic. For example, composition of C^r functions is not differentiable. To see this, if f is only C^r, then:

$$\delta(f(g(x)) = \delta f(g(x)) + Df(g(x)).\delta g(x) \qquad (2.3.2.2)$$

is only C^{r-1}. If N acts on functions which are analytic in some domain, however, and Nf is analytic on a larger domain, then N is C^∞

Under a compactness condition (§2.3.8), the analysis of DN is much the same as in finite dimensional spaces. Firstly, one finds its eigenvalues, i.e. $\lambda \in \mathbb{C}$ such that there exists $\delta\varphi$ s.t.

$$DN.\delta\varphi = \lambda\ \delta\varphi. \qquad (2.3.2.3)$$

Eigenvalues outside the unit circle correspond to unstable directions under N. Then one finds the eigenvectors corresponding to the unstable eigenvalues, in order to identify whether the unstable directions are in important directions or not. The ones that are important I call *essential.*

If one has found a linearly stable fixed point, an important question is to find its basin of attraction. For a general fixed point, the stable eigenspaces at the fixed point can be extended nonlinearly to give the set of points attracted to the fixed point. This is called its stable manifold. So one would like to know how far this extends. Such problems are known as determining the "universality class" of the fixed point. In the 1-D period doubling problem, for example, it appears (Chang et al., 1981) that the stable manifold of the fixed point described in §2.2.2 is bounded by the stable manifold of another fixed point with two unstable directions (see figure 2.3.2.1), corresponding to a map with a quartic maximum.[9]

[9] Sullivan (1991) has made great progress in determining the universality class of the Feigenbaum quadratic fixed point. MacKay and van Zeijts (1988) found that the quartic point is but one point of a whole horseshoe for period-doubling renormalisation.

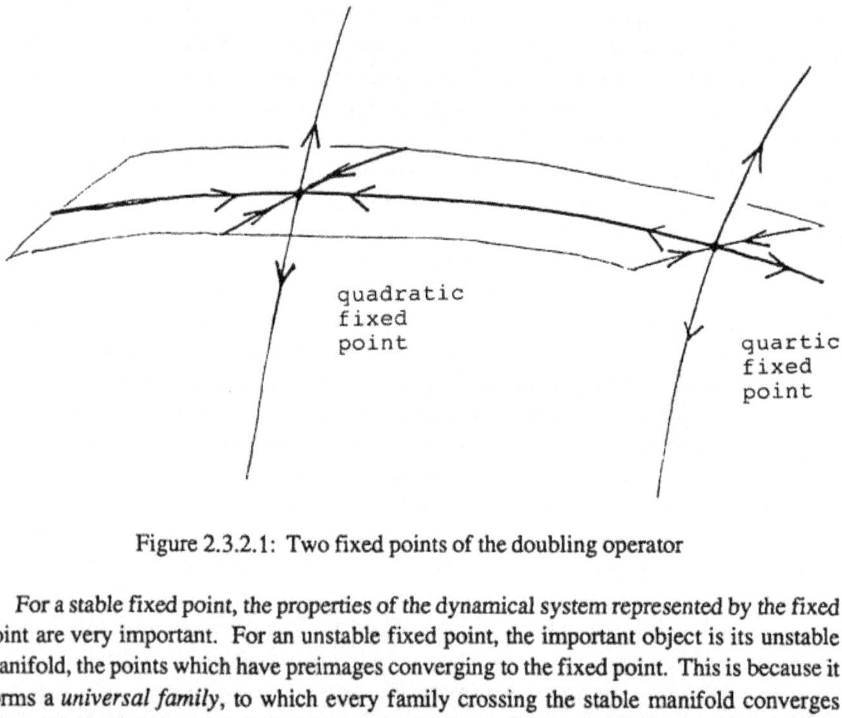

Figure 2.3.2.1: Two fixed points of the doubling operator

For a stable fixed point, the properties of the dynamical system represented by the fixed point are very important. For an unstable fixed point, the important object is its unstable manifold, the points which have preimages converging to the fixed point. This is because it forms a *universal family*, to which every family crossing the stable manifold converges after rescaling in parameters. So it is very interesting to know where the unstable manifold goes. In §4.5.4, I will discuss a case where, in one direction, the unstable manifold of one fixed point ends on a stable fixed point (figure 4.5.4.1).

Throughout §2.3 I will consider a general renormalisation:

$$N(f) = B\,P(f)\,B^{-1} \tag{2.3.2.4}$$

where f could represent a system with more than one map, if desired, and $P(f)$ could be any compositions of its elements. B could be any sort of coordinate change, chosen to normalise $P(f)$ in some way.

§2.3.3 **Normal forms for rescaling.** The coordinate change involved in the renormalisation (2.2.2.5) is linear. In general one could consider more complicated coordinate changes. If (f, B) is a fixed point of the general renormalisation N of (2.3.2.4), then so is:

$$(CfC^{-1},\ CBC^{-1}) \tag{2.3.3.1}$$

for any coordinate change C. So one might as well simplify things by finding normal forms for the possible coordinate changes B. I consider normal forms for the coordinate changes B, rather than the systems f, because I expect the coordinate changes to shrink down on some point, say 0, whereas f typically moves points around in a complicated way. After a few rescalings, the first few coordinate changes look like their linearisations at 0. In fact, Sternberg's theorem (Nelson, 1969, Belitskii, 1978) shows that provided there are no resonance relations:

$$\lambda_1^{i_1} \dots \lambda_n^{i_n} = \lambda_j \tag{2.3.3.2}$$

among the eigenvalues λ_k of the linearisation (counted according to multiplicity), then B is conjugate to its linearisation in a neighbourhood of 0. So B can be chosen to be linear diagonal. In (2.3.3.2) the i_k are non-negative integers with:

$$i_1 + \dots + i_n \geq 2. \tag{2.3.3.3}$$

Collet et al. (1981a) examined the resonant case when $\lambda_2 = \lambda_1^2$, in 2-D. They showed conjugacy to either a linear diagonal normal form, or the normal form:

$$B(x, y) = (\lambda_1 x,\ \lambda_1^2(y + x^2)). \tag{2.3.3.4}$$

They also showed that even if the conjugacies in this and the general case were non-area preserving, f an area preserving fixed point of N implies CfC^{-1} is too.

Having chosen a normal form for B, there remains the problem of a prescription for its evaluation. One of the simplest ways is as in (2.2.2.6), where it was chosen to enforce a normalisation. This is the method I will use throughout this thesis. An alternative, however, was suggested by Greene, and will be discussed in §2.3.9.

§2.3.4 **Coordinate change eigenfunctions.** Given a fixed point of a renormalisation N, an important question is what is its stability under N. In this section, I pick out some of the eigenvectors of DN which are common to any renormalisation problem in dynamical systems. They are the *coordinate change* eigenvectors, associated with coordinate transforms of the fixed point. Thus in a coordinate free representation they would be irrelevant, but whenever one is working in a specific coordinate system, they will turn up.

They were first identified by Feigenbaum (1979), in the 1-D period doubling problem. Incidentally, as there are infinitely many, he conjectured that together with δ, they give the whole spectrum. This is wrong, however, because it would imply that all coordinate free properties, like the multipliers of all the periodic orbits, are dependent on each other, for small perturbations of the fixed point, which is false. Furthermore, I found the next non-coordinate change eigenvalue in this problem after δ. The multipliers of the orbits of period 2^n converge in the critical case to a universal value of -1.601191328074. The convergence is asymptotically geometric at rate:

$$\delta' = -0.1237 \tag{2.3.4.1}$$

so I identify this as the largest non-coordinate eigenvalue after δ.

I generalise again to the renormalisation (2.3.2.4). Since the coordinate transforms of a fixed point form an invariant space under N, one expects N to induce a renormalisation on coordinate changes. This is indeed the case. Suppose f is a fixed point, i.e.:

$$f = B_0 P(f)\, B_0^{-1} \tag{2.3.4.2}$$

where B_0 is the rescaling evaluated at f. Then for a coordinate transform CfC^{-1} of f:

$$N(CfC^{-1}) = B_1\, C\, P(f)\, C^{-1} B_1^{-1} = B_1\, CB_0^{-1} f B_0\, C^{-1} B_1^{-1} \tag{2.3.4.3}$$

where B_1 is evaluated at CfC^{-1}. So N induces:

$$N_c(C) = B_1\, CB_0^{-1} \tag{2.3.4.4}$$

on coordinate changes C. For infinitesimal ones:

$$C(\underline{x}) = \underline{x} + \delta\sigma(\underline{x}) \tag{2.3.4.5}$$

this takes the form:

$$DN_c : \delta\sigma' = DB_{B^{-1}}.\delta\sigma\, B^{-1} + \delta B_{B^{-1}} \tag{2.3.4.6}$$

where δB is the variation in B_1 with C.

If B has linear diagonal normal form:

$$B(\underline{x}) = \underline{a}:\underline{x} \equiv (a_1x_1,\ldots, a_nx_n) \tag{2.3.4.7}$$

and we ignore the contribution from δB, then we can diagonalise DN_c easily, with eigenvectors $x_1^{p_1}\ldots x_n^{p_n}\underline{e}_i$, where $\underline{e}_i$ are the unit vectors, and eigenvalues $\alpha_i\,\alpha_1^{-p_1}\ldots\alpha_n^{-p_n}$. If B is chosen to enforce certain scales, then the effect of δB will be to change the scale change eigenvalues to 0, and to add certain amounts of scale change vector to each of the othereigenvectors.

Note that the infinitesimal coordinate change (2.3.4.5) generates the perturbation:

$$\delta f = CfC^{-1} - f = -\delta\sigma f + Df.\delta\sigma. \tag{2.3.4.8}$$

So the eigenvectors of DN_c induce eigenvectors of DN, with the same eigenvalues. In 2-D, one can easily find which are area preserving and which are symmetric.

§2.3.5 **Scaling-coordinates and scaling-parameters.** When there are unstable coordinate change eigenfunctions, it is necessary to choose coordinates carefully for a system in order that it have no component in their direction, if one wants to see the self-similarity. For example, one always has to find the correct point to rescale about. This is equivalent to killing the components in the "shift of origin" directions. The way I go about killing other components, when necessary, is to evaluate the positions of some points $\underline{x}^{(n)}$ which I expect to scale like $\underline{\alpha}_n$, and write them in the form:

$$x_i^{(n)} \sim \sum_j L_{ij}\,\alpha_j^n\,\xi_j + a_{ij}\,\delta\sigma_j(\underline{\alpha}^n : \xi)) \tag{2.3.5.1}$$

where : is as defined in (2.3.4.7). The sum over $\delta\sigma$ is over the unstable coordinate change eigenfunctions, and the object is to solve (2.3.5.1) for the unknown coefficients L_{ij}, a_{ij}. It is usually best to choose various $\underline{x}^{(n)}$, fitting them to different ξ. I do not have any very systematic way for solving for the coefficients. I use symmetry wherever possible. I usually solve approximately for the dominant ones first, and then refine the estimates later. The resulting coordinates ξ, I call scaling-coordinates, because all features scale asymptotically like α in them. For specific examples of the procedures of this section, see §§4.4.1, 4.7.6.

To obtain faster convergence to the self-similar behaviour, it may be worth removing some more coordinate change components, even though they are stable. Typically, one can find the largest non-coordinate eigenvalue δ' inside the unit circle by looking at the convergence rate of some coordinate-free property in critical cases (e.g. see (2.3.4.1)). This is the limiting factor in the convergence. So one might as well remove any coordinate change components which die away slower than δ'.

So far, I have discussed only the critical case. The practical significance of a fixed point of a renormalisation lies in its unstable manifold. I like to regard this as a universal family, satisfying:

$$f^*_{\delta.\mu} = N(f^*_\mu). \tag{2.3.5.2}$$

Usually, I use constant scaling in the above, using its value at the fixed point, though it is not necessary.[10] To see the universal family in a given family, one has to determine scaling-parameters as well as scaling coordinates. For example, one has to at least find the critical parameter value. If the unstable manifold has more than one dimension, one will have to determine more. The scaling parameters can be chosen without finding scaling coordinates, by looking at coordinate-free properties. So choose some sequence $\underline{\mu}^{(n)}$ of parameter values which one expects to scale like $\underline{\delta}$. Writing the analogous expression to (2.3.5.1), where the $\delta\sigma$ are replaced by unstable parameter changes, one can determine scaling-parameters.

For parameter values off critical, the scaling coordinates determined at critical may not be good enough. There can be unstable parameter dependent coordinate changes, and these have to be killed too. Again, in the determination of scaling-parameters and scaling-coordinates, it may be worth killing stable eigenvectors which die slower than δ'.

§2.3.6 **Restricting to invariant subspaces.** The first thing to do in looking for a fixed point is to consider whether one expects it to belong to some restricted class, preserved by the renormalisation. For example, the renormalisation (2.2.2.5) preserves the space of even maps. For any analytic map with a quadratic maximum there is locally a coordinate change to make it even (I learnt this from Eckmann). Thus it is worth restricting attention to the class of even maps.

Incidentally, to find the above coordinate change, consider the map:

$$x' = f(x) = a_0 + a_2 x^2 b(x) \tag{2.3.6.1}$$

with $a_2 \neq 0$, $b(0) > 0$. Then $b(x)$ has a square root (in fact precisely two) $g(x)$ analytic in some neighbourhood of 0, with $g(0) = \pm\sqrt{b(0)} \neq 0$. This can be shown by series expansion and comparison of coefficients.[11] Thus:

[10] The universal family can be regarded as a fixed point of the operator on one-parameter families $\tilde{f} : \tilde{N}(\tilde{f})_\mu = N(\tilde{f}_{\mu/\delta})$ with δ a rescaling chosen to enforce some normalisation condition. The operator $\tilde{N}$ is usually referred to as the Sinai operator, although it already appeared in Greene et al (1981), and Vul & Khanin, 1981).

[11] Or by the implicit function theorem.

$$x' = a_0 + a_2 x^2 \, (g(x))^2. \tag{2.3.6.2}$$

Define a new coordinate y by:

$$y = xg(x). \tag{2.3.6.3}$$

This is an invertible transformation, since $g(0)$ is non-zero.[12] Then the map becomes:

$$y' = (s_0 + s_2\, y_2) \;\; g(a_0 + a_2 y^2) \tag{2.3.6.4}$$

which is even in y.

§2.3.7 **Finding approximate fixed points.** One can often find remarkably good approximate fixed points by hand. It is generally best to consider coordinate-free properties. As an example, I will determine an approximate fixed point for 1-D period doubling, and its δ.

The map:

$$f(x) = \mu - x^2 \tag{2.3.7.1}$$

has fixed points at:

$$x = -\frac{1}{2} \pm \left(\frac{1}{4} + \mu \right)^{1/2} \tag{2.3.7.2}$$

with stability:

$$f'(x) = -2x = 1 \mp \sqrt{1 + 4\mu}. \tag{2.3.7.3}$$

So there is a tangent bifurcation at $\mu = -\frac{1}{4}$, and period doubling at $\mu = \frac{3}{4}$. The 2-cycle is given by:

$$\begin{aligned} x_1 &= \mu - x_0^2 \\ x_0 &= \mu - x_1^2. \end{aligned} \tag{2.3.7.4}$$

Writing

$$\Sigma = x_1 + x_0, \;\; \Delta = x_1 - x_0 \tag{2.3.7.5}$$

[12] In fact, invertible up to the next critical points of f, if any.

one finds:

$$\Sigma = 2\mu - \frac{1}{2}(\Sigma^2 + \Delta^2), \quad \Delta = \Delta\Sigma. \tag{2.3.7.6}$$

The 2-cycle has $\Delta \neq 0$. Therefore $\Sigma = 1$, and so:

$$\Delta^2 = 4\mu - 3. \tag{2.3.7.7}$$

Its stability is given by:

$$f'(x_0)f'(x_1) = \Sigma^2 - \Delta^2 = 4(1-\mu). \tag{2.3.7.8}$$

So it period doubles at $\mu = \frac{5}{4}$. At the accumulation point we expect all the orbits to have asymptotically the same multipliers, because the multipliers of a periodic orbit are invariant under coordinate change. So the accumulation point is given approximately by:

$$4(1-\mu) = 1 - \sqrt{1+4\mu} \tag{2.3.7.9}$$

i.e.

$$\mu = \frac{7+\sqrt{17}}{8} = 1.390388. \tag{2.3.7.10}$$

Compare the empirical result:

$$\mu_\infty = 1.40115518909205060. \tag{2.3.7.11}$$

The eigenvalue δ can be evaluated as:

$$\frac{\frac{\partial}{\partial\mu}\, 4(1-\mu)}{\frac{\partial}{\partial\mu}\, 1 - \sqrt{1+4\mu}} = 2\sqrt{1+4\mu} = \frac{\sqrt{17}+1}{2} = 5.123 \tag{2.3.7.12}$$

which is reasonably close to the empirical value (2.2.2.2). This sort of scheme is due to Derrida and Pomeau (1980). One can also find an approximation for α.

§2.3.8 **Domains and compactness.** In the previous section, I determined a quadratic approximation to the fixed point. If one wants to carry the above sort of thing to higher degree, one has to worry about domains of definition for the functions. Otherwise one has no guarantee that the effect of truncation will go to zero as the degree goes to infinity.

Generally the renormalisation involves a composition. So domains have to be chosen such that evaluation of the renormalised functions on the domains requires knowledge of the original ones only on the domains. If one wants to show that a fixed point is analytic, one has to consider complex domains. Thus one should find an approximate fixed point, and search for good domains for it. Then it will have a neighbourhood for which the same domains will work, hopefully large enough to contain the fixed point.

One can be misled, however. For example, in §4.4.2 I tried a domain which was safe for a degree 14 approximate fixed point, but the results got worse as the degree was increased, indicating that in the norm defined by the domain, I was nowhere near the fixed point. In fact the fixed point probably was not even in the space, as it probably had a singularity in my domain. I will discuss how to be sure this does not happen in §2.3.10. In fact the estimates required essentially amount to an existence proof for the fixed point.

The determination of domains is also very important for relating the spectrum of DN to that for truncations of DN. The spectrum of arbitrarily high degree truncations of a linear operator does not necessarily have much to do with the actual spectrum. An example brought to my attention by Similon is the translation operator t, whose action on functions f is:

$$t(f)(x) = f(x+1). \tag{2.3.8.1}$$

It is linear, and leaves the spaces of polynomials of any finite degree invariant. On any of these spaces its Jordan normal form consists of one big Jordan block with eigenvalue 1. But on the space of entire functions, it has eigenvectors for any non-zero eigenvalue, namely, $e^{\lambda x}$, with eigenvalue e^{λ}!

Under a simple condition, such pathology can be excluded. Suppose we are working in a Banach space B, i.e. a vector space with a norm, such that every Cauchy sequence converges. For example, the space of functions analytic on the unit disc is a Banach space, when equipped with the ℓ_1 norm:

$$\| \Sigma a_n x^n \| = \Sigma | a_n |. \tag{2.3.8.2}$$

The norm of a linear operator L on B is defined by:

$$\| L \| = \sup_{\|v\| = 1} \| Lv \| \tag{2.3.8.3}$$

and the operator is said to be *bounded* if its norm is finite. Then a bounded linear operator is said to be *compact* if the closure of the image of the unit ball is compact. Equivalently, if the image of every bounded sequence has a convergent subsequence. Compact operators behave like finite dimensional ones (e.g. Krasnosel'skii et al., 1972, §18). Specifically,

(i) Apart from 0, the spectrum i.e. $\{\lambda : L \quad \lambda I$ is not invertible$\}$ consists only of eigenvalues i.e. $\{\lambda : \exists v \neq 0$ s.t. $Lv = \lambda v\}$, and they have finite multiplicity and are isolated.

(ii) If L is compact and $\| L' - L \| \to 0$, then the spectrum of L' converges to that of L, apart from at 0. With care in the case of multiple eigenvalues, the eigenvectors converge too.

If N has good domains, it follows that DN is compact with respect to an appropriate norm.

§2.3.9 **Finding arbitrarily good fixed points.** If the fixed point has no unstable directions, straightforward iteration from a reasonably close starting point will converge to it. So one could work in a space of polynomials, say, and truncate the renormalisation at some degree with respect to some domains, and expect convergence to a reasonable fixed point.

If it has only a few unstable directions, a technique suggested to me by Kruskal can be used to kill them. Consider the operator N_K, a polynomial in N:

$$N_K = \sum_{i=0}^{n} a_i N^i. \tag{2.3.9.1}$$

Then, if

$$\Sigma a_i = 1 \tag{2.3.9.2}$$

the fixed points of N are also fixed points of N_K (though not always vice versa). Furthermore, each eigenvector of DN with eigenvalue λ, is an eigenvector of DN_K with eigenvalue:

$$\Sigma a_i \lambda^i. \tag{2.3.9.3}$$

So if, for example, DN has all eigenvalues within the unit circle apart from one at δ, then a choice (with $n = 2$):

$$\begin{gathered} a_0 + a_1 + a_2 = 1 \\ | a_0 + a_1 x + a_2 x^2 | < 1 \text{ for } | x | < 1 \\ | a_0 + a_1 \delta + a_2 \delta^2 | < 1 \end{gathered} \tag{2.3.9.4}$$

will make DN_K attracting.[13] Note that after the first evaluation of N_K, further iteration requires only one evaluation of N at each step.

[13] I am grateful to Keith Briggs for correcting an error here.

I tried this for the $1\text{-}D$ period doubling problem. Noting that the renormalisation preserves the space of even maps, I iterated N_K on an even polynomial, using MACSYMA and a truncation at some degree, and got results close to those of Feigenbaum (1979, app.A1), apart from his printing error (g_1 should be negative!). My results are given in table 2.3.9.1. Here and elsewhere, the symbol *"e"* or *"d"* stands for "times 10 to the power of". I also tried removing the restriction to even maps. This creates a new unstable eigenvalue of α, corresponding to shift of origin. Increasing n to 3, and choosing the coefficients to kill them both off, I got convergence to the fixed point under N_K again.

Table 2.3.9.1: Approximate $1\text{-}D$ fixed point of the doubling operator

$f^*(x) =$	1	
	-1.52763308	x^2
	+0.10481532	x^4
	+0.02670563	x^6
	-3.527422e-3	x^8
	+8.1617e-5	x^{10}
	+2.5275e-5	x^{12}
	-2.551e-6	x^{14}
	-9.9e-8	x^{16}
	+2.91e-8	x^{18}
	+6.e-10	x^{20}

$\alpha = -2.5029078.$

The above technique could be turned into an existence proof for a fixed point (though one would have to check that it really is a fixed point of N, not just N_K). One would need to choose a norm, and show that N_K mapped some bounded set to itself, and was a contraction on that set. Then it has a unique fixed point on that set.

It has two slight disadvantages. One is that the convergence is only geometric, being given by the largest eigenvalue. The other is that it does not give you the spectrum, apart from the fact that it lies in the set where:

$$|\Sigma a_i \lambda^i| \leq 1. \tag{2.3.9.5}$$

A more effective way to find a fixed point is to use Newton's method. Given an operator N, one considers the operator N':

$$N'f = f - J.(Nf - f) \tag{2.3.9.6}$$

where

$$J = (DN - 1)^{-1} \tag{2.3.9.7}$$

provided $(DN - 1)$ is invertible. Then N' has the same fixed points as N. Moreover:

$$DN' = 1 - J.(DN-1) = 0 \tag{2.3.9.8}$$

so its fixed points attract faster than exponentially. In fact the convergence is quadratic. The derivative DN can be evaluated with little more work than N, so applying a truncation, one can find an approximate fixed point without too much difficulty.

In fact, it is not necessary to choose J to satisfy (2.3.9.7) exactly. Any invertible J such that all the eigenvalues of:

$$1 - J.(DN - 1) \tag{2.3.9.9}$$

lie inside the unit circle will do. This is guaranteed, for example, if it has norm less than 1, in some norm. The convergence is no longer quadratic, but this can save evaluation of DN.

Newton's method leads to an alternative determination of the rescaling B, suggested by Greene (private communication). If the effects of variations in B are left out from DN, then $DN - 1$ is not invertible, as DN has scale change eigenvectors, with eigenvalue +1 (§2.3.4). But there will be special values of the rescaling for which $Nf - f$ lies in the range of $DN - 1$. This determines the rescaling. Then $(DN - 1)^{-1}(Nf - f)$ in (2.3.9.6) can be defined as one of the preimages of $(Nf - f)$. In my opinion, the choice should still be made to preserve some normalisation, otherwise one might drift off to infinity.

§2.3.10 **Existence proofs.** Newton's method can be extended to a proof of existence of a fixed point, as was first done by Lanford (1981) for the 1-D period doubling problem, and more recently by Eckmann et al. (1982) for period doubling in area preserving maps (see §3.2.3). Incidentally, there is a completely different proof in the 1-D case by Campanino et al. (1981).

The strategy is as follows. Find a norm (e.g. ℓ_1 norm (2.3.8.2) on power series in certain domains), and an approximate fixed point f_0 such that:

$$\| Nf_0 - f_0 \| < \varepsilon. \tag{2.3.10.1}$$

Find an invertible bounded linear operator J, $\| J \| = \kappa$, such that:

$$\| DN' \| = \| 1 - J.(DN - 1) \| \le \lambda < 1 \text{ on } \| f - f_0 \| < \eta \tag{2.3.10.2}$$

with

$$\varepsilon\kappa \leq \eta(1 - \lambda). \tag{2.3.10.3}$$

To make the estimate (2.3.10.2) will necessitate finding domains such that $\| f - f_0 \| \leq \eta$, f analytic on its domain, implies Nf analytic on a larger domain, so that the column norms of DN will decrease geometrically. Then N' (i) leaves $\| f - f_0 \| \leq \eta$ invariant, and (ii) is a contraction on it. To see this:

(i)
$$\begin{aligned} \| N'f - f_0 \| &\leq \| N'f - N'f_0 \| + \| N'f_0 - f_0 \| \\ &\leq \| DN' \| \ \| f - f_0 \| + \| J \| \ \| Nf_0 - f_0 \| \\ &\leq \lambda\eta + \kappa\varepsilon \leq \eta \end{aligned} \tag{2.3.10.4}$$

by (2.3.10.3).

(ii)
$$\| N'f_1 - N'f_2 \| \leq \| DN' \| \ \| f_1 - f_2 \| \leq \lambda \| f_1 - f_2 \| \tag{2.3.10.5}$$

for $\| f_i - f_0 \| \leq \eta$, $i = 1,2$.

Hence, by contraction mapping principle, there is a unique fixed point of N' on $\| f - f_0 \| \leq \eta$, given by $\lim_{n \to \infty} N'^n f$, for any f in the ball. As the estimates on the derivatives, in particular, are quite involved, the proofs so far, using this method, have used computers to do the tedious grind. One can also get estimates on the spectrum.

It struck me that the above idea can be modified for spaces in which M is not differentiable, like spaces of C^r functions on some domain. Choose a norm, e.g. absolute values of the first few derivatives at zero plus the supremum of the rth derivative over the domain. Find an approximate fixed point f_0 as in (2.3.10.1). Find an invertible bounded linear operator J, $\| J \| = \kappa$, such that:

$$\| N'f_1 - N'f_2 \| \leq \lambda \| f_1 - f_2 \| \text{ for } \| f_i - f_0 \| \leq \eta, \ i = 1,2 \tag{2.3.10.6}$$

for some $\lambda < 1$, such that (2.3.10.3) is satisfied. Then N' leaves invariant the ball $\| f - f_0 \| \leq \eta$, and is a contraction on it. These estimaates can be done by hand. Sometime soon, I hope to use this idea to prove the existence of the fixed point for 1-D period doubling for C^3 maps.[14] Essentially the same idea will be used in §4.7.4 to give a C^3 neighbourhood attraction for a fixed point of a renormalisation.

---

[14] I have not done this yet, but Lanford and Davie have addressed the more important question of whether allowing C^r maps introduces new unstable directions (talks given at Warwick in March 1992).

Chapter 3
Period doubling in area preserving maps

Closely parallel to the one dimensional case, universally self-similar period doubling sequences have been found in area preserving maps. They can be explained by a fixed point of a renormalisation. I describe various properties of the resulting universal one parameter family. The chapter concludes with some appendices on other bifurcation sequences and period doubling in other settings.

§3.1 PERIOD DOUBLING SEQUENCES

I begin by analysing period doubling bifurcations in reversible maps, leading to the notion of a dominant symmetry. Then I describe results indicating that there are infinite period doubling sequences with asymptotic self-similarity. The behaviour appears to be the same for most one parameter families of area preserving maps after a period doubling bifurcation.

§3.1.1 **Dominant symmetry for period doubling bifurcations.** In §1.2.4.4 I described the generic period doubling bifurcations for one parameter families of area preserving maps. As a preliminary to this chapter, I want to study the effect of imposing symmetry. For generic period doubling of a symmetric periodic orbit in reversible maps or reversible area preserving maps, I will show that the daughter orbit is also symmetric. As it has even period, it can belong to only one half-family of symmetries (§§1.2.3.1, 1.1.4.2), which I call the *dominant* half-family. Similar results in this direction were found independently by Rimmer (1979). Note that this is a different notion of dominant symmetry from that in §1.2.3.4. In fact, twist implies that a fixed point on the dominant half-line of §1.2.3.4 must period double across that line (see below).

Without loss of generality, consider a symmetric fixed point 0 of a reversible map T. It is a fixed point of all the symmetries. Let us choose one, and call it S. I already showed in (1.2.3.2.3) that, writing:

$$DT_0 = \begin{vmatrix} A & B \\ C & D \end{vmatrix} \tag{3.1.1.1}$$

in symmetry coordinates for S (1.1.4.2.1), then $A = D$ and $AD - BC = 1$. If the fixed point has residue 1, then $A = D = -1$, and $BC = 0$. So at least one of B and C has to be zero. Typically they will not both be zero, as one can find arbitrarily small (reversible or reversible area preserving) perturbations of this case to make one of them non-zero, a situation which is stable to perturbation, because B and C are continuous functions of T.

First, I will consider the case of $C = 0, B \neq 0$. The only modification that symmetry imposes on the period doubling bifurcation analysis of §1.2.4.4 is that $b = 0$. Thus, we still typically get precisely one 2-cycle, bifurcating for one sign of the parameter ε, as in (1.2.4.4.16), with coordinates:

$$x_{\pm} = \left(- \frac{\varepsilon}{d} \right)^{1/2} + 0(\varepsilon)$$
$$y_{\pm} = 0(\varepsilon^{3/2}). \tag{3.1.1.2}$$

So it is within an angle of $0(\varepsilon^{3/2})$ of the symmetry line of S. But this 2-cycle must be its own reflection by S, as there are no other 2-cycles locally. Thus it actually lies on the symmetry line for small enough ε.

The case $B = 0,\ C \neq 0$ reduces to this one if one changes to symmetry coordinates for TS. To see this, I first find a coordinate change H to reduce the linear part of TS to the standard form:.

$$D(TS) = \begin{vmatrix} -1 & 0 \\ C & -1 \end{vmatrix} \begin{vmatrix} 1 & 0 \\ 0 & -1 \end{vmatrix} = \begin{vmatrix} -1 & 0 \\ C & 1 \end{vmatrix} = H \begin{vmatrix} 1 & 0 \\ 0 & -1 \end{vmatrix} H^{-1} \tag{3.1.1.3}$$

so

$$H = \begin{vmatrix} 0 & -1 \\ 1 & \frac{C}{2} \end{vmatrix}. \tag{3.1.1.4}$$

Then evaluating DT in the new coordinates:

$$H^{-1}\, DT\, H = \begin{vmatrix} -1 & -C \\ 0 & -1 \end{vmatrix} \tag{3.1.1.5}$$

as claimed.

In summary, for generic period doubling of a symmetric fixed point, the daughter lies on Fix(S) if $B \neq 0$, and Fix(TS) if $C \neq 0$, at bifurcation and in symmetry coordinates for S. One could express this in the following coordinate free way. At a fixed point DS has eigenvalues ± 1. Call the corresponding eigenvectors $v_{\pm}$. At a period doubling point, DT has two eigenvalues of -1. At least one of $v_{\pm}$ is also an eigenvector of DT. If v_- is not, then S is dominant. If v_+ is not, then S is subdominant. If both are, we get the special case of diagonal Jordan normal form.

For a symmetric fixed point of a DeVogelaere map (§1.1.4.3), it is easy to see that S_1 is dominant, because

$$DT = \begin{vmatrix} f'(x) & -1 \\ f'(x')f'(x) - 1 & -f'(x') \end{vmatrix}. \tag{3.1.1.6}$$

Similarly, the simple symmetry of McMillan maps is subdominant. Also, for any map with an action generating function with the symmetry (1.1.4.4.1):

$$\tau(x', x) = \tau(x, x') \tag{3.1.1.7}$$

the corresponding symmetry is dominant, because:

$$B = -\frac{1}{\tau_{12}} \neq 0. \tag{3.1.1.8}$$

But with the symmetry (1.1.4.4.2) (also (1.2.3.4.1)):

$$\tau(x, x') = \tau(-x', -x) \tag{3.1.1.9}$$

that symmetry must be subdominant, because:

$$C = \frac{1}{\tau_{12}} \neq 0 \tag{3.1.1.10}$$

justifying the remark made in the opening paragraph of this section.

Next, I consider the case of a symmetric periodic orbit of even period n. It belongs to one half-family of symmetries (§1.2.3.1), so let S be a symmetry such that two of the points lie on Fix(S). This is because, from the preceding analysis, each must double generically along either Fix(S) or Fix(T^nS). But it is impossible to have four points of the same orbit on one symmetry line. I call the point which doubles along Fix(S), the *good* point, and the other the *bad* point. Equivalently, the good point is the one whose dominant symmetry is the same as that of its parent.

So if there is a sequence of period doublings starting from a symmetric periodic point, the sequence can be followed by first finding the dominant symmetry for the point, and then finding the sequence of good points on that symmetry line. Bountis (1981) called this the "symmetry road".

In fact, the bifurcation analysis of §1.2.4.4 can be extended to give a fair indication of which daughter is likely to be good. It can only be an indication because bifurcation analysis is only local, and at least at present I only define "goodness" at residue 1, which is far from birth. This proceeds as follows. Write the tangent maps for the two daughters as:

$$DT^2 = \begin{vmatrix} A_\pm & B_\pm \\ C_\pm & D_\pm \end{vmatrix} \tag{3.1.1.11}$$

To the order at which I was working in (1.2.4.4.17), I got $B_\pm = -2$, but this is insufficient to distinguish between them. So it is necessary to include some higher order terms in the analysis. Evaluation of $B_\pm$ to order $\varepsilon^{\frac{1}{2}}$ requires x'' to order $\varepsilon^{3/2}$. So write:

$$x' = -x + y + ax^2 + ex^3 + fxy + g\varepsilon x + O(\varepsilon^2) \tag{3.1.1.12}$$

and use y' as before in (1.2.4.4.11). Note that symmetry implies $b = 0$, and some relations between the coefficients, but I will work it out in the general case. Then:

$$\begin{aligned} x'' = & -(-x + y + ax^2 + ex^3 + fxy + g\varepsilon x + O(\varepsilon^2)) \\ & + (-y + bx^2 + cxy + dx^3 + \varepsilon x + O(\varepsilon^2)) \\ & + a(-x + y + ax^2)^2 + e(-x)^3 + f(-x)(-y) + g\varepsilon(-x) + O(\varepsilon^2) \end{aligned} \tag{3.1.1.13}$$

$$= x - 2y + bx^2 + (d - 2e - 2a^2)x^3 + (c - 2a)xy + (1 - 2g)\varepsilon x + O(\varepsilon^2). \tag{3.1.1.14}$$

So in the end the new terms do not contribute, and one gets:

$$B_\pm = -2 + (c - 2a)\, x_\pm + O(\varepsilon). \tag{3.1.1.15}$$

Thus, $B_\pm$ is heading away from 0 for one, and towards 0 for the other. So the former is likely to be good, the latter bad. Furthermore, the behaviour of $C_\pm$ is tied to $B_\pm$ by $BC = A^2 - 1$. This indicator turns out to be good for all the cases that I have looked at (e.g. see figure 3.3.2.1).

§3.1.2 **Self-similar period doubling sequences.** The remarkable discoveries of universally self-similar period doubling sequences in one dimensional maps (§2.2.2) inspired several people (Benettin et al. 1980a,b, Bountis, 1981, Derrida, Eckmann and Koch (unpublished), van Zeijts (referred to by Helleman, 1980), Greene et al., 1980, 1981) to look for period doubling sequences in one parameter families of area preserving maps. They found closely parallel behaviour to the 1-*D* case, but with different δ and α. As area preserving maps live on a 2-*D* space, one expects scaling in two directions. I found the second scaling factor and its direction.

Period doubling sequences do indeed occur in one parameter families of area preserving maps, and are very common. I looked mainly at the quadratic map. The dominant symmetry for a symmetric fixed point is easily found when the map is expressed in DeVogelaere form:

$$\begin{aligned} x' &= f(x) - y \\ y' &= x - f(x') \\ f(x) &= px - (1-p)x^2. \end{aligned} \tag{3.1.2.1}$$

Thus, $y = 0$ is a dominant symmetry line for its symmetric fixed points. The new form for $f(x)$ is a historical accident. Figure 3.1.2.1 shows the fixed point at $x = 0$ undergoing period doubling, and figure 3.1.2.2 shows the daughter doubling. One can follow as many doublings as one pleases (I have followed up to 17). Table 3.1.2.1 gives the parameter values and positions for the first 12 period doubling bifurcations of this sequence.

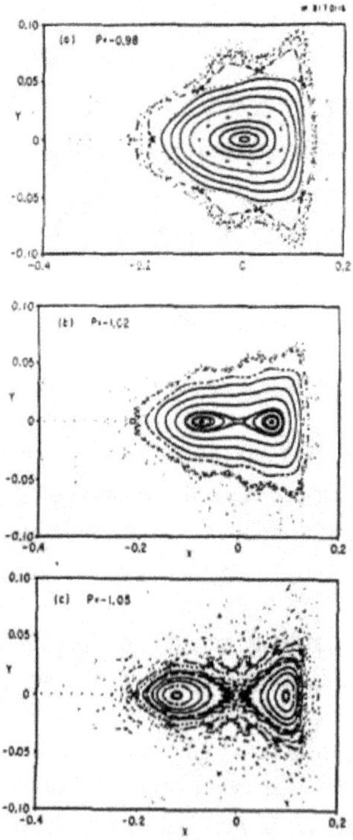

Figure 3.1.2.1: Period doubling of the fixed point of the quadratic area preserving map

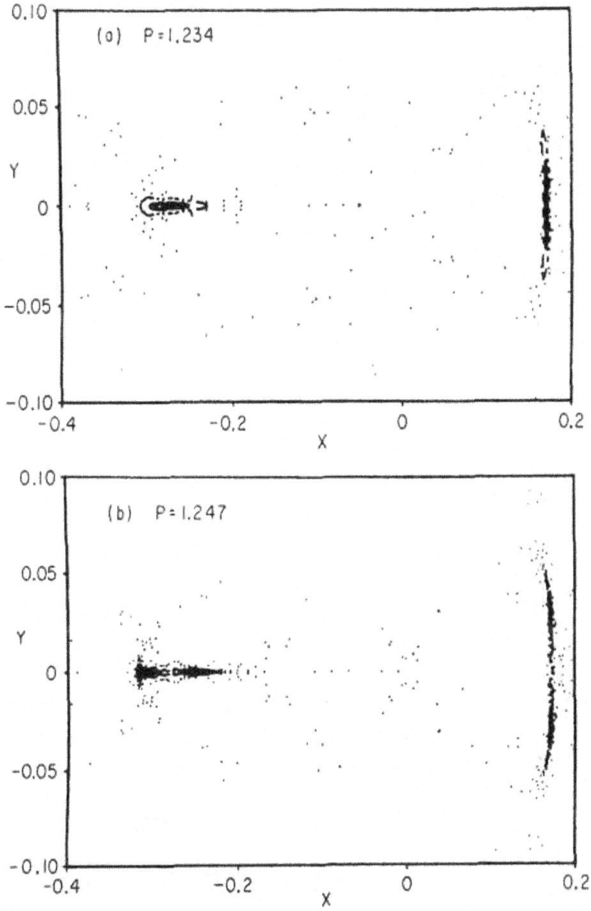

Figure 3.1.2.2: Period doubling of the 2-cycle of the quadratic area preserving map

Plotting the positions of the periodic points lying on the dominant line against parameter gives figure 3.1.2.3. The parameter values for successive doublings accumulate at a value p^*, and the positions of the good points at a point x^*. For this particular case:

$$\begin{aligned} p^* &= -1.26631127692210\ldots \\ x^* &= -0.2360060924260\ldots \end{aligned} \tag{3.1.2.2}$$

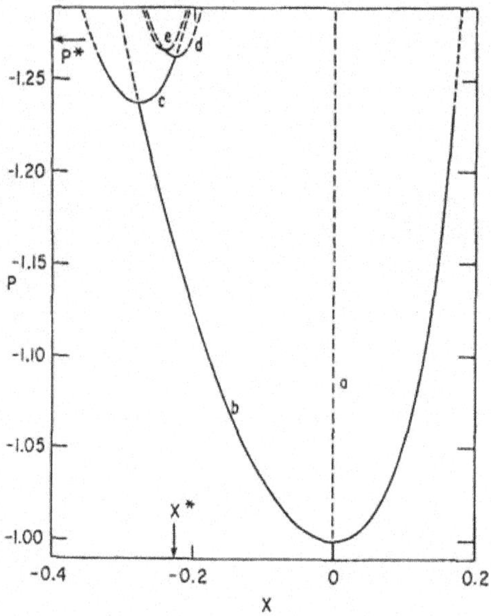

Figure 3.1.2.3: The part of the period doubling tree of the fixed point of the quadratic map, that lies on the dominant line

The important thing to see in figure 3.1.2.3 is that the period doubling tree repeats itself on successive doubling of the periods and rescaling about the accumulation point. Asymptotically, the self-similarity becomes exact, with scale factors of:

$$\delta = 8.712097200\ldots \text{ in parameter}$$
$$\alpha = -4.018076704\ldots \text{ along the symmetry line.} \tag{3.1.2.3}$$

For example, the parameter values p_n for the nth bifurcation converge asymptotically geometrically with ratio δ:

$$r_n = \frac{p_{n+1} - p_n}{p_{n+2} - p_{n+1}} \to \delta. \tag{3.1.2.4}$$

This is shown in table 3.1.2.2, but in fact for a different period doubling sequence. Table 3.1.2.2 and subsequent tables in this section are for the period doubling sequence of one of the 3-cycles of the quadratic map, because I happened to look at it in more detail. They begin at period 6, however, because at that time I had written my program to find even cycles only!

Table 3.1.2.1: Parameter values and positions for the first 12 period doubling bifurcations of the sequence beginning from the fixed point of the quadratic map (3.1.2.1)

PERIOD	RESIDUE
1	.999999999999999999999999999987
2	.999999999999999999999999999849
4	1.000000000000000000000000001363
8	.999999999999999999999999993020
16	1.000000000000000000000001759475
32	1.000000000000000000000021222428
64	1.000000000000000000000312047853
128	.999999999999999999996739854115
256	1.000000000000000000111880640226
512	1.000000000000000000999880526949
1024	1.000000000000000001228905643955
2048	1.000000000000000369955228675900
	PARAMETER VALUE
1	-1.000000000000000000000000000000
2	-1.236067977499789696409173668747
4	-1.262841686313575492599739636232
8	-1.265913483003443894905301377840
16	-1.266265663774486328063119022899
32	-1.266306046722604021356864916458
64	-1.266310677203981419114320091466
128	-1.266311208155739610282145458380
256	-1.266311269037040844938329668542
512	-1.266311276017963076103742053063
1024	-1.266311276818426832943421088066
2048	-1.266311276910211586449112033808
	POSITION ON SYMMETRY LINE
1	-.000000000000000000000000000000
2	-.276393202250021030359082633140
4	-.224612021927501036463879628863
8	-.238675841208976215156703479960
16	-.235323099797104402768963024429
32	-.236173934088673034778807461866
64	-.235964075973725644525324400804
128	-.236016521203697637581302694041
256	-.236003493741647217237398922097
512	-.236006738805204053403892410437
1024	-.236005931515893372675095720885
2048	-.236006132467727762172612857047

Table 3.1.2.2: Asymptotically geometric convergence of the parameter values for successive period doublings from a 3-cycle of the quadratic map.

Period	Parameter value p_n	r_n
6	-.510401575553307655713307333041	
12	-.511644183605752915716245801797	
24	-.511786718529065077379005694018	8.717919956527
48	-.511803072043921728203803757739	8.715858612755
96	-.511804947223579622235238090174	8.721038961683
192	-.511805162241835102434910666382	8.721025355294
384	-.511805186896791880193292408490	8.721096427724
768	-.511805189723839779154468059397	8.721096231449
1536	-.511805190048001732277978029172	8.721097191452
3072	-.511805190085171591316369821085	8.721097187608
6144	-.511805190089433653552787651757	8.721097200502
12288	-.511805190089922360665948562535	8.721097200429
24576	-.511805190089978398008718546989	8.721097200609

Similarly, the positions x_n of the good points at bifurcation, converge with ratio α. Table 3.1.2.3 shows the difference Δx_n between the positions of the good and bad points on the dominant symmetry line, and their convergence rate.

Table 3.1.2.3: Asymptotically geometric convergence of the differences in position of the good and bad points on the dominant line.

period	Δx_n	ratio
6	.679750818040771681083650194360	
12	-.165036332415921861586956853752	-4.118794983
24	.041660536550037927582510647267	-3.961454798
48	-.010329954653922757134373167568	-4.032983487
96	.002573453851940660825676378052	-4.014043090
192	-.000640306494160258929763967359	-4.019096909
384	.000159366695098805067310625599	-4.017818740
768	-.000039661796123800129233725186	-4.018141150
1536	.000009870880494516428057514895	-4.018060612
3072	-.000002456615782232220834052969	-4.018080713
6144	.000000611391113979441236488527	-4.018075706
12288	-.000000152160130600866163915344	-4.018076953
24576	.000000037868897017703774548184	-4.018076642

There is self-similarity across the symmetry line too. Figure 3.1.2.4 shows the orbits in the period doubling sequence at the accumulation parameter value p^*. It repeats itself on successive doubling of the periods and rescaling about $(x^*, 0)$. Asymptotically the self-similarity becomes exact, with rescaling factors of α along the symmetry line, as before, and

$$\beta = 16.363896879... \text{ across the symmetry line.} \quad (3.1.2.5)$$

For example, defining (x_n, y_n) to be the position of the point one quarter of the way round the nth orbit from its good point (one of the daughters of the previous bad point), then x_n converges to x^* at rate α, and y_n converges to zero at rate β. This is shown in table 3.1.2.4, except that I have taken the positions at period doubling rather than at the accumulation parameter value. The table is for the period doubling sequence of the 3-cycle of the quadratic map, but begins at period 12 because "one quarter of the way round" can be defined only when the period is a multiple of 4. Note that $\beta \neq \alpha^2$, although they differ by only 1%. This was a point of confusion to some workers, who were not using symmetry coordinates. If the symmetry line is bent, then one sees results which look like α^2 scaling across the line. Why β is so close to α^2, however, is still a mystery.

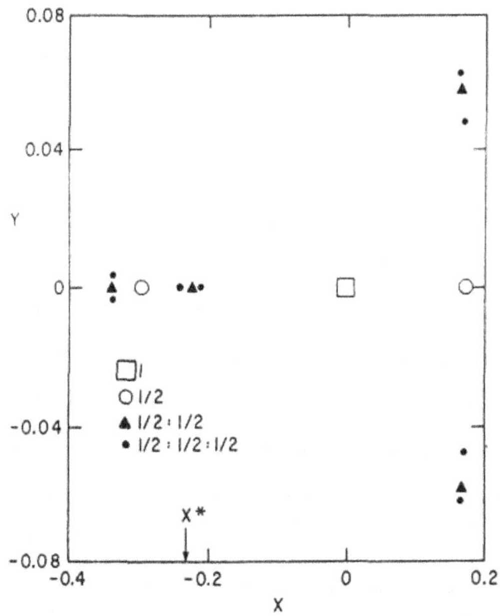

Figure 3.2.2.4: Positions of some of the orbits of the period doubling sequence of the fixed point in the quadratic map, at the accumulation parameter value

Table 3.1.2.4: Asymptotically geometric scaling across the dominant line

period	y_n	ratio
12	.011797697777518476974994007162	
24	.000838833885484369559068451797	14.06440295
48	.000049305475214230393123962230	17.01299666
96	.000003042294305194558514837386	16.20667505
192	.000000185470960384173134987420	16.40307624
384	.000000011340908407531420588807	16.35415380
768	.000000000692941868154258923896	16.36632007
1536	.000000000042347333959867571949	16.36329382
3072	.000000000002587827699981952077	16.36404694
6144	.000000000000158142869336487795	16.36385953
12288	.000000000000009664127113446582	16.36390617
24576	.000000000000000590576227084841	16.36389456

Another way to see the second scaling β is to evaluate the tangent maps:

$$\begin{vmatrix} A_n & B_n \\ C_n & D_n \end{vmatrix} \tag{3.1.2.6}$$

to the periodic orbits, at the good point. One finds (both at p^*, and at p_n) that:

$$\frac{B_{n+1}}{B_n} \rightarrow \frac{\beta}{\alpha} \text{ and } \frac{C_{n+1}}{C_n} \rightarrow \frac{\alpha}{\beta} \tag{3.1.2.7}$$

except that at $p_n, C_n = 0$ (see §3.1.1). For example, table 3.1.2.5 shows B_n for the good points of the period doubling sequence of the 3-cycle, at the parameter values where they period double. This is consistent with scaling in x by α and in y by β for each doubling of the period, because

$$\begin{vmatrix} 1/\alpha & 0 \\ 0 & 1/\beta \end{vmatrix} \begin{vmatrix} A_n & B_n \\ C_n & D_n \end{vmatrix} \begin{vmatrix} \alpha & 0 \\ 0 & \beta \end{vmatrix} = \begin{vmatrix} A_n & \frac{\beta}{\alpha} B_n \\ \frac{\alpha}{\beta} C_n & D_n \end{vmatrix} \tag{3.1.2.8}$$

Benettin et al. (1980b) found β in essentially this way, independently, though they did not make the connection between the scaling directions and symmetry.

Table 3.1.2.5: B_n for the period doubling sequence of the 3-cycle of the quadratic map, measured at period doubling.

period	B_n	ratio
6	.58580504655397961240189l952D+02	
12	-.24225314973224807396659773 3D+03	-4.135388576
24	.98302058058801211532140464 6D+03	-4.057823733
48	-.40069260881859086398261752 5D+04	-4.076136519
96	.16314912518757066739425896 0D+05	-4.071677929
192	-.66447264213832267257495695 8D+05	-4.072793166
384	.27060735065441611003472597 9D+06	-4.072513050
768	-.11020711244245596434909365 2D+07	-4.072583844
1536	.44882573308739523791753456 9D+07	-4.072565945
3072	-.18278744289839039283864190 2D+08	-4.072570474
6144	.74441453369599139984638156 4D+08	-4.072569329
12288	-.30316800138308534189030437 3D+09	-4.072569618
24576	.12346727696539201364012586 3D+10	-4.072569545

Also consistent with the scaling, A_n (= D_n, by symmetry) has a limit, evaluating at p^*:

$$A_n \to -1.27176... . \tag{3.1.2.9}$$

Alternatively, in terms of the residues R_n at p^*,

$$R_n \to R^* = 1.135877549336... \tag{3.1.2.10}$$

Incidentally, the convergence is again asymptotically geometric, at a rate

$$\delta' = -0.117... . \tag{3.1.2.11}$$

Note that in order to extract the limits mentioned in this section, I typically "superconverged" the results to several levels. By superconverging a sequence x_n, $n = 1,..., N$, I means estimating the accumulation point from each triple x_{i-1}, x_i, x_{i+1}, $i = 2,..., N-1$. To do this, set:

$$r_i = \frac{x_{i-1} - x_i}{x_i - x_{i+1}} \tag{3.1.2.12}$$

and

$$a_i = \frac{r_i x_{i+1} - x_i}{r_i - 1}. \tag{3.1.2.13}$$

The a_i form a new sequence which should have the same limit, and be close to it. Roundoff errors, however, eventually lead to divergence, so one has to pick the value that looks best.

§3.1.3 **Universality**. The above behaviour appears to be the same for most periodic orbits of most one parameter families. If an orbit loses stability by direct period doubling (§1.2.4.4), then it is likely to have an infinite sequence of period doublings, with the same scalings as above. Besides the period doubling sequences of the fixed point and 3-cycle of the quadratic map, I followed that of the 1/5-orbit bifurcating from the fixed point, which gave the same scalings. Other people have found aspects of these results for several periodic orbits of other maps, such as some generalised standard maps (Benettin et al., 1980a, b), and even for continuous time conservative systems like a forced Duffing oscillator (Vivaldi, private communication) and a particle in a standing wave (Bialek, private communication). Note that numerical errors are much larger in continuous time, than in discrete time, so it is not possible to follow as many period doublings.

In summary, it looks as if there are coordinate changes B_n, commuting with the dominant symmetry, such that for a map T at the accumulation of a period doubling sequence from initial period m:

$$B_n(T^m)^{2^n} B_n^{-1} \tag{3.1.3.1}$$

converges to some universal map T^*, with:

$$B_{n+1} \sim B\, B_n, \;\; B(x, y) = (\alpha x, \beta y) \tag{3.1.3.2}$$

$$\begin{aligned} \alpha &= -4.018076704... \\ \beta &= 16.363896879.... \end{aligned} \tag{3.1.3.3}$$

Furthermore, for a one parameter family T_p passing through accumulation of period doubling, there appear to be a reparametrisation μ and parameter dependent coordinate changes B_n satisfying the same scaling as above, such that:

$$B_n \left(T^m_{\mu\delta^{-n}} \right)^{2^n} B_n^{-1} \tag{3.1.3.4}$$

where

$$\delta = 8.710297200... \tag{3.1.3.5}$$

converges to some universal one parameter family T^*_μ. The scale in μ can be chosen, for example, so that the fixed point nearest the origin period doubles at $\mu = -1$.

All the above mentioned systems are reversible, but reversibility is probably not necessary to see the effects of the universal one parameter family. Greene (private communication) followed the first five bifurcations of a period doubling sequence in Rannou's map (1.1.4.3.11), believed to have no symmetry. He found the same δ to 4 figures, though he did not accurately determine the spatial scalings. Note that this would involve finding scaling parameters/coordinates, as discussed in §2.3.5.

§3.2 RENORMALISATION

The universally self-similar period doubling sequences can be explained by a fixed point of a renormalisation. I derive a simple approximation for the fixed point, and then discuss various approaches towards finding arbitrarily good approximations, and to proving its existence.

§3.2.1 **Outline.** The universal self-similarity discussed in §3.1 suggests that one consider the following renormalisation:

$$N(T) = B\, T^2\, B^{-1} \tag{3.2.1.1}$$

where B is a coordinate change which may depend on T, chosen to "renormalise" the map T^2 in some sense. B should have constant Jacobian so that T area preserving implies $N(T)$ is too, and for reversible T, B should commute with a dominant symmetry. For example, in symmetry coordinates one could choose B to be linear and diagonal:

$$B(x, y) = (\alpha x, \beta y) \tag{3.2.1.2}$$

with

$$\left(\frac{1}{\alpha}, \frac{1}{\beta}\right) = T(T(0, 0)). \tag{3.2.1.3}$$

The results of the previous section suggest that N has a fixed point T^* in the space of reversible area preserving maps, with essentially only one unstable direction, eigenvalue δ. Also, we expect the next largest essential eigenvalue after δ to be δ' (3.1.2.11).

To test out this idea, I used MACSYMA to perform N a few times on the quadratic map in scaling coordinates and at p^*, using a constant B as above. Although it was not possible to apply N more than 4 times without introducing a truncation, there appeared to be reasonable convergence. Table 3.2.1.1 contains the approximate fixed point so produced.

Table 3.2.1.1: (x_1, y_1) are the coefficients of the DeVogelaere quadratic map shifted to its accumulation point. Powers of x increase downwards and powers of y increase across. (x_2, y_2), (x_4, y_4) are its first two iterates under N, using the empirical scalings, and truncating at the levels indicated.

x_1	0.4086	-1.0			
	-0.19658				
	-2.2663				
y_1	0.0501	-2.0487	2.2663		
	0.59724	0.891044			
	-4.555	10.2723			
	2.01938				
	11.64015				
x_2	0.402836	-1.0061	0.06801337		
	-0.19449	-0.1089037			
	-2.26752	0.31246			
	-0.2501569				
	0.3588674				
y_2	0.050245	-2.056	2.3967	-0.314	0.0106
	0.596378	0.7846	0.4426	-0.034	
	-4.5912	10.906	-2.1255	0.0976	
	1.76336	2.011	-0.23449		
	12.411	-4.788	0.336		
	2.28	-0.538			
	-3.59	0.515			
	-0.412				
	0.296				
x_4	0.40349	-1.00566	0.0625	1.289e-3	-1.013e-4
	-0.19470	-0.09866	-4.807e-3	1.557e-4	
	-2.26760	0.28838	8.636e-3	-9.43 e-4	
	-0.2240	-0.0219	1.078e-3		
	0.33267	0.01928	-3.29 e-3		
	-0.02499	2.489e-3			
	0.0143	-5.104 e-3			
	1.91 e-3				
	-2.96 e-3				
y_4	0.050225	-2.05546	2.383	-0.283	2.53e-3
	0.59641	0.7952	0.394	-6.64e-3	
	-4.5870	10.839	-1.923	0.027	
	1.7940	1.778	-0.047		
	12.326	-4.344	0.106		
	2.007	-0.1108			
	-3.271	0.1815			
	0.0871				
	0.1151				

§3.2.2 **Simple approximation for the fixed point.** Several people have generated approximate fixed points by hand. Greene and I (Greene et al., 1981) worked out one involving truncation to DeVogelaere quadratic form. Others were produced by Derrida and Pomeau (1980), Collet et al. (1981b), and Helleman (1980). I made some corrections to Helleman's scheme, and extended it to give the second scaling (Appendix to Helleman, 1983). As it is cleaner than that of Greene and myself, it is the one I will present here.

Consider the quadratic DeVogelaere map, which I choose to write in the following form:

$$\begin{aligned} x' &= -y + f(x) \\ x &= y' + f(x') \end{aligned} \tag{3.2.2.1}$$

with

$$f(x) = Cx + x^2. \tag{3.2.2.2}$$

The parameter C is the same as p in (3.1.2.1), but the scales in x are different. It has a fixed point at the origin, and a 2-cycle at:

$$\begin{aligned} \hat{x} &= a \pm b \\ \hat{y} &= 0 \end{aligned} \tag{3.2.2.3}$$

where

$$a = -\frac{1+C}{2} \tag{3.2.2.4}$$

$$b^2 = \frac{(C+1)(C-3)}{2}.$$

Writing

$$\begin{aligned} \Delta x &= x - \hat{x} \\ \Delta y &= y - \hat{y} \end{aligned} \tag{3.2.2.5}$$

we get

$$\begin{aligned} \Delta x' &= -\Delta y + (C + 2\hat{x})\Delta x + \Delta x^2 \\ \Delta x &= \Delta y' + (C + 2\hat{x}')\Delta x' + \Delta x'^2. \end{aligned} \tag{3.2.2.6}$$

Iterating twice gives:

$$
\begin{aligned}
\Delta x'' &= -\Delta y' + (C + 2\hat{x}')\Delta x' + \Delta x'^2 \\
&= -\Delta x + 2(C + 2\hat{x}')\Delta x' + 2\Delta x'^2 \\
&= -\Delta x + 2(C + 2\hat{x}')(-\Delta y + (C + 2\hat{x}')\Delta x + \Delta x^2) + 2(-\Delta y + (C + 2\hat{x})\Delta x + \Delta x^2)^2 \\
&= -\frac{\beta}{\alpha}\Delta y + C'\Delta x + \alpha\Delta x^2 + 0(\Delta x^3)
\end{aligned}
\tag{3.2.2.7}
$$

where

$$
\begin{aligned}
C' &= 2(C + 2\hat{x}')(C + 2\hat{x}) - 1 \\
a &= 2(C + 2\hat{x}') + 2(C + 2\hat{x})^2 \\
\beta &= 2\alpha(C + 2\hat{x}').
\end{aligned}
\tag{3.2.2.8}
$$

In the above, I have ordered Δy as Δx^2. By symmetry, the other equation gives:

$$\Delta x = \frac{\beta}{\alpha}\Delta y'' + C'\Delta x'' + \alpha\Delta x''^2 + 0(\Delta x^3). \tag{3.2.2.9}$$

Thus, truncating at $0(\Delta x^2)$, we see that we have a rescaled version of the original map, rescaled by α in x and β in y, with C replaced by C':

$$C' = -2C^2 + 4C + 7. \tag{3.2.2.10}$$

This transformation on C has a fixed point at:

$$C_\infty = \frac{(3 - \sqrt{65}\,)}{4} = -1.2656 \tag{3.2.2.11}$$

(compare p^* of (3.1.2.2)), with multiplier:

$$\delta = -4C_\infty + 4 = 1 + \sqrt{65} = 9.062. \tag{3.2.2.12}$$

At C_∞ the rescaling factors are:

$$\alpha = \left(-2 - 2\sqrt{(C+1)(C-3)}\right) + 2\left(-2 + 2\sqrt{(C+1)(C-3)}\right)^2 = -4.0955 \tag{3.2.2.13}$$

$$\beta = \left(-2 - 2\sqrt{(C+1)(C-3)}\right)\alpha = -4.1286\,\alpha = 16.909. \tag{3.2.2.14}$$

Compare the empirical values of §3.1.2.

§3.2.3 **Existence of the fixed point.** One would like to extend the above sort of analysis to arbitrary truncation level. Collet et al. (1981b) tried iterating a Newton method to find the fixed point in a space of polynomials, truncating at some degree. They had trouble, however, with the coordinate change eigenvalue of $\beta/\alpha^2 = 1.0136$, because it is too close to 1. This can be eliminated by restricting to maps with symmetry, since it corresponds to a coordinate change which does not preserve symmetry. Note that N preserves symmetry, provided S and B commute. The problem was to find how to impose the restriction.

One scheme I thought of, but never tried, was to consider a modification of the renormalisation:

$$N_S: T' = B\, S\, T^{-1}\, S\, T\, B^{-1}. \tag{3.2.3.1}$$

For any T, T' has symmetry S, provided S and B commute, since it can be written as a product of S with another involution:

$$T' = S.\, BT^{-1}STB^{-1}. \tag{3.2.3.2}$$

The second factor is an involution, because it is a coordinate transform of S. Furthermore, if T has symmetry S, then

$$N_S(T) = N(T). \tag{3.2.3.3}$$

So N_S restricts itself to symmetric maps, and acts just the same as N on them. This scheme has been superceded, however, by the symmetric representation (§1.1.4.4) (Koch, private communication). Note that although composition is difficult in general in the symmetric representation, composition of a map with itself is easy.

Another way to restrict to symmetric maps is to use generating functions. As mentioned in §1.1.4.4, symmetry is easy to impose on Poincaré's generating function, but I could not find an easy algorithm for their composition. Meanwhile, Widom and Kadanoff (1981) hit on the action generating function (§1.1.3.3). The symmetry constraint is easy to apply, and composition relatively straightforward. They obtained an extremely good fixed point, using truncation at some degree.

Recently, Eckmann et al. (1982) proved existence of the fixed point, in the action representation. They did this along the same lines as Lanford's (1981) proof for the one dimensional case.

Note that there are lots of other fixed points, for example, that of Zisook and Shenker (1982), and a whole bunch of linear ones whose interpretation is not clear to me.

§3.3 THE UNIVERSAL ONE PARAMETER FAMILY

The significant object for any unstable fixed point of a renormalisation is its unstable manifold, which gives a universal family. I describe properties of the universal one parameter family for the period doubling fixed point. First I describe the universal bifurcation tree. It looks as if there is no region of stability at the accumulation point. I show, however, that there are universal islets of stability beyond period doubling. Period doubling is significant because it leads to disintegration of the stable region associated with an initially stable periodic orbit.

§3.3.1 **Introduction.** The significance of this renormalisation and fixed point lies in the unstable manifold of the fixed point, which can be parametrised naturally to give the universal one parameter family. Any one parameter family crossing the stable manifold transversally (including non-zero speed), will behave the same as the universal one, on a small enough scale in space and long enough time scale.

In this section I describe some properties of the universal one parameter family T^*_μ. I use the normalisation that $\mu = -1$ when the fixed point nearest the origin period doubles. Note that the self-similarity implies that there are infinitely many fixed points further away (Greene et al., 1981). This is because, if T has a fixed point x, then x is also a fixed point of T^2. For the universal family:

$$T_{\mu.\delta} = BT^2_\mu B^{-1} \tag{3.3.1.1}$$

so that Bx is a fixed point of $T_{\mu.\delta}$. Its residue R' is related to the residue R of x by:

$$R' = 4R(1 - R) \tag{3.3.1.2}$$

Similarly, existence of a periodic orbit of odd period implies existence of another one. Existence of a periodic orbit of even period implies existence of two periodic orbits of half the period and the same residue.

§3.3.2 **The bifurcation tree.** I start with the stability of the orbits of the period doubling sequence of the fixed point nearest the origin. Their residues increase almost linearly from 0 at birth, pass through 1 at period doubling, and through R^* at $\mu = 0$. The dependence is not quite linear, however, as

$$R^* \neq \frac{\delta}{\delta - 1} \tag{3.3.2.1}$$

showing that the following points (μ, R) for the 2-cycle:

$$(-1, 0),\ (-\frac{1}{\delta}, 1),\ (0, R^*) \tag{3.3.2.2}$$

are not collinear. Note, from the bifurcation analysis of §1.2.4.4, that one expects the residue to be locally linear near birth.

The residue is related to the component A (= D, by symmetry) of the tangent map:

$$DT = \begin{vmatrix} A & B \\ C & D \end{vmatrix} \tag{3.3.2.3}$$

by

$$R = \frac{1 - A}{2}. \tag{3.3.2.4}$$

One can also follow the other components of the tangent map. They are all shown in figure 3.3.2.1 for the good and bad points. They have been scaled so that $B = 1$ at birth. The pictures in this section were actually produced for the fourth power of the quadratic map:

$$\begin{aligned} Q : x' &= f(x) - y \\ y' &= x - f(x') \end{aligned} \tag{3.3.2.5}$$

with

$$f(x) = px - (1-p)x^2. \tag{3.3.2.6}$$

The fourth power was sufficiently large that deviations from exact self-similarity were below the resolution of the pictures. Note that the behaviour of B for the good point when it period doubles can be predicted to be:

$$\frac{\beta}{2\alpha} = 2.0363... \tag{3.3.2.7}$$

as its daughters must be born with:

$$B' = \frac{\beta}{\alpha} \tag{3.3.2.8}$$

by scaling, and the daughters of a point with tangent map:

$$\begin{vmatrix} -1 & B \\ 0 & -1 \end{vmatrix} \tag{3.3.2.9}$$

are born with tangent map:

$$\begin{vmatrix} -1 & B \\ 0 & -1 \end{vmatrix}^2 = \begin{vmatrix} 1 & -2B \\ 0 & 1 \end{vmatrix}. \tag{3.3.2.10}$$

So $B' = -2B$. Hence result.

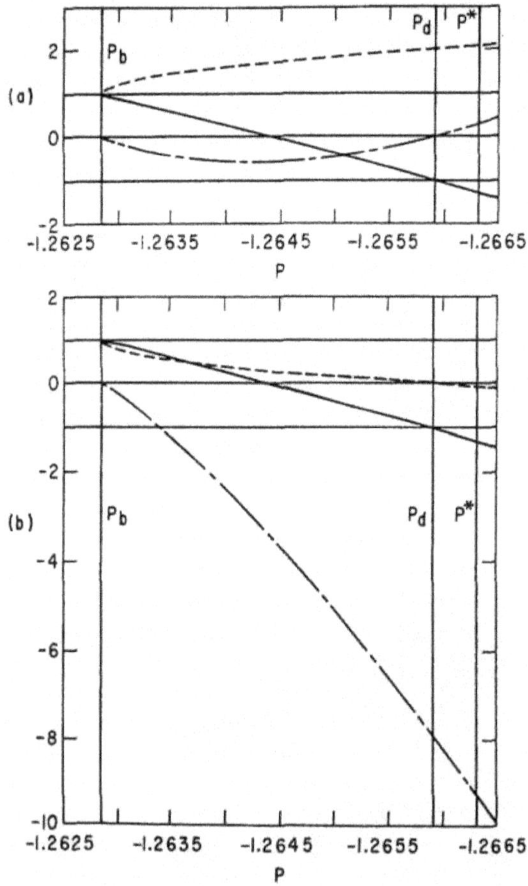

Figure 3.3.2.1: Behaviour of the elements A (continuous line), B (dashed) and C (chain-dashed) of the tangent map to an orbit of the period doubling sequence in the universal map, at (a), its good point, (b) its bad point. The orbit is born at parameter value p_b, period doubles at p_d, and the accumulation of period doubling is at p^*.

The universal family has many more periodic orbits than those of the period doubling sequence. Some of them are shown at $\mu = 0$ in figure 3.3.2.2. The orbits I looked at are all attached to the original fixed point by bifurcations. This is shown in figure 3.3.2.3. I checked that the residues all had the correct local parameter dependence at birth (cf. §1.2.4.4).

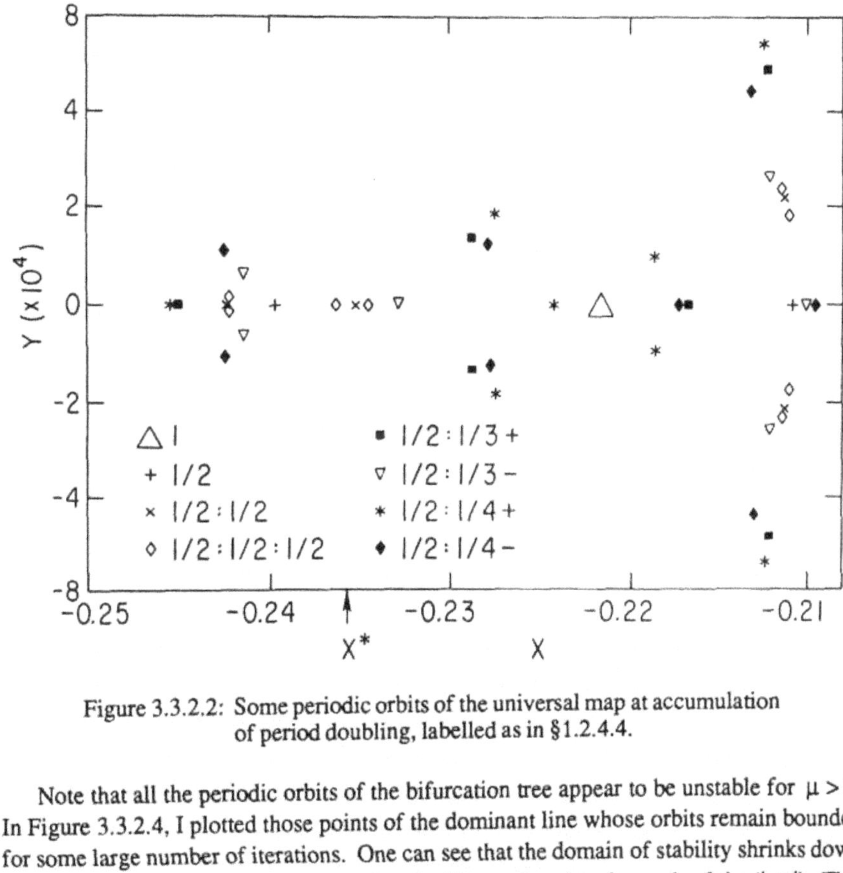

Figure 3.3.2.2: Some periodic orbits of the universal map at accumulation of period doubling, labelled as in §1.2.4.4.

Note that all the periodic orbits of the bifurcation tree appear to be unstable for $\mu > 0$. In Figure 3.3.2.4, I plotted those points of the dominant line whose orbits remain bounded for some large number of iterations. One can see that the domain of stability shrinks down to nothing at $\mu = 0$ ($p = p^*$). Note also the big cut forming the neck of the "cat". This corresponds to third order resonance, as is made clear by superposing figures 3.3.2.3, 3.3.2.4 to get figure (iii) of the abstract. Stability analysis (§1.2.1.4) shows that a point with multipliers $e^{\pm 2\pi i/3}$ is typically unstable. Of course, that is just a local result, so there could be other regions of stability. These figures, however, indicate that they must be very small.

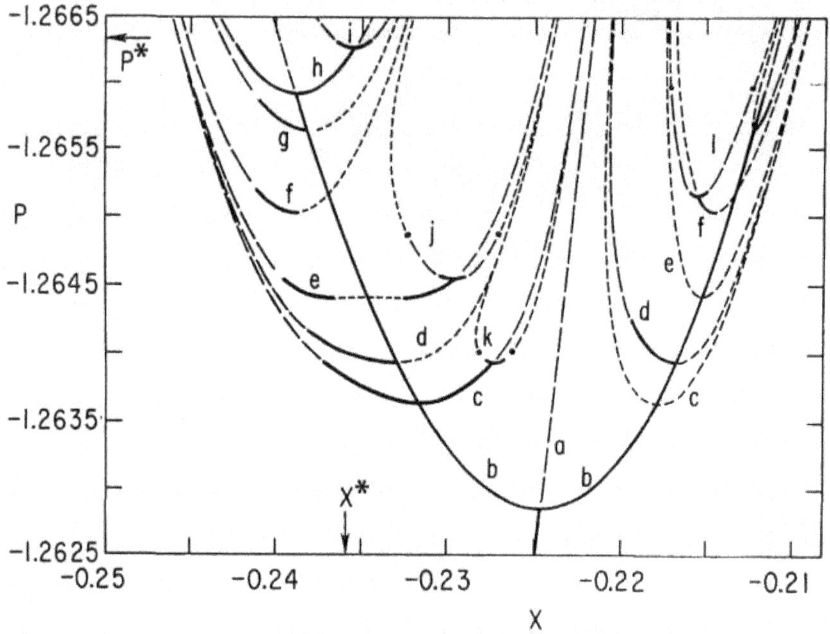

Figure 3.3.2.3: Some of the branches of the bifurcation tree of the universal map: a) the fixed point 1; b) 1/2; c) 1/2:1/6; d) 1/2:1/5; e) 1/2:1/4; f) 1/2:1/3; g) 1/2:2/5; h) 1/2:1/2; i) 1/2:1/2:1/2; j) 1/2:1/4:1/2; k) 1/2:1/6:1/2; l) 1/2:1/3:1/2. Thick line indicates elliptic, long dahses, inversion hyperbolic, and short dashes, regular hyperbolic.

To calculate the domain of stability, the quadratic map has a simple escape criterion, namely, for $p < 1$, an orbit is unbounded in some direction in time iff it has a point (x, y) with $x < -1$. Iterating this criterion one step forward and backward shows that it is sufficient to search only a finite area for bounded orbits, viz.:

$$|y| \leq 1 + f(x) = (1 + x)(1 - (1 - p)x). \qquad (3.3.2.11)$$

To derive the escape criterion $x < -1$, write the map in second difference form:

$$x'' - 2x' + x = 2f(x') - 2x' = -2(1 - p)x(1 + x). \qquad (3.3.2.12)$$

Then, writing $x = -1 - \xi$, if $\xi' > 0$ the second difference is negative, and less than:

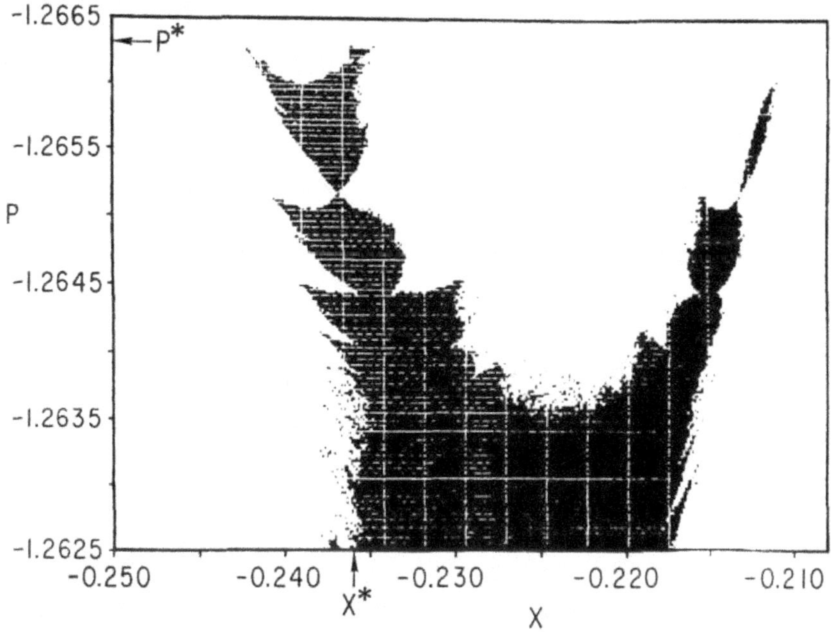

Figure 3.3.2.4: Points of the dominant line whose orbits remain bounded

$$-2(1 - p) \max(\xi', \xi'^2). \tag{3.3.2.13}$$

Thus, at least one of ξ, ξ'' is greater than $(1 - p) \max(\xi', \xi'^2)$, so by induction ξ_n goes to $+\infty$ in at least one direction in time. In fact the divergence is quadratic. Alternatively, see §1.4.3.

In figure 3.3.2.5, I show the total trapped area as a function of parameter, for the quadratic map. I iterated points until they escaped, or for a maximum of 5000 iterations. To check that this was sufficiently large, I checked how many points remained after 2500 iterations. On average, fewer than 5% of these escaped in the next 2500 iterations. I took points in the region (3.3.2.11) on a grid with spacing depending on parameter in such a way as to make the number remaining trapped on the order of 100, to give picture accuracy.

The trapped area shrinks to zero at the accumulation point, but not monotonically. In fact, it appears to come down to zero every time the multipliers of an orbit of the period doubling sequence pass through $e^{\pm 2\pi i/3}$ (third order resonance), as previously discussed. Presumably the trapped area has a dense set of jump discontinuities as outermost invariant

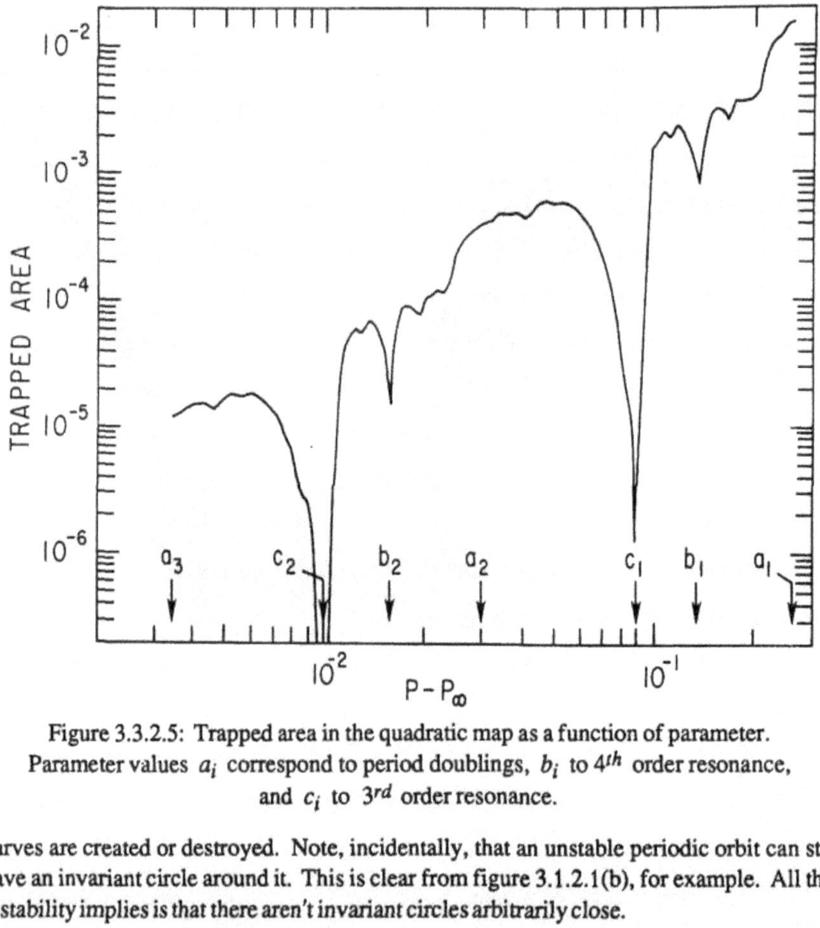

Figure 3.3.2.5: Trapped area in the quadratic map as a function of parameter. Parameter values a_i correspond to period doublings, b_i to 4th order resonance, and c_i to 3rd order resonance.

curves are created or destroyed. Note, incidentally, that an unstable periodic orbit can still have an invariant circle around it. This is clear from figure 3.1.2.1(b), for example. All that instability implies is that there aren't invariant circles arbitrarily close.

The area decreases by a factor of about 33 for each factor of δ that p comes closer to p^*. This is consistent with the universal scaling, which would give a factor of:

$$\left| \frac{\alpha\beta}{2} \right| = 32.876\ldots . \tag{3.3.2.14}$$

The factor of 2 comes from the fact that we are looking at the *total* trapped area, rather than just that associated with the good point of period 2^n. The trapped area of a neighbourhood of the good point is repeated 2^n times by its images. The total trapped area can not be meaningfully defined for the universal map, as it could only take the values 0 and ∞, but it would scale like $|\alpha\beta|$ with parameter.

§3.3.3 **Islets of stability beyond period doubling.** For p at and beyond the accumulation point, I found that all points of a grid with a spacing of 0.005 in x, y and p, escaped. Note that it is not necessary to check beyond $p = -\sqrt{5} = -2.236...$, because an argument of Newhouse (1980) shows that then the only bounded orbits are those of a horseshoe, which always has zero area (see §1.4.3).

The above search on a grid, however, was misleading. There are in fact, universally, islets of stability beyond the accumulation of period doubling. These arise by tangent bifurcation, and are closely analogous to the stable windows for one dimensional maps. The existence of such tangent bifurcations was suggested by counting periodic orbits of the quadratic map (see §1.2.3.5). Newhouse's horseshoe contains 2^n fixed points of Q^n, so in particular it has 6 symmetric 5-cycles. But only 4 5-cycles occur in the bifurcation tree of the fixed point. So I looked for a symmetric 5-cycle tangent bifurcation. The approximate location is easiest found by looking for a tangency of $\text{Fix}(Q^5S) = Q^2\text{Fix}(QS)$ and $\text{Fix}(S)$ (see figure 3.3.3.1). Then a 2-D root finder for a point of $\text{Fix}(QS)$ with second iterate on $\text{Fix}(S)$ and residue 0, gives the tangent bifurcation to arbitrary accuracy.

The period doubling sequence of the fixed point accumulates from above at $p^* = -1.2663113...$, and I find a period 5 tangent bifurcation at $p_0 = -1.5596292...$. The 5-cycle of positive residue is elliptic only up to $p_0' = -1.5596327...$ .that is for an interval of 3.5×10^{-6}, so it is not surprising that it did not show up in the grid search. It is reproduced as an elliptic $5 \times 2n$-cycle from parameters p_n to p_n', converging to p^* at rate δ (to 8 figures), and similar scaling in x (and presumably y too). The results are given in table 3.3.3.1. Thus I claim that T_μ^* has a period 5 tangent bifurcation too.

Similarly, there are two symmetric 6-cycles (belonging to QS) not attached to the bifurcation tree. They pop up by tangent bifurcation at $p = -1.584447622...$ and the 6-cycle of positive residue is elliptic for an interval of 5.2×10^{-8}. I found several other symmetric tangent bifurcations beyond p^* (of periods 7 and 8). Table 3.3.3.2 shows how many to expect, up to period 8, based on the counting arguments of §1.2.3.5, and the possible bifurcations from the fixed point. I expect there are infinitely many of higher period, and non-symmetric ones too. Table 3.3.3.3 gives the values for the 8-cycles belonging to S.

The least period for which there is a tangent bifurcation not attached to the bifurcation tree of the fixed point is 5 in the quadratic map, by counting, but the universal map even possesses elliptic *fixed* points beyond $\mu = 0$. For example, corresponding to the period 8 points above, there will be fixed points three factors of δ, α, β away.

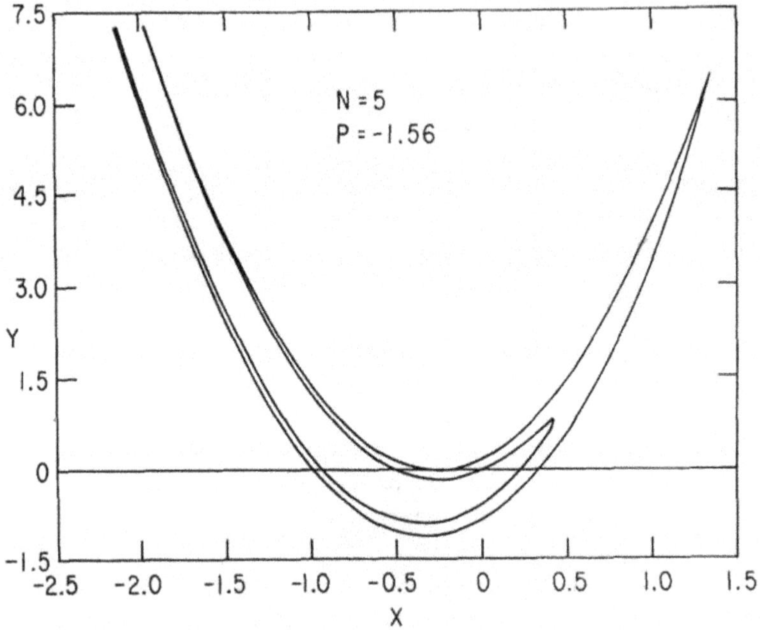

Figure 3.3.3.1: Locating a tangent bifurcation by tangency of $Q^2\mathrm{Fix}(QS)$ to $\mathrm{Fix}(S)$.

Table 3.3.3.1: Tangent bifurcations and residue 1 points for periods 5×2^n.

period	residue	position	parameter value
5	0	-.27103079557742647046834	-1.5596291604160276065317213 2
5	1	-.27040841382230190569308	-1.55963268760887242100353062
10	0	-.24027175376777103002527	-1.30339358643678358888651144
10	1	-.24047263139613939392434	-1.30339415005115888371599158
20	0	-.23666039440256712796129	-1.27053278607569382758888846
20	1	-.23661041360593196549231	-1.27053284942501821274441080
40	0	-.23604072895859858338452	-1.26679575056859787675632768
40	1	-.23605320262007689731744	-1.26679575785016304944273126
80	0	-.23602015265080155665430	-1.26636682360656341539773171
80	1	-.23601704859668992340028	-1.26636682444127938400649745
160	0	-.23600519363581487110222	-1.26631764622815342582303580
160	1	-.23600596622354358503696	-1.26631764632386883600086454
320	0	-.23600661429630931254676	-1.26631200725427723341593459
320	1	-.23600642202280357065655	-1.26631200726525235066507405
640	0	-.23600599673618426491317	-1.26631136066526561571231289
640	1	-.23600604458884413059588	-1.26631136066652407229049124
1280	0	-.23600612016132843644199	-1.26631128652446486196654565
1280	1	-.23600610825204283147338	-1.26631128652460916221767720
2560	0	-.23600608597293199986358	-1.26631127802314930823452467
2560	1	-.23600608893686566694018	-1.26631127802316585434663485

Table 3.3.3.2: Numbers of tangent bifurcations to expect in the quadratic map, unattached to the period 1 tangent bifurcation. They are calculated by subtracting off the orbits arising by bifurcation from the fixed point (using the notation of §1.2.4.4), from the totals given in table 1.2.3.5.1. For even period, the numbers are subdivided into orbits belonging to S and QS.

period	periodic orbits attached to the period 1 tangent bifurcation	# of n-cycles attached	#of unattached symmetric tangent bifurcations
1	$1\pm$	2	0
2	$\frac{1}{2}$	1,0	0
3	$\frac{1}{3}\pm$	2	0
4	$\frac{1}{4}\pm, \frac{1}{2}, \frac{1}{2}$	2,1	0
5	$\frac{1}{5}\pm, \frac{2}{5}\pm$	4	1
6	$\frac{1}{6}\pm, \frac{1}{3}:\frac{1}{2}, \frac{1}{2}:\frac{1}{3}\pm$	3,2	1,0
7	$\frac{1}{7}\pm, \frac{2}{7}\pm, \frac{3}{7}\pm$	6	4
8	$\frac{1}{8}\pm, \frac{1}{4}:\frac{1}{2}, \frac{3}{8}\pm, \frac{1}{2}:\frac{1}{4}\pm, \frac{1}{2}:\frac{1}{2}:\frac{1}{2}$	6,2	3,2

Table 3.3.3.3: Parameter values p and positions x_0, x_1 of period 8 tangent bifurcations in the quadratic map.

p	x_0	x_1
-1.38583245784912311	0.129071252737934835	-0.927276281248210995
-1.55714135496325422	0.308871563599705721	-0.27043587000868346
-1.58822574651785372	-0.273938204616775725	-0.99131424421208461

Following Karney et al. (1982), the appropriate power of a map in the neighbourhood of a tangent bifurcation is well approximated by the quadratic map, though with the more natural choice of coordinates and parametrisation:

$$x' = p - y - x^2$$
$$y' = x. \tag{3.3.3.1}$$

Thus, the whole picture of period doubling followed by islets of stability will be repeated in miniature. I say this choice of parametrisation/coordinates is more natural, because it gives generic bifurcation for the fixed points at multipliers +1, unlike (3.1.2.1).

The trapped area beyond p^* presumably also scales like $|\frac{\alpha\beta}{2}|$ with each factor of δ in parameter, but with a different universal function from that of figure 3.3.2.5. Although it is possible that the set of parameter values beyond p^* for which there is an elliptic periodic orbit is dense, I expect its measure to be small.

§3.3.4 **Significance.** Period doubling is significant in that it is associated with the disintegration of the stable region around an initially stable periodic orbit. It is replaced by some sort of stochastic region (or possibly, by escape in an unbounded system). Where one is, in the transition, can be measured by the residue of the periodic orbit. The disintegration is essentially complete when the residue exceeds R^* (3.1.2.10). I expect that T^*_μ has universal horseshoes, but I did not check it. It would be very interesting to check this, and to find the value of the residue in the universal family for formation of a period 1 horseshoe.[1]

I conclude this section with a caution. In any particular example, the universal map only exerts its influence close enough in space and parameter to the accumulation point for period doubling. Thus there can be large, non-universal regions of stability further away. For example, the period doubling sequence of the 5-cycle of positive residue bifurcating from the fixed point at $p = \cos\left(\frac{2}{5}\pi\right) = 0.30901699...$, accumulates at $p = 0.119353761903...$, leaving plenty invariant curves around the fixed point, but destroying completely the fifth-order island chain. As another example, the period doubling sequence of the fixed point (0,0) of the standard map accumulates at $k = 2\pi \times 1.05597806$ (Benettin et al., 1980a), but at $k = 2\pi$ an islet of stability pops up by tangent bifurcation (Chirikov, 1979). Note that the above period doubling sequence has anomalous behaviour because the standard map has a simple commutator (§1.2.4.7). The fixed point of positive index of the tangent bifurcation is elliptic until

$$k = \left((2\pi)^2 + 16\right)^{1/2} = 2\pi \times 1.1854471.... \tag{3.3.4.1}$$

[1] Which would be associated with a certain homoclinic tangency (cf. Davis et al, 1991, where intervals of hyperbolicity are found in the parameter space for the quadratic map).

§3.4 APPENDICES

This section contains a few appendices on other bifurcation sequences, period doubling in integrable maps, and in 4-D symplectic maps.

§3.4.1 **Other bifurcation sequences**. One can follow other bifurcation sequences than period doubling. For example, one can follow successive period triplings. I did this for two initial periodic orbits in the quadratic map, in the form (3.3.3.1). Table 3.4.1.1 gives the results for the tripling sequence of the fixed point. It shows self-similarity again. The parameter values for successive triplings converge at rate:

$$\delta = 20.187. \tag{3.4.1.1}$$

The self-similarity in position, however, is only between alternate triplings. I found two-step scalings of:

$$\begin{aligned} \alpha &= -43.99 \text{ along the symmetry line} \\ \beta &= -186.8 \text{ across the symmetry line.} \end{aligned} \tag{3.4.1.2}$$

The implication is that there is a 2-cycle of the renormalisation:

$$T \to B\, T^3 B^{-1}. \tag{3.4.1.3}$$

I found the same scalings for the tripling sequence of the 2-cycle of the quadratic map.

Bak and Jensen (1981) have followed some other bifurcation sequences and found self-similarity again.[2] Note, however, that there is no need for a fixed point (or periodic orbit) of a renormalisation to have an associated bifurcation sequence. In 1-D maps, for example, Eckmann and Koch (Collet and Eckmann, 1980) found fixed points of the tripling and quadrupling operators, despite the fact that the only bifurcations in 1-D are tangent and period doubling (§1.2.4.2). These fixed points correspond to the map at the accumulation point of a sequence of stable windows of periods 3^n, 4^n respectively.

---

2 Meiss (1986) has pursued this further.

Table 3.4.1.1: Data for a period tripling sequence

period	parameter value for tripling	ratio
1	1.25	
3	1.18649724154936992	
9	1.18177677838815665	13.45265
27	1.18157966907744623	23.94845
81	1.18156992911447644	20.23717
243	1.18156942485775178	20.26892
729	1.18156942477345777	20.18719
2187	1.18156942359425795	20.18662

period	position on Fix(QS)	ratio
1	0.5	
3	1.43185326390959455	
9	1.33276208445687517	-9.403998
27	1.31363225672015903	5.179931
81	1.31585729121086753	-8.597542
243	1.31629080260096078	5.132586
729	1.31624019349691551	-8.565878
2187	1.31623033566259329	5.133897

period	off-diagonal element of DT^n	ratio
1	-1.0	
3	0.25401103380 25204	-0.254011
9	-3.89936849657123559	-15.351178
27	1.11598042520536447	-0.286195
81	-16.56955318008927650	-14.847530
243	4.74089141824891269	-0.286121
729	-70.34753352550348770	-14.838461
2187	20.12831713630986920	-0.286127

§3.4.2 **Period doubling in integrable maps.** Integrable systems have a lot of invariant curves (§1.3.1.2). As I showed that the universal period doubling sequence leads to destruction of closed invariant curves, it cannot occur in integrable systems.

Finite sequences of period doublings can occur. For example:

$$T: x' = y \qquad y' = -x + f(y) \tag{3.4.2.1}$$

with

$$f(y) = \frac{2ky}{1+y^2} \tag{3.4.2.2}$$

in area preserving, and has the integral (Laslett, 1978):

$$I(x, y) = x^2 + y^2 + x^2 y^2 - 2kxy. \tag{3.4.2.3}$$

Its fixed points are given by:

$$2y = f(y) \text{ i.e. } y = 0 \text{ or } 1 + y^2 = k. \tag{3.4.2.4}$$

The fixed points of T^2 are given by:

$$y(1 + y^2)^2 = k^2 y. \tag{3.4.2.5}$$

So there is a 2-cycle $1 + y^2 = -k$. Thus as k decreases through -1, the fixed point at 0 loses stability and gives birth to a stable 2-cycle. But there are no more period doublings, as non-parabolic points must be stationary points of I, and the stationary points of I are given by:

$$I_x = I_y = 0 \text{ i.e. } y(1 + y^2)^2 = k^2 y \tag{3.4.2.6}$$

the same equation as for the fixed points of T^2. Note that integrable maps can have uncountably many parabolic periodic points, corresponding to closed invariant curves with rational rotation number. Bountis and Helleman (Helleman, 1980) followed period doubling numerically in other integrable systems, but always found only a finite sequence.

Indeed, I think it is impossible for an integrable map to possess an infinite period doubling sequence, provided it takes place in a bounded domain and produces non-parabolic points, because the invariant would have to have infinitely many isolated stationary points in a bounded domain, which is not possible, by compactness.

§3.4.3 **Period doubling in 4-D symplectic maps.** The universality results for period doubling in 1-D maps extend to higher dimensional dissipative systems (§2.2.3). An interesting question is whether the period doubling sequences of area preserving maps carry over to higher dimensional symplectic maps in any form. One should compare the situation with that of adding dissipation, which pushes the area preserving period doubling sequences into dissipative ones. Bountis and I looked for period doubling sequences in 4-D symplectic maps. As a periodic orbit in 4-D has two pairs of multipliers, we considered two parameter families, so that each pair could be moved independently. Although there are period doubling bifurcations, we never found anything looking like a self-similar infinite sequence. Still, there is a lot of space in two parameter families of 4-D maps, so we may just have missed it!

Period doubling typically occurs whenever a periodic orbit has a pair of multipliers passing through -1. The first thing I will do is to find the connection between the stability parameters (§1.2.1.6) of the daughter and parent. Suppose the parent orbit has stability parameters A, B, and multipliers r, r^{-1}, s, s^{-1}. Then:

$$A = r + r^{-1} + s + s^{-1} \tag{3.4.3.1}$$

and

$$B = 2 + rs + \frac{1}{rs} + \frac{r}{s} + \frac{s}{r}. \tag{3.4.3.2}$$

At birth, the multipliers of the daughter are the squares of those of its parent, so it is born with stability parameters A', B':

$$A' = r^2 + r^{-2} + s^2 + s^{-2} \tag{3.4.3.3}$$

and

$$B' = 2 + r^2 s^2 + \frac{1}{r^2 s^2} + \frac{r^2}{s^2} + \frac{s^2}{r^2}. \tag{3.4.3.4}$$

By considering the expressions for A^2 and $(B-2)^2$, one finds that:

$$\begin{aligned} A' &= A^2 - 2B \\ B' &= (B-2)^2 - 2 - 2A' = B^2 + 2 - 2A^2. \end{aligned} \tag{3.4.3.5}$$

Note that:

$$R' = B' + 2 + 2A' = (B-2)^2 \tag{3.4.3.6}$$

and

$$S' = B' + 2 - 2A' = RS. \tag{3.4.3.7}$$

Also the Krein signature of a pair in the left half plane changes, and of a pair in the right half does not. The signature of a pair at $\pm i$ goes to 0, on doubling.

I considered a simple class of symplectic maps:

$$\begin{aligned} T: \underline{x}' &= -\underline{y} + \underline{\nabla}\varphi(\underline{x}), \quad (\underline{x}, \underline{y}) \in \mathbb{R}^2 \times \mathbb{R}^2 \\ \underline{y}' &= \underline{x}. \end{aligned} \tag{3.4.3.8}$$

They are reversible, with symmetries:

$$\begin{aligned} S: \underline{x}' &= \underline{x} \quad \text{symmetry plane } \underline{y} = \tfrac{1}{2}\underline{\nabla}\varphi(\underline{x}) \\ \underline{y}' &= -\underline{y} + \underline{\nabla}\varphi(\underline{x}) \end{aligned} \tag{3.4.3.9}$$

and

$$\begin{aligned} TS: \underline{x}' &= \underline{y} \quad \text{symmetry plane } \underline{y} = \underline{x} \\ \underline{y}' &= \underline{x}. \end{aligned} \tag{3.4.3.10}$$

Its fixed points are given by:

$$2\underline{x} = \underline{\nabla}\varphi(\underline{x}) \tag{3.4.3.11}$$

and their stability by

$$DT = \begin{vmatrix} M & -I \\ I & 0 \end{vmatrix}, \quad M = \underline{\nabla\nabla}\varphi. \tag{3.4.3.12}$$

So

$$A = \operatorname{Tr} M \tag{3.4.3.13}$$

and

$$(DT)^2 = \begin{vmatrix} M^2 & I & -M \\ & M & -I \end{vmatrix}. \tag{3.4.3.14}$$

So

$$B = \det M + 2. \tag{3.4.3.15}$$

Thus

$$R = 2A + B + 2 = 2\mathrm{Tr}\, M + \det M + 4$$

$$S = 2A - B - 2 = 2\,\mathrm{Tr}\, M - \det M - 4 \qquad (3.4.3.16)$$

$$U = \frac{A^2}{4} - B + 2 = \frac{1}{4}(\mathrm{Tr}\, M)^2 - \det M.$$

Note that since M is symmetric, $U \geq 0$, with equality iff M is a multiple of the identity. This reflects the fact that the Krein signatures are always the same.

Period doubling occurs when $R = 0$. Analysis of the 2-cycles $\underline{x}, \underline{x}'$ yields:

$$R_2 = \det MM'$$

$$S_2 = 4\,\mathrm{Tr}\, MM' - 16 - \det MM' \qquad (3.4.3.17)$$

$$U_2 = \frac{1}{4}(\mathrm{Tr}\, MM')^2 - \det MM'.$$

Note that U_2 is not necessarily positive as MM' need not be symmetric. This analysis gives one a good idea of where to start looking for period doubling sequences.

I looked at the specific case:

$$\varphi(\underline{x}) = -\frac{1}{2}ax_1^2 + bx_1x_2 + \frac{1}{3}x_1^3 + \frac{c}{2}x_2^2. \qquad (3.4.3.18)$$

When $b = 0$, it decouples into two area preserving maps, one linear and the other quadratic. So I found the usual area preserving period doubling sequence from the quadratic map. In order to get self-similarity for the whole map, one has to follow the line in parameter space where c is such that the multipliers of the linear map are at $e^{\pm 2\pi i/3}$, so that they simply interchange on doubling. This also preserves the Krein signature. Thus I followed the straight line in the stability plane from $(A, B) = (1,0)$ to $(-3,4)$.

Next I took small coupling $b = 0.01$, so that if the period doubling sequence persisted in any form, it should be close to the uncoupled case. This enabled me to know where to look, in the parameter plane (a, c). There were period doublings up to period 32, but the 32-cycle never attained $R = 0$. It is not clear what happened to it. Thus unlike the effect of adding dissipation, coupling to more degrees of freedom appears to annihilate the area preserving period doubling sequences. It is still an open question, however, whether 4-D symplectic maps have their own period doubling sequences of a completely different form.[3]

[3] Such have been found subsequently by Mao and Helleman (1986).

Chapter 4
Renormalisation for invariant circles

In this chapter, I introduce a renormalisation operator relevant to invariant circles. Restricting attention to noble rotation numbers, it has two important fixed points, a simple one and a critical one. Before discussing them, I discuss features of fixed point analysis common to both. The simple fixed point corresponds to an integrable twist map and is essentially attracting, giving a new view on KAM theory. The critical fixed point corresponds to maps with a non-smooth noble circle. Its unstable manifold gives a universal one parameter family describing the breakup of invariant circles. I discuss possible extensions of this work, and give details on one in particular, to the question of when a map on a circle is conjugate to rotation.

§4.1 RENORMALISATION

I introduce a renormalisation operator which is well adapted to questions involving rotation number. It necessitates, however, a generalisation of periodic maps to commuting pairs of maps.

§4.1.1 **Motivation**. The proofs of persistence of invariant circles depend crucially on the type of irrationality of the rotation number (§1.3.1.5). I will now motivate a renormalisation operator which has the important feature that it is well adapted to the continued fraction expansion (1.3.1.5.10) of the rotation number. Remember that I generalised rotation number to other orbits than those on an invariant circle (1.3.1.3.4).

Without loss of generality, consider the orbit of the origin $\underline{0}$. If it has rotation number ω, then:

$$\pi_1 F^q R^p(\underline{0}) = q\omega - p + o(q) \text{ as } q \to \infty \tag{4.1.1.3}$$

where R is back-rotation through one period:

$$R(\theta, z) = (\theta - 1, z). \tag{4.1.1.4}$$

If $\underline{0}$ belongs to a circle on which F is topologically conjugate to rotation, then we have the stronger statement:

$$\pi_1 F^{q_n} R^{p_n}(\underline{0}) \to 0 \text{ if } q_n\omega - p_n \to 0. \tag{4.1.1.5}$$

In the case of differentiable conjugacy to rotation, one can say even more:

$$\pi_1 F^{q_n} R^{p_n}(0) \sim K(q_n\omega - p_n) \text{ as } q_n\omega - p_n \to 0 \tag{4.1.1.6}$$

where K is the derivative of the conjugacy at 0. This suggests that we consider a sequence of maps:

$$B_n F^{q_n} R^{p_n} B_n^{-1} \tag{4.1.1.7}$$

where the B_n are coordinate changes, expanding the scale in the θ direction. It will be convenient to expand the scale in z too, so as to keep the twist roughly constant.

A choice of p_n, q_n for which $q_n\omega - p_n$ is particularly small is given by the convergents of ω (see 1.3.2.1.4):

$$\frac{p_n}{q_n} = [m_0,\ldots, m_n] \tag{4.1.1.8}$$

where

$$\omega = [m_0, \ldots]. \tag{4.1.1.9}$$

For this choice, there is a systematic way to generate the sequence (4.1.1.7). For each integer m, define the *renormalisation operator* N_m, acting on pairs of maps (U, T), by:

$$\begin{aligned} N_m\colon U' &= BTB^{-1} \\ T' &= BT^mUB^{-1} \end{aligned} \tag{4.1.1.10}$$

B is a coordinate change, chosen to *renormalise* the pair (T, T^mU) in some sense. There is a lot of freedom in the precise specification of B, so we will not make a specific choice here. Then it follows from the definition of convergents that:

$$\begin{aligned} p_i &= m_i p_{i-1} + p_{i-2} \\ q_i &= m_i q_{i-1} + q_{i-2} \end{aligned} \tag{4.1.1.11}$$

with initial conditions:

$$\begin{aligned} p_{-2} &= 0,\ p_{-1} = 1 \\ q_{-2} &= 1,\ q_{-1} = 0. \end{aligned} \tag{4.1.1.12}$$

So by induction:

$$N_{m_n} \dots N_{m_0}(U,T) = (B_n\, U^{q_{n-1}}\, T^{p_{n-1}}\, B_n^{-1},\, B_n\, U^{q_n}\, T^{p_n}\, B_n^{-1}) \quad (4.1.1.13)$$

where B_n is the composition of the successive coordinate changes B. So the sequence (4.1.1.7) is generated by successive applications of N_{m_i} on the pair (F, R), for m_i following the continued fraction expansion of ω.

This is essentially the same renormalisation as that introduced by Kadanoff (1981a) and Shenker, and applied to the dissipative case by Feigenbaum et al. (1982) and Rand et al. (1982). It is also closely related to the approximate renormalisation of Escande and Doveil (1981) for Hamiltonians.

An apparent problem with the renormalisation is that in looking on successively smaller scales one loses sight of the periodicity of the map in θ. In the next section, I will show how the essence of periodicity can be saved, by generalising the class of periodic maps to that of commuting pairs of maps.

§4.1.2 **Commuting Pairs**. To say that F is periodic in θ is equivalent to saying that F commutes with R (4.1.1.4). So let us generalise the important concepts for periodic maps, in particular, rotation number and Calabi invariant, to commuting pairs (U, T). I will use coordinates $\underline{x} = (x, y)$ in place of (θ, z) to indicate that there is not necessarily any periodicity in x.

Firstly, I generalise the notions of orbits and invariant circles.

Definitions: The *orbit* of $\underline{x}$ under a commuting pair (U, T) is $\{U^q T^p \underline{x} : p, q \in \mathbb{Z}\}$.

A point $\underline{x}$ is *periodic* if $\exists (p, q) \in \mathbb{Z}^2 \backslash \{\underline{0}\}$ s.t. $U^q T^p \underline{x} = \underline{x}$.

It has *type* $(\underline{p}, \underline{q})$ if these are the smallest such integers $(q \geq 0)$.

An *invariant curve* is a curve from $x = -\infty$ to $+\infty$, invariant under both U and T.

Next I generalise rotation number. If (θ, z) has rotation number ω under a periodic map F, then:

$$\frac{\pi_1 F^q R^p (\theta, z)}{q} = \omega - \frac{p}{q} + o(1) \text{ as } q \to \infty. \quad (4.1.2.1)$$

This tends to zero for a sequence p_n, q_n iff $p_n/q_n \to \omega$. So let us generalise (and also allow $\omega = \infty$, i.e. I consider rotation number as an element of the projective line $\mathbb{R}P$):

$\underline{x}$ has *rotation number* $\omega \in \mathbb{R}P$ under (U, T) if for all sequences

$p_n, q_n \in \mathbb{Z}$ s.t. $r_n = \max(|p_n|, |q_n|) \to \infty$, then

$$\frac{\pi_1 U^{q_n} T^{p_n} \underline{x}}{r_n} \to 0 \text{ iff } \frac{p_n}{q_n} \to \omega. \qquad (4.1.2.2)$$

Poincaré's theorem (§1.3.1.3) generalises to invariant curves of a commuting pair under the condition that:

$$\exists\, m, n \in \mathbb{Z},\ K > 0 \text{ s.t. } \pi_1 U^m T^n \underline{x} \leq x - K \qquad (4.1.2.3)$$

for all points $\underline{x}$ on the curve (cf. $\pi_1 R(\theta, z) = \theta - 1$). I prove this in §4.7.7. So an invariant curve has a rotation number.

I will say an invariant curve of rotation number μ/ν is *smooth* if there is a sufficiently differentiable coordinate function $\varphi(\underline{x})$ (with differentiable inverse) on the curve, such that

$$\varphi U^q T^p \varphi^{-1}(t) = t + q\mu - p\nu. \qquad (4.1.2.4)$$

For the *twist condition*, I want both U and T to have action generating functions, i.e. $\partial x'/\partial y$ should have constant sign. This is probably more restrictive than necessary, and slightly unfortunately so, as R does not satisfy the twist condition.[1]

If U, T commute and have generating functions υ, τ then the generating functions for UT and TU can differ by only a constant, so I call it the *Calabi invariant* $C(\upsilon, \tau)$. This is not a direct generalisation from the periodic case (1.1.3.3.4), because R does not have a generating function, but it is essentially the same. It would be nice to work out a direct generalisation of the definition (1.1.3.1.5).[2]

I call the extension of class A to commuting pairs *class AA*. Presumably the Poincaré-Birkhoff theorem (§1.2.2.2), Moser's twist theorem (§1.3.1.5), Mather's theorem (§1.3.2.2), and other results like Denjoy's theorem (§1.3.1.3) would generalise under suitable conditions.[3]

---

[1] But we could always start from the pair (FR^{m_0}, FR^{m_0+1}) and use a slightly different form of the renormalisation operator to generate the sequence (4.1.1.7) (cf. Rand et al, 1982). In any case, after two renormalisations of a pair (F, R) not too far from integrable, one obtains a pair of twist maps.

[2] In October 1982 I realised how to do this: choose any point $\underline{x}$ and curves C_1, C_2 connecting it to $U\underline{x}$, $T\underline{x}$ respectively. Take their images under T, U, respectively. If U and T commute then the resulting four curves form one closed curve. The Calabi invariant is the (algebraic) area enclosed.

[3] I never worked through these generalisations, but Rand (1992) showed that one can always glue a commuting pair back into an annulus map, so commuting pairs are not really a generalisation, just a useful coordinate system on the space of annulus maps.

§4.1.3 **Renormalisation and rotation number.** Based on the preceding discussion of rotation number, I can now make some nice statements:

> $\underline{x}$ has rotation number ω under (U, T) iff
> $B\underline{x}$ has rotation number ω' under (U', T'),
> where ω, ω' are related by $\omega = m + \frac{1}{\omega'}$
>
> $\underline{x}$ has rotation number $\omega_0 = [m_0, m_1, \ldots]$ under (U, T) iff $B_{n-1} \ldots B_0\underline{x}$ has rotation number $\omega_n = [m_n, \ldots]$ under $N_{m_{n-1}} \ldots N_{m_0}(U, T)$, where the B_j are the successive coordinate changes.
>
> (U, T) has an invariant curve of rotation number ω_0 iff $N_{m_{n-1}} \ldots N_{m_0}(U, T)$ has an invariant curve of rotation number ω_n.

In particular, a periodic map F has an invariant circle of rotation number ω_0 iff $N_{m_{n-1}} \ldots N_{m_0}(F, R)$ has an invariant curve of rotation number ω_n.

Restricting B to linear diagonal scale changes $B(x, y) = (\alpha x, \beta y)$, N_m induces the following renormalisation on action generating functions:

$$\upsilon'(x, x') = \alpha\beta\, \tau\left(\frac{x}{\alpha}, \frac{x'}{\alpha}\right)$$

$$\tau'(x, x') = \alpha\beta\, \upsilon \oplus \tau \oplus \ldots \oplus \tau\left(\frac{x}{\alpha}, \frac{x'}{\alpha}\right). \tag{4.1.3.1}$$

Note that this implies that the Calabi invariant renormalises as follows:

$$C(\upsilon', \tau') = -\alpha\beta\, C(\upsilon, \tau). \tag{4.1.3.2}$$

In particular, zero Calabi invariant is preserved.

The idea of renormalisation is not new to KAM theory. Most proofs consist in finding successive coordinate changes to make the system look more like a linear twist, restricting attention each time to a smaller annulus, in such a way as to reduce the deviations from a linear twist faster than exponentially (see, for example, Moser, 1973, Herman, 1981, Rüssmann, 1981, Gallavotti, 1982). So scale changes are made in the z direction. I believe that the freedom I have to make scale changes in the θ direction too will make this renormalisation more powerful, because I think it will lead to a global KAM theorem, namely, it will give the boundary in class A of the set of maps with an invariant circle of given rotation number. The only expense is that at each step one has to change the generators for the group $\{F^q R^p : p, q \in \mathbb{Z}\}$.

§4.2 FIXED POINT ANALYSIS

In the rest of this chapter, I restrict attention to noble rotation numbers, as the simplest and most important case. The renormalisation for nobles has two important fixed points. In this section, I discuss features of their analysis which are common to both. First I discuss stability of a fixed point, and then the significance of convergence to a fixed point. The renormalisation leaves invariant several important spaces. Its action in some of them is simple enough to give easy eigenvectors.

§4.2.1 **Stability.** Quadratic irrationals have eventually periodic continued fraction expansion. So for maps with an orbit of quadratic irrational rotation number $[b_0,...,b_j, (c_1,...,c_k,)^\infty]$, this suggests that one might find asymptotic behaviour under $N_{c_k} \ldots N_{c_1}$, after removing the aperiodic head by applying $N_{b_j} \ldots N_{b_0}$. The simplest form of asymptotic behaviour is convergence to a fixed point of the renormalisation. This whole chapter is essentially about two important fixed points, so in this section I will give the analysis which is common to any fixed point, and treat them individually later.

I will look at the simplest case, namely, nobles, for which the repeat pattern is $c = [1]$. Any other quadratic irrational could be tackled in the same way. So I consider the operator:

$$N_1 : U' = BTB^{-1} \qquad T' = BTUB^{-1}. \tag{4.2.1.1}$$

Suppose that (U^*, T^*) is a fixed point. Then the first thing one would want to find is its stability under N_1. This is given by the derivative:

$$DN_1 : \delta U' = DB_{TB^{-1}}.\delta T\, B^{-1} \qquad \delta T' = DB_{TUB^{-1}}.\delta T\, U\, B^{-1} + DB_{TUB^{-1}}.DT_{UB^{-1}}.\delta U\, B^{-1}. \tag{4.2.1.2}$$

Here I have ignored contributions to $\delta U', \delta T'$ due to variation of B with (U, T), which depend on the particular prescription for renormalising (T, TU). Specifically δB produces:

$$\delta U' = \delta B\, TB^{-1} + DB_{TB^{-1}}.DT_{B^{-1}}.\delta B^{-1} \tag{4.2.1.3}$$

and similarly for $\delta T'$, with T replaced by TU. Now,

$$B^{-1}B = I \text{ so } \delta B^{-1}B + (\delta B^{-1})_B\,.\delta B = 0. \tag{4.2.1.4}$$

So

$$\delta B^{-1} = -DB^{-1}.\delta B\ B^{-1}. \tag{4.2.1.5}$$

Thus

$$\delta U' = \delta B\ TB^{-1} - DB_{TB^{-1}}.DT_{B^{-1}}.DB^{-1}.\delta B\ B^{-1}. \tag{4.2.1.6}$$

But the first term can be written as $\delta B\ B^{-1}U'$, and for the second, compare:

$$DU' = DB_{TB^{-1}}.DT_{B^{-1}}.DB^{-1}. \tag{4.2.1.7}$$

Thus the correction due to δB is:

$$\delta U' = \delta B\ B^{-1}U' - DU'.\delta B\ B^{-1} \tag{4.2.1.8}$$

$\delta T'$ is given by the same formula with U' replaced by T'. These contributions are only in the direction of coordinate changes (§2.3.4), so they will have no essential effect. If B is chosen to preserve some normalisation on the scales, however, they will change the scale change eigenvalues from 1 to 0, which is important for making the fixed point problem well-posed.

To restrict to area preserving maps I will often use the action representation. N_1 induces the following renormalisation on pairs (υ, τ) of actions, in the case that B is a linear diagonal scale change, $B(x, y) = (\alpha x, \beta y)$:

$$\upsilon'(x, x') = \alpha\beta\, \tau\left(\frac{x}{\alpha}, \frac{x'}{\alpha}\right)$$

$$\tau'(x, x') = \alpha\beta \left\{ \upsilon\left(\frac{x}{\alpha}, \bar{x}\left(\frac{x}{\alpha}, \frac{x'}{\alpha}\right)\right) + \tau\left(\bar{x}\left(\frac{x}{\alpha}, \frac{x'}{\alpha}\right), \frac{x'}{\alpha}\right)\right\} \tag{4.2.1.9}$$

where $\bar{x}$ is chosen to make the sum stationary. The derivative DN_1 is given by:

$$\delta\upsilon'(x, x') = \alpha\beta\, \delta\tau\left(\frac{x}{\alpha}, \frac{x'}{\alpha}\right)$$

$$\delta\tau'(x, x') = \alpha\beta \left\{ \delta\upsilon\left(\frac{x}{\alpha}, \bar{x}\left(\frac{x}{\alpha}, \frac{x'}{\alpha}\right)\right) + \delta\tau\left(\bar{x}\left(\frac{x}{\alpha}, \frac{x'}{\alpha}\right), \frac{x'}{\alpha}\right)\right\}. \tag{4.2.1.10}$$

Note that variations in $\bar{x}$ due to $\delta\upsilon$, $\delta\tau$ do not contribute, by stationarity. The contributions from variations in α, β, have been left out once more. They are as follows:

$$\delta\upsilon'(x,x') = \delta(\alpha\beta)\,\tau\left(\frac{x}{\alpha}.\frac{x'}{\alpha}\right) + \alpha\beta\left(x\tau_1\left(\frac{x}{\alpha},\frac{x'}{\alpha}\right) + x'\tau_2\left(\frac{x}{\alpha},\frac{x'}{\alpha}\right)\right)\delta\left(\frac{1}{\alpha}\right)$$

$$= \frac{\delta(\alpha\beta)}{\alpha\beta}\upsilon'(x,x') + \alpha\,\delta\left(\frac{1}{\alpha}\right)(x\upsilon_1'(x,x') + x\upsilon_2'(x,x')) \qquad (4.2.1.11)$$

and the same for $\delta\tau'$ with υ' replaced by τ'.

§4.2.2 **Significance of convergence to a fixed point.** In this section, I show how Mather's theorem (§1.3.2.2) can probably be used to show that for F area preserving and periodic, if $N_1^n(F,R)$ converges to any fixed point of N_1 with $\alpha\beta > 1$, then F has a golden circle. If Mather's theorem generalises to commuting pairs, I could similarly deduce a golden curve for (U,T) whenever $N_1^n(U,T)$ converges to a fixed point. In particular, the fixed point would have a golden curve. This section requires some improvement, however. In particular I need to discuss existence of Birkhoff orbits for commuting pairs, and whether they have a point in the domain of convergence.[4]

Write $\Delta W_{p/q}^{(n)}$ for $\Delta W_{p/q}$ for $N_1^n(F,R)$. I will also use the continued fraction notation (1.3.2.1.4) for rationals. Convergence to the fixed point (in C^1 so actions exist and are continuous), implies that:

$$\Delta W_{[1]}^{(n)} \to \Delta W_{[1]}^{*} \qquad (4.2.2.1)$$

But

$$\Delta W_{[1]}^{(n)} = (\alpha\beta)_n \dots (\alpha\beta)_1\, \Delta W_{[(1,)^{n+1}]}^{(0)} \qquad (4.2.2.2)$$

from (4.2.1.9), where the $(\alpha\beta)_j$ are the successive values of $\alpha\beta$. Convergence to the fixed point also implies that:

$$(\alpha\beta)_j \to \alpha\beta > 1.$$

[4] I never worked out this approach to constructing invariant curves, as another approach due to David Rand turned out to be simpler. He developed it in 1983 in the context of dissipative annulus maps (eventually published in 1992), but I adapted it to the area-preserving case in April 1984 (unpublished). The idea formed the basis for part of Haydn's PhD thesis, eventually published in 1990. Stirnemann (1992) has independently worked out a similar approach to Rand's, but with the hypotheses formulated more topologically.

So

$$\Delta W^{(0)}_{[(1,)^n]} \to 0$$

and F has a golden circle, by Mather's theorem (§1.3.2.2). Note that the convergence need only be C^1 for Mather's theorem to apply.

The periodic orbits of rotation number [1] might not be in the domain of definition. But those of rotation number $[(1,)^n]$ are, for large enough n, so the same arguments could be applied to them.

§4.2.3 **Invariant spaces under the renormalisation.** I want to analyse the stability of a fixed point. Before doing this it is worth considering the following important spaces, which N_1 leaves invariant:

(i) commuting pairs
(ii) area preserving pairs
(iii) commuting area preserving pairs with zero Calabi invariant (class AA)
(iv) symmetric commuting pairs
(v) coordinate transforms of any fixed point
(vi) all intersections of the above.

So if the fixed point lies in one or more of these spaces, perturbations from it should decompose into components in the separate tangent spaces and a complementary component.

Note with regard to (iv) above, that I say T is *symmetric* if it is reversible (§1.1.4), with the particular symmetry $S(x, y) = (-x, y)$ (1.1.4.4.2). Remember from §1.1.4.2 that up to coordinate change this is locally the most general symmetry. Although the space of symmetric commuting pairs is preserved, symmetry alone is not preserved. In fact, for U, T symmetric, UT is symmetric iff U, T commute. To see this, U, T symmetric implies that:

$$(SUT)^2 = SUTSUT = U^{-1}SST^{-1}UT = U^{-1}T^{-1}UT. \qquad (4.2.3.1)$$

§4.2.4 **Some easy eigenvectors.** Some of the eigenvectors are easy to obtain for a general fixed point. In particular, I will identify eigenvectors for coordinate changes, non-commuting directions, non-zero Calabi invariant, constant Jacobian not equal to 1, and constants in the actions. In this section I assume that I can expand in power series.

(i) coordinate changes

Following §2.3.4, if B has the normal form (see §2.3.3):

$$B(x, y) = (\alpha x, \beta y) \tag{4.2.4.1}$$

then we obtain coordinate change eigenvectors

$$\delta\sigma(x, y) = (x^p y^q, 0), (0, x^p y^q), \; p, q \geq 0 \tag{4.2.4.2}$$

with eigenvalues

$$\alpha^{1-p} \beta^{-q}, \; \alpha^{-p} \beta^{1-q}. \tag{4.2.4.3}$$

Throughout this section, p and q will be non-negative integers. One has to check, incidentally, whether these coordinate changes give independent perturbations. Note that there is a natural degeneracy between $\delta\sigma(x, y) = (x^{p+1}y^q, 0)$ and $(0, x^p y^{q+1})$. There is precisely one area preserving combination (up to scale), viz. $((q+1)x^{p+1}y^q, -(p+1)x^p y^{q+1})$. The eigenvectors $(y^q, 0)$ and $(0, x^p)$, neither of which suffer natural degeneracies, are also area preserving. Lastly, the magnification:

$$\delta\sigma(x, y) = (x, y) \tag{4.2.4.4}$$

is not area preserving, but transforms area preserving maps to area preserving maps.

One might also want to know what perturbations are produced in the action representation by coordinate changes. Area preserving coordinate changes $C: (x, y) \to (X, Y)$ can be generated by the mixed generating function (§1.1.3.4):

$$\begin{aligned} x &= X + f_2(X, y) \\ Y &= y + f_1(X, y). \end{aligned} \tag{4.2.4.5}$$

If $T: (x, y) \to (x', y')$ is generated by the action generating function $\tau(x, x')$, i.e.

$$\begin{aligned} y' &= \tau_2(x, x') \\ y &= -\tau_1(x, x') \end{aligned} \tag{4.2.4.6}$$

then I would like to know the action generating function for CTC^{-1}. I will do this in two steps. First let us find the action for

$$TC^{-1}: X, Y \overset{C^{-1}}{\rightarrow} x, y \overset{T}{\rightarrow} x', y'. \tag{4.2.4.7}$$

We have to assume C is close to the identity. Then:

$$\begin{aligned} y' &= \tau_2(X + f_2(X, y), x') \\ &\sim \tau_2(X, x') + \tau_{21}(X, x') f_2(X, y) \end{aligned} \tag{4.2.4.8}$$

$$\begin{aligned} Y &= -\tau_1(X + f_2(X, y), x') + f_1(X, y) \\ &\sim -\tau_1(X, x') - \tau_{11}(X, x') f_2(X, y) + f_1(X, y). \end{aligned} \tag{4.2.4.9}$$

But this could be generated by:

$$\bar{\tau}(X, x') = \tau(X, x') - f(X, -\tau_1(X, x')). \tag{4.2.4.10}$$

Using the same result again, we get the action for CTC^{-1} to be:

$$\tau(X, X') - f(X, -\tau_1(X, X')) + f(X', \tau_2(X, X')). \tag{4.2.4.11}$$

Now we easily see that the effect of the renormalisation on a perturbation of the above form is:

$$f'(X, y) = \alpha\beta f\left(\frac{X}{\alpha}, \frac{y}{\beta}\right). \tag{4.2.4.12}$$

Hence the coordinate change eigenvectors are generated by $f(X, y) = X^p y^q$, with eigenvalue $\alpha^{1-p}\beta^{1-q}$. Note also that $\delta\tau = \tau$, $\delta\upsilon = \upsilon$ is a scale change eigenvector, eigenvalue +1, related to the magnification (4.2.4.4).

(ii) <u>non-commuting</u>

Rand et al. (1982) identified two non-commuting eigenvectors for the parallel problem for maps on a circle. In fact, one can identify them all. I show how this is done for the two dimensional area preserving case we are considering.

If U, T are close to being commuting, then

$$U'T' - T'U' = BT\,TU\,B^{-1} - BT\,UT\,B^{-1} \sim D(BT)_{TUB^{-1}}.(TU - UT)\,B^{-1}. \tag{4.2.4.13}$$

The fixed point equation implies that $D(BT)_{TUB^{-1}(0)}$ has eigenvalues α, β. This is because at a commuting fixed point:

$$BTTUB^{-1} = UT = B^{-1}TB \text{ and } TUB^{-1} = B^{-1}T. \tag{4.2.4.14}$$

So

$$D(BT)_{TUB^{-1}}.D(B^{-1}T) = D(B^{-1}T)_B.DB \tag{4.2.4.15}$$

i.e.

$$D(BT)_{TUB^{-1}} = D(B^{-1}T)_B.DB.[D(B^{-1}T)]^{-1} \tag{4.2.4.16}$$

which at $\underline{0}$ is a similarity transform of DB_0.

Call its eigenvectors (a_ρ, b_ρ), $\rho = \alpha, \beta$. Then if a perturbation has commutator:[5]

$$(UT - TU)(x, y) = (a_\rho x^p y^q, b_\rho x^p y^q) \tag{4.2.4.17}$$

its image under DN_1 has a commutator of:

$$-\rho\alpha^{-p}\beta^{-q}(a_\rho x^p y^q, b_\rho x^p y^q) + \text{terms of higher degree.} \tag{4.2.4.18}$$

One has to check that these commutators are all independently attainable, and then one has non-commuting (possibly generalised) eigenvectors for each. In summary, the non-commuting spectrum is the negative of the spectrum for coordinate changes.

Note from (4.2.3.1), that there can be no symmetric non-commuting eigenvectors of DN_1.

One might also want to find the non-commuting eigenvectors in the action representation. I define the commutator of actions (υ, τ) by:

$$C(\upsilon, \tau)(x_0, x_2) = \upsilon(x_0, x_1) + \tau(x_1, x_2) - \tau(x_0, x_1') - \upsilon(x_1', x_2) \tag{4.2.4.19}$$

where x_1, x_1' are chosen by stationarity. Then

$$C(\upsilon', \tau')(x_0, x_3) = \alpha\beta\left\{\tau\left(\frac{x_0}{\alpha}, \frac{x_1}{\alpha}\right) + \upsilon\left(\frac{x_1}{\alpha}, \frac{x_2}{\alpha}\right) + \tau\left(\frac{x_2}{\alpha}, \frac{x_3}{\alpha}\right)\right.$$

$$\left. - \upsilon\left(\frac{x_0}{\alpha}, \frac{x_1'}{\alpha}\right) - \tau\left(\frac{x_1'}{\alpha}, \frac{x_2'}{\alpha}\right) - \tau\left(\frac{x_2'}{\alpha}, \frac{x_3}{\alpha}\right)\right\} \tag{4.2.4.20}$$

where x_1, x_2, x_1', x_2' are chosen by stationarity. If υ, τ are close to commuting and to a fixed point, then $x_2 \sim x_2' \sim \hat{x}(x_0, x_3)$, the intermediate point for the composition $\tau \oplus \upsilon$

[5] Usually the word commutator would be used for $UTU^{-1}T^{-1}$, which would be close to the identity here, but it turned out easier to use $UT-TU$.

(note that this is the opposite order from usual). Thus infinitesimally close to the fixed point:

$$\delta C'(x, x') = -\alpha\beta\, \delta C\left(\frac{x}{\alpha}, \frac{1}{\alpha}\, \hat{x}(x, x') \right). \tag{4.2.4.21}$$

So we see that a commutator of x^p grows like $-\alpha^{1-p}\beta$.

Greene found the rest of the commutator eigenfunctions, viz.

$$\delta C(x, x') = x^p\left(-\upsilon_1(\alpha x, \alpha x')\right)^q \tag{4.2.4.22}$$

with eigenvalue $-\alpha^{1-p}\beta^{1-q}$. In fact he showed that the eigenfunctions in actions are $\frac{1}{2}x^p y^q$ + a symmetric or antisymmetric term, according to p, q.

(iii) Calabi invariant

The particular case of a constant commutator in actions corresponds to perturbations with non-zero Calabi invariant. Clearly, from the above, this gives an eigenvector with eigenvalue $-\alpha\beta$. It is non-commuting in the action representation, but since the commutator is only a constant, it gives commuting maps. Note that it has to be non-symmetric, as symmetry implies zero Calabi invariant.

(iv) Constants in the actions

The action representation also introduces two new eigenvectors, corresponding to changes in the constant terms.[6] For constant perturbations,

$$\begin{aligned} \delta\upsilon' &= \alpha\beta\, \delta\tau \\ \delta\tau' &= \alpha\beta\, (\delta\upsilon + \delta\tau). \end{aligned} \tag{4.2.4.23}$$

Hence we obtain eigenvalues $\gamma\alpha\beta$, $-\alpha\beta/\gamma$. Apart from these two additions and the non-zero Calabi invariant eigenvalue above, the spectrum in the action representation is the same as the area preserving part of the spectrum for maps.

(v) non-area preserving

Following Zisook's example for the case of period doubling (1982), we can find two non-area preserving eigenvectors, those which have constant Jacobian different from 1. If B has constant Jacobian, then:

[6] Remarkably, these turn out to have physical significance in the solid-state physics context (MacKay, 1991).

$$\begin{aligned} \operatorname{Det} DU' &= \operatorname{Det} DT_{B^{-1}} \\ \operatorname{Det} DT' &= DT_{UB^{-1}} \operatorname{Det} DU_{B^{-1}}. \end{aligned} \tag{4.2.4.24}$$

So

$$\begin{aligned} \operatorname{Det} DU &= 1+\varepsilon \\ \operatorname{Det} DT &= 1+\eta \end{aligned} \quad \text{implies} \quad \begin{aligned} \varepsilon' &= \eta \\ \eta' &= \varepsilon+\eta. \end{aligned} \tag{4.2.4.25}$$

Hence we obtain eigenvalues $\gamma, -1/\gamma$. I could not find any more, however, apart from the non-area preserving coordinate changes about which we already know.[7]

§4.3 SIMPLE FIXED POINT

The renormalisation has a simple fixed point corresponding to an integrable twist map with a golden curve. Its stability is analysed, both in the space of maps and in the action representation. DN_1 is compact at the simple fixed point, in suitable domains. I decompose its spectrum and show that the simple fixed point is essentially attracting in class A. This implies persistence of noble circles for small enough perturbations of the integrable twist.

§4.3.1 **Simple fixed point.** N_1 has a simple fixed point:

$$T: \begin{aligned} x' &= x+y+1 \\ y' &= y \end{aligned} \quad U: \begin{aligned} x' &= x+\frac{y}{\gamma}-\gamma \\ y' &= y \end{aligned} \quad B: \begin{aligned} x' &= -\gamma x \\ y' &= -\gamma^2 y. \end{aligned} \tag{4.3.1.1}$$

It corresponds to a linear shear, with $y=0$ as a golden curve.

I can write out DN_1 in full at the simple fixed point:

[7] Davie and Dutta (1992) have found the complete non-area preserving spectrum from the area-preserving period-doubling fixed point in terms of the area-preserving spectrum. I have applied their method to solve the corresponding problem for the renormalisation under consideration here (MacKay, 1992d).

$$\delta U_x'(x,y) = -\gamma\,\delta T_x\left(-\frac{x}{\gamma}, -\frac{y}{\gamma^2}\right) \tag{4.3.1.2}$$

$$\delta T_x'(x,y) = -\gamma\,\delta U_x\left(-\frac{x}{\gamma}, -\frac{y}{\gamma^2}\right) - \gamma\,\delta T_x\left(-\frac{x}{\gamma} - \frac{y}{\gamma^3} - \gamma, -\frac{y}{\gamma^2}\right) - \gamma\,\delta U_y\left(-\frac{x}{\gamma}, -\frac{y}{\gamma^2}\right)$$

$$\delta U_y'(x,y) = -\gamma^2\,\delta T_y\left(-\frac{x}{\gamma}, -\frac{y}{\gamma^2}\right)$$

$$\delta T_y'(x,y) = -\gamma^2\,\delta U_y\left(-\frac{x}{\gamma}, -\frac{y}{\gamma^2}\right) - \gamma^2\,\delta T_y\left(-\frac{x}{\gamma} - \frac{y}{\gamma^3} - \gamma, -\frac{y}{\gamma^2}\right).$$

Here I am ignoring the effects of variation of B with (U, T), which are only in the direction of coordinate changes as already discussed (§4.2.1).

Let us define an order on monomials:

$$1 < y < x < y^2 < xy < x^2 < \ldots \tag{4.3.1.3}$$

and say a polynomial is of *rank* p,q if its largest monomial, with respect to this order, is $x^p y^q$. Then observe that at the simple fixed point, DN_1 never increases the rank of a polynomial perturbation. Thus, we can put DN_1 into Jordan normal form on the space of polynomial perturbations, with polynomial eigenvectors or generalised eigenvectors. What happens on the rest of the space, I will discuss in §4.3.4. Expanding

$$\delta U_x(x,y) = \Sigma\, a_{pq} x^p y^q \tag{4.3.1.4}$$

$$\delta T_x(x,y) = \Sigma\, b_{pq} x^p y^q$$

$$\delta U_y(x,y) = \Sigma\, c_{pq} x^p y^q$$

$$\delta T_y(x,y) = \Sigma\, d_{pq} x^p y^q$$

and ordering the coefficients by rank, and in the above order within rank, we see that DN_1 is block upper triangular, with 4×4 diagonal blocks:

$$\begin{vmatrix} 0 & -\gamma & 0 & 0 \\ -\gamma & -\gamma & -\gamma & 0 \\ 0 & 0 & 0 & -\gamma^2 \\ 0 & 0 & -\gamma^2 & -\gamma^2 \end{vmatrix} \times (-\gamma)^{-p} (-\gamma^2)^{-q} \tag{4.3.1.5}$$

for rank p, q. The diagonal block has eigenvalues:

$$-\gamma^2,\ 1,\ -\gamma^3,\ \gamma \times (-\gamma)^{-p} (-\gamma^2)^{-q} \tag{4.3.1.6}$$

with respective eigenvectors:

$$\begin{vmatrix} 1 \\ \gamma \\ 0 \\ 0 \end{vmatrix},\ \begin{vmatrix} \gamma \\ -1 \\ 0 \\ 0 \end{vmatrix},\ \begin{vmatrix} 1/2\gamma \\ \gamma/2 \\ 1 \\ \gamma \end{vmatrix},\ \begin{vmatrix} \gamma \\ -\gamma \\ \gamma \\ -1 \end{vmatrix} \tag{4.3.1.7}$$

$$B_{pq},\quad D_{pq},\quad A_{pq},\quad C_{pq}$$

For each of these eigenvectors for the block one can determine coefficients of lower rank to make eigenvectors or generalised eigenvectors for DN_1. They span the space of polynomial perturbations. As there is a lot of degeneracy, the coefficients need not be uniquely determined. I used the freedom to choose the eigenvectors to respect the invariant spaces (§4.2.3), and the result is shown in table 4.3.1.1, where they are labelled according to their terms of maximal rank as in (4.3.1.7) above. The decomposition will be proved in §4.3.3. Note that in the above, and much subsequent analysis in §4.3, I use relations between powers of the golden ratio to simplify the expressions. The relevant relations are given in §4.3.7.

§4.3.2 **Simple fixed point in actions.** The simple fixed point can be written in actions:

$$\tau(x, x') = \frac{1}{2}(x' - x - 1)^2$$

$$\upsilon(x, x') = \frac{\gamma}{2}(x' - x + \gamma)^2 \tag{4.3.2.1}$$

with

$$\alpha = -\gamma,\ \beta = -\gamma^2 \tag{4.3.2.2}$$

and intermediate point $\bar{x}$ for the composition $\upsilon \oplus \tau$ given by:

Table 4.3.1.1: Decomposition of the spectrum of DN_1 at the simple fixed point, according to area preservation (a.p.), commutativity (comm), coordinate transforms of the fixed point (c.c.), non-zero Calabi invariant (C.I) and symmetry (s). Prefix n- means non-

classification				eigenvector	eigenvalue	eigenvalues at least 1 in modulus
a.p.	comm	c.c.	s	B_{pq}, p even	$-\gamma^2(-\gamma)^{-p}\,(-\gamma^2)^{-q}$	$-\gamma^2,\ 1,\ -1$
				D_{0q}	$(-\gamma^2)^{-q}$	1
			ns	B_{pq}, p odd	$-\gamma^2(-\gamma)^{-p}\,(-\gamma^2)^{-q}$	γ
		C.I.	ns	A_{00}	$-\gamma^3$	$-\gamma^3$
	n-comm		ns	D_{pq}, $p \geq 1$	$(-\gamma)^{-p}\,(-\gamma^2)^{-q}$	
				C_{p0}	$\gamma(-\gamma)^{-p}$	$\gamma,\ -1$
				A_{p0}, $p \geq 1$	$-\gamma^3(-\gamma)^{-p}$	$\gamma^2,\ -\gamma,\ 1$
n-a.p.	comm	c.c.	s	A_{pq}, $q \geq 2$, p odd	$-\gamma^3(-\gamma)^{-p}\,(-\gamma)^2)^{-q}$	
			ns	A_{pq}, $q \geq 2$, p even	$-\gamma^3(-\gamma)^{-p}\,(-\gamma^2)^{-q}$	
		n-c.c.	ns	A_{p1}	$\gamma(-\gamma)^{-p}$	$\gamma,\ -1$
	n-comm		ns	C_{pq}, $q \geq 1$	$\gamma(-\gamma)^{-p}\,(-\gamma^2)^{-q}$	

$$\bar{x}(x, x') = \frac{x}{\gamma} + \frac{x'}{\gamma^2} \quad 1 - \frac{1}{\gamma^2}. \tag{4.3.2.3}$$

I can diagonalise DN_1 in actions at the simple fixed point. To do this I change to sum and difference variables:

$$\begin{aligned} \sigma &= x' + x \\ \delta &= x' - x \end{aligned} \tag{4.3.2.4}$$

and write:

$$t(\sigma, \delta) = \tau(x, x'),\ u(\sigma, \delta) = \upsilon(x, x'). \tag{4.3.2.5}$$

Then the simple fixed point is:

$$t(\sigma, \delta) = \frac{1}{2}(\delta - 1)^2$$

$$u(\sigma, \delta) = \frac{\gamma}{2}(\delta + \gamma)^2 \tag{4.3.2.6}$$

with intermediate point for the composition:

$$\bar{x}(\sigma, \delta) = \frac{\sigma}{2} - \frac{\delta}{2\gamma^3} - 1 - \frac{1}{\gamma^2}. \tag{4.3.2.7}$$

In sum and difference variables, the composition $u \oplus t$ is given by:

$$u \oplus t(\sigma, \delta) = u(\bar{x} + \frac{\sigma - \delta}{2}, \ \bar{x} - \frac{\sigma - \delta}{2}) + t(\frac{\sigma + \delta}{2} + \bar{x}, \ \frac{\sigma + \delta}{2} - \bar{x}). \tag{4.3.2.8}$$

Thus the derivative DN_1 at the simple fixed point is:

$$\delta u'(\sigma, \delta) = \gamma^3 \, \delta t(-\frac{\sigma}{\gamma}, -\frac{\delta}{\gamma})$$

$$\delta t'(\sigma, \delta) = \gamma^3 \Big(\delta u(-\frac{\sigma}{\gamma} + \frac{\delta}{\gamma^2} + \frac{1}{\gamma} + \frac{1}{\gamma^3}, \frac{\delta}{\gamma^3} + \frac{1}{\gamma} + \frac{1}{\gamma^3}) \tag{4.3.2.9}$$

$$+ \delta t(-\frac{\sigma}{\gamma} - \frac{\delta}{\gamma^3} + \frac{1}{\gamma} + \frac{1}{\gamma^3}, -\frac{\delta}{\gamma^2} - \frac{1}{\gamma} - \frac{1}{\gamma^3})\Big).$$

Hence expanding:

$$\delta u(\sigma, \delta) = \Sigma \, a_{pq} \sigma^p \delta^q$$

$$\delta t(\sigma, \delta) = \Sigma \, b_{pq} \sigma^p \delta^q \tag{4.3.2.10}$$

and ordering by rank $1 < \delta < \sigma < \delta^2 < \sigma\delta < \sigma^2 < ...$, we get DN_1 in block upper triangular form, with 2×2 diagonal blocks:

$$\begin{vmatrix} 0 & (-\gamma)^{-q} \\ (-\gamma^3)^{-q} & (-\gamma^2)^{-q} \end{vmatrix} \times \gamma^3(-\gamma)^{-p}. \tag{4.3.2.11}$$

Hence eigenvalues

$$\gamma^4, -\gamma^2 \times (-\gamma)^{-p}(-\gamma^2)^{-q} \tag{4.3.2.12}$$

with eigenvectors

$$\begin{vmatrix} \gamma^q \\ \gamma \end{vmatrix}, \begin{vmatrix} \gamma^q \\ -\dfrac{1}{\gamma} \end{vmatrix}. \tag{4.3.2.13}$$

$$I_{pq} \qquad II_{pq}$$

A perturbation $\delta u, \delta t$ of the simple fixed point in actions gives the following perturbation in maps:

$$\delta U_x(x, y) = \frac{1}{\gamma}(u_1 - u_2)\left(2x + \frac{y}{\gamma} - \gamma, \frac{y}{\gamma} - \gamma\right)$$
$$\delta T_x(x, y) = (t_1 - t_2)(2x + y + 1, y + 1)$$

$$\delta U_y(x, y) = 2u_1\left(2x + \frac{y}{\gamma} - \gamma, \frac{y}{\gamma} - \gamma\right)$$
$$\delta T_y(x, y) = 2t_1(2x + y + 1, y + 1). \tag{4.3.2.14}$$

So a perturbation $\sigma^p \, \delta^q$ in actions gives a perturbation $x^p y^{q-1}$ in maps, unless $q = 0$, when it gives x^{p-1}, unless $p = 0$ too, when it gives zero. In fact, one finds the following correspondence:

$$I_{pq} = B_{p,q-1} \quad q \neq 0$$
$$II_{pq} = D_{p,q-1} \quad q \neq 0$$
$$I_{p0} = A_{p-1,0} \quad p \neq 0$$
$$II_{p0} = C_{p-1,0} \quad p \neq 0 \tag{4.3.2.15}$$

I_{00} and II_{00}, being constants in the actions, generate the zero perturbation in maps.

§4.3.3 **Decomposition of the spectrum**. In this section I derive the decomposition of the spectrum of DN_1 at the simple fixed point, given in table 4.3.1.1. A minor uncertainty is that I did not check that all the Jordan blocks are simple (i.e. one dimensional). p, q will denote non-negative integers throughout this section.

(i) coordinate changes

First I determine the coordinate change eigenvectors.

$\delta\sigma(x, y) = (x^p y^q, 0)$ generates

$$\begin{aligned} \delta U_x(x, y) &= -(x + \frac{y}{\gamma} - \gamma)^p y^q + x^p y^q \\ \delta T_x(x, y) &= -(x + y + 1)^p y^q + x^p y^q \\ \delta U_y(x, y) &= 0 \\ \delta T_y(x, y) &= 0 \end{aligned} \tag{4.3.3.1}$$

whose terms of maximal rank give a $B_{p-1,\, q+1}$ eigenvector for $p \neq 0$, and zero for $p = 0$.

$\delta\sigma(x, y) = (0, x^p y^q)$ generates

$$\begin{aligned} \delta U_x(x, y) &= \frac{1}{\gamma} x^p y^q \\ \delta T_x(x, y) &= x^p y^q \\ \delta U_y(x, y) &= -(x + \frac{y}{\gamma} - \gamma)^p y^q + x^p y^q \\ \delta T_y(x, y) &= -(x + y + 1)^p y^q + x^p y^q \end{aligned} \tag{4.3.3.2}$$

whose terms of maximal rank give a B_{pq} eigenvector.

This double covering of B_{pq}, $q \geq 1$ can be resolved by considering the following combinations, for which terms of maximal rank cancel:

$$\delta\sigma(x, y) = (x^{p+2} y^{q-2}, (p + 2) x^{p+1} y^{q-1}) \tag{4.3.3.3}$$

for $q \geq 2$, giving an A_{pq} eigenvector, and

$$\delta\sigma(x, y) = (x y^q, y^{q+1}) \tag{4.3.3.4}$$

giving a D_{0q} eigenvector. So I define A_{pq}, $q \geq 2$, and D_{0q} to be these particular coordinate changes. The best choice for B_{pq} is the area preserving coordinate changes:

$$\begin{aligned} \delta\sigma(x, y) &= (q x^{p+1} y^{q-1}, -(p + 1) x^p y^q), & q \geq 1 \\ \delta\sigma(x, y) &= (0, x^p), & q = 0. \end{aligned} \tag{4.3.3.5}$$

Note that the coordinate changes $(y^q, 0)$ do not affect the map at the fixed point, since y is an invariant of this map, and thus the eigenfunctions they attempt to generate vanish. By the same token, the D_{0q} are area preserving, despite the fact that the coordinate changes generating them are not.

(ii) area preservation

The analysis in actions (§4.3.2) shows that we can take $B_{pq}, D_{pq}, A_{p0}, C_{p0}$ to be area preserving, and the rest can not be for any choice of lower rank coefficients. Alternatively,

$$\det D(T+\delta T) = \det \begin{vmatrix} 1+\delta T_{x,x} & 1+\delta T_{x,y} \\ \delta T_{y,x} & 1+\delta T_{y,y} \end{vmatrix}, \tag{4.3.3.6}$$

where subscripts after a comma denote partial derivatives. So δT is area preserving iff

$$\delta T_{x,x} + \delta T_{y,y} - \delta T_{y,x} = 0. \tag{4.3.3.7}$$

Similarly, δU is area preserving iff

$$\delta U_{x,x} + \delta U_{y,y} - \frac{1}{\gamma}\,\delta U_{y,x} = 0.$$

Applied to the terms of maximal rank for the eigenvectors, this shows that:

A_{pq}, $q \neq 0$, and C_{pq}, $q \neq 0$ are non-area preserving.

Counting area preserving dimensions by using the action representation shows that the rest can be chosen to be area preserving. There are $2(d+1)$ dimensions for actions of degree d. At the simple fixed point a perturbation of degree d in actions gives a perturbation of degree $d-1$ for maps. So we get $2(d+2)$ area preserving dimensions of degree d. The total number of dimensions of degree d is $4(d+1)$, so there are $2d$ non-area preserving dimensions. The non-area preserving eigenvectors found above already span them, so the rest I will restrict to be area preserving.

(iii) commutation

Now I consider commutation. $(\delta U, \delta T)$ preserves commutation iff

$$\delta U\, T + DU_T.\delta T\, U - DT_U.\delta U = 0. \tag{4.3.3.8}$$

At the simple fixed point the x and y components (I will call them the x- and y-commutators) are respectively:

$$\delta U_x(x+y+1, y) + \delta T_x(x, y) + \frac{1}{\gamma}\delta T_y(x, y) - \delta T_x(x + \frac{y}{\gamma} - \gamma, y) - \delta U_x(x, y) - \delta U_y(x, y) \tag{4.3.3.9}$$

$$\delta U_y(x+y+1, y) + \delta T_y(x, y) - \delta T_y(x + \frac{y}{\gamma} - \gamma, y) - \delta U_y(x, y).$$

We see that the x-commutator is non-zero at maximal rank for C_{pq}, so these eigenvectors are non-commuting.

Next we show that $D_{pq}, p \neq 0$, are non-commuting, by evaluating to the next rank. D_{pq} has eigenvalue $(-\gamma)^{-p}(-\gamma^2)^{-q}$, so its next coefficients $c_{p-1,\, q+1}$ and $d_{p-1,\, q+1}$ are determined by:

$$\begin{vmatrix} 0 & -\gamma & 0 & 0 \\ -\gamma & -\gamma & 0 & 0 \\ 0 & 0 & 0 & -\gamma \\ 0 & 0 & -\gamma & -\gamma \end{vmatrix} \begin{vmatrix} c \\ d \\ \gamma \\ -1 \end{vmatrix} = \begin{vmatrix} c \\ d \\ \gamma \\ -1 \end{vmatrix}. \tag{4.3.3.10}$$

So

$$c = -\gamma d \tag{4.3.3.11}$$

but d is arbitrary. The arbitrariness can be removed, however, because I am requiring the D_{pq} to be area preserving. This implies

$$d = \frac{p}{q+1}. \tag{4.3.3.12}$$

The x-commutator, however, has a term in $x^{p-1}y^{q+1}$ of

$$\gamma p + \frac{d}{\gamma} + \frac{p}{\gamma} - c \neq 0 \text{ for } d \neq -p. \tag{4.3.3.13}$$

So D_{pq}, $p \neq 0$ are non-commuting.

Next I show that $A_{p0}, p \geq 1$ are non-commuting. To do this requires knowledge of all coefficients of degree p, since the commutator will turn out to be of lower degree. Note, however, that terms of given rank in the y-commutator depend only on terms of strictly higher rank, so I will not need to evaluate further. One can check that in actions there is an eigenvector with eigenvalue $-\gamma^3(-\gamma)^{-p}$, with terms of maximal degree $p+1$ as follows:

$$\delta\tau(x, x') = \sum_{i+j=p+1} x^i x'^j = \frac{x^{p+2} - x'^{p+2}}{x - x'} \tag{4.3.3.14}$$

$$\delta\upsilon(x, x') = \frac{1}{\gamma}\,\delta\tau(x, x'). \tag{4.3.3.15}$$

It gives an A_{p0} eigenvector. But there is only one eigenvalue $-\gamma^3(-\gamma)^{-p}$ for degree p, so A_{p0} is uniquely determined to terms of maximal degree. The terms of maximal degree for the y-components are:

$$\delta T_y(x, y) = f(x, x + y)$$

$$\delta U_y(x, y) = \frac{1}{\gamma} f\left(x, x + \frac{y}{\gamma}\right) \tag{4.3.3.16}$$

where

$$f(x, x') = (p + 2)\,\frac{x^{p+1} - x'^{p+1}}{x - x'}. \tag{4.3.3.17}$$

So, correct to rank $(p - 1, 0)$, the y-commutator is:

$$\frac{1}{\gamma} f(x+y+1,\ x+\gamma y+1) + f(x, x+y) - f\left(x + \frac{y}{\gamma} - \gamma,\ x+\gamma y-\gamma\right) - \frac{1}{\gamma} f\left(x, x + \frac{y}{\gamma}\right)$$

$$= (p+2) \sum_{i+j=p} \frac{1}{\gamma}(x+y+1)^i (x+\gamma y+1)^j + x^i(x+y)^j - \left(x + \frac{y}{\gamma} - \gamma\right)^i (x+\gamma y-\gamma)^j - \frac{1}{\gamma} x^i\left(x + \frac{y}{\gamma}\right)^j \tag{4.3.3.18}$$

It is easy to check (in the action representation) that the commutator vanishes to leading degree. But evaluating the coefficient of x^{p-1}, for $p \neq 0$, in the y-commutator above, gives

$$p(p + 2)\left(\frac{1}{\gamma} + \gamma\right) \neq 0 \tag{4.3.3.19}$$

A_{00} is commuting in maps, but not in actions:

$$\delta\tau(x, x') = x + x' - \gamma - \frac{1}{\gamma}$$

$$\delta\upsilon(x, x') = \frac{x}{\gamma} + \frac{x'}{\gamma} + \gamma + \frac{1}{\gamma}. \qquad (4.3.3.20)$$

The commutator of the actions is $2 - 5\gamma \neq 0$, so they commute apart from a constant term. Thus A_{00} corresponds to change of the Calabi invariant.

Next I have to count non-commuting directions. There are two commutators of each rank, one an x-commutator, the other a y-commutator. C_{pq} has commutator $(x^p y^q, 0)$ + lower rank. $D_{pq}, p \geq 1$ has y-commutator $x^{p-1}y^{q+1}$ + lower rank, and $A_{p0}, p \geq 1$ has y-commutator x^{p-1} + lower rank, so we have got them all, and the rest can be chosen to be commuting. This leaves only the A_{p1} unidentified. By elimination they can be chosen to be non-area preserving commuting non-coordinate changes. They will be discussed further under (iv) symmetry, below. Incidentally, A_{01} is one of the eigenvectors (4.2.4.25). The other is non-commuting.

Alternatively, I can get all the area preserving non-commuting eigenvectors from the action representation. At the simple fixed point

$$\hat{x}(x, x') = \frac{x'}{\gamma} + \frac{x}{\gamma^2} + 1 + \frac{1}{\gamma^2}. \qquad (4.3.3.21)$$

So from (4.2.3.20)

$$C'(x, x') = -\gamma^3 C\left(-\frac{x}{\gamma}, -\frac{x'}{\gamma^2} - \frac{x}{\gamma^3} - \frac{1}{\gamma} - \frac{1}{\gamma^3}\right). \qquad (4.3.3.22)$$

Thus a commutator of $x^p x'^q$ generates

$$-\gamma^3(-\gamma)^{-p}(-\gamma^2)^{-q}\, x^p x'^q + \text{terms of lower rank}. \qquad (4.3.3.23)$$

Note that I am using the reverse ordering for rank here.

(iv) <u>symmetry</u>

A perturbation δT preserves symmetry iff

$$(S(T + \delta T))^2 = \text{identity}$$

$$\text{i.e. } S\,\delta T\, ST + D(STS).\delta T = 0 \qquad (4.3.3.24)$$

i.e.

$$-\delta T_x(-x - y - 1, y) + \delta T_x(x, y) - \delta T_y(x, y) = 0 \qquad (4.3.3.25)$$

$$\delta T_y(-x - y - 1, y) + \delta T_y(x, y) = 0 \qquad (4.3.3.26)$$

and similarly for δU. Applied to the terms of maximum rank, we see that the following do not have symmetry:

C_{pq}, for all p; A_{pq}, p even; B_{pq}, p odd; D_{pq}, p odd.

It is easy to check that a coordinate change $\delta\sigma$ is symmetric if $\delta\sigma_x$ is odd in x, and $\delta\sigma_y$ is even in x. Hence B_{pq}, p even, D_{0q}, A_{pq}, $q \geq 2$, p odd, preserve symmetry. Non-commuting eigenvectors can not be symmetric, as mentioned in §4.2.4(ii) (though generalised ones could?), so the remaining non-commuting eigenvectors are automatically non-symmetric. This leaves only A_{p1}, p odd, to be considered. They are also non-symmetric, as I will now show.

To maximum degree, A_{p1} is given by:

$$\begin{aligned} \delta T_x(x, y) &= -(x + y)^{p+2}y^{-1} + x^{p+2}y^{-1} + (p + 2)x^{p+1} \\ \delta T_x(x, y) &= -(p + 2)(x + y)^{p+1} + (p + 2)x^{p+1}. \end{aligned} \qquad (4.3.3.27)$$

This was motivated by evaluating the coordinate changes A_{pq}, $q \geq 2$. Note that it is not singular, despite the y^{-1} terms. This is the only form for A_{p1} to maximum degree, because all other eigenvalues from degree $p + 1$, rank $\leq x^p y$, are less than $\gamma(-\gamma)^p$, the eigenvalue of A_{p1}. It is symmetric to maximum degree. So next I consider the terms of degree p, starting at the maximum rank.

There is one degree of freedom in choosing the x^p terms, as C_{p0} has the same eigenvalue. One choice is zero, since DN acting on something of the form $f(x, y)y$ always produces something of the same form. But in principle, I could add a multiple of C_{p0}:

$$\begin{aligned} \delta T_x &= -a\gamma\, x^p \\ \delta T_y &= -a\, x^p. \end{aligned} \qquad (4.3.3.28)$$

Checking the x-symmetry (4.3.3.25) at rank x^p, however, shows that A_{p1} is non-symmetric if $a \neq 0$.

So now evaluate δT_y to rank $x^{p-1}y$:

$$\gamma(-\gamma)^p\, \delta U_y(x, y) = -\gamma^2\, \delta T_y\left(-\frac{x}{\gamma}, -\frac{y}{\gamma^2}\right)$$

(4.3.3.29)

$$\gamma(-\gamma)^{-p}\, \delta T_y(x, y) = -\gamma^2\, \delta U_y\left(-\frac{x}{\gamma}, -\frac{y}{\gamma^2}\right) - \gamma^2\, \delta T_y\left(-\frac{x}{\gamma} - \frac{y}{\gamma^3} - \gamma, -\frac{y}{\gamma^2}\right).$$

So

$$\gamma^2(-\gamma)^{-2p}\, \delta T_y(x, y) = \gamma^4\, \delta T_y\left(\frac{x}{\gamma^2}, \frac{y}{\gamma^4}\right) - \gamma^3(-\gamma)^{-p}\, \delta T_y\left(\frac{x}{\gamma} - \frac{y}{\gamma^3} - \gamma, -\frac{y}{\gamma^2}\right). \quad (4.3.3.30)$$

Try

$$\delta T_y(x, y) = (p+2)\left(-(x+y)^{p+1} + x^{p+1} + bx^{p-1}y\right) + \text{lower rank.} \quad (4.3.3.31)$$

Then one finds:

$$b = -\gamma p(p+1). \quad (4.3.3.32)$$

Now check y-symmetry (4.3.3.26) to rank $x^{p-1}y$. It turns out to be non-symmetric.

§4.3.4 **Compactness of DN_1 at the simple fixed point.** Next I show that DN_1 is a compact operator at the simple fixed point, in a suitable norm, and that the error in truncating it at finite degree goes to zero as the degree goes to infinity. Thus, standard results in functional analysis (e.g. Krasnosel'skii et al., §18, 1972) imply that the diagonalisation of table 4.3.1.1 is complete, apart from a component with eigenvalue 0. It is clear, incidentally, that table 4.3.1.1 does not cover the whole space, as there are arbitrarily small perturbations of the simple fixed point, in class AA, which are not coordinate transforms of the simple fixed point. For example, for $k = 0$ the standard map is equivalent to the simple fixed point, and has a whole circle of points of type (0, 1), but for $k \neq 0$ there are only two.

To show compactness of DN_1, I show that N_1 is analyticity improving in a neighbourhood of the simple fixed point, on suitable domains. Specifically, if T is analytic on the product of discs $|x| \le X$, $|y| \le Y$, for some $X, Y > 0$ with

$$X > \gamma^3 + \frac{Y}{\gamma} \quad (4.3.4.1)$$

and U is analytic on $|x| \le X'$, $|y| \le Y'$, with

$$\frac{X}{\gamma} < X' < \gamma X, \ \frac{Y}{\gamma^2} < Y' < \gamma^2 Y \tag{4.3.4.2}$$

then (U', T') is analytic on larger discs, for (U, T) close enough to the simple fixed point. Close enough means with respect to the ℓ_1 norm (2.3.8.2) for power series expansions in these discs (i.e. sum of the absolute values of the Taylor coefficients, normalised so the discs have unit radius).

The error in truncating DN_1 at degree d goes to zero as $d \to \infty$. In fact,

$$\| DN_1 - DN_1^{(d)} \| \le C \lambda^{d+1} \tag{4.3.4.3}$$

where

$$\lambda = \max\left(\frac{X}{\gamma X'}, \frac{X'}{\gamma X}, \frac{Y}{\gamma Y'}, \frac{Y'}{\gamma Y}, \frac{\frac{X}{\gamma} + \frac{Y}{\gamma^3} + \gamma}{X} \right) < 1 \tag{4.3.4.4}$$

and C depends on the relative weights assigned to δU_x etc. To see this, the norm of a bounded linear operator A is defined by:

$$\| A \| = \sup_{x \ne 0} \frac{\| Ax \|}{\| x \|}. \tag{4.3.4.5}$$

Now $DN_1 - DN_1^{(d)}$ gives zero acting on vectors of degree d or lower, and on monomials of degree larger than d, it is just DN_1. For a monomial vector, $x^p y^q$, we get that:

$$\| \delta U_x' \| \le \gamma \left(\frac{X'}{\gamma X} \right)^p \left(\frac{Y'}{\gamma^2 Y} \right)^q \| \delta T_x \| \tag{4.3.4.6}$$

$$\| \delta T_x' \| \le \gamma \left(\frac{X}{\gamma X'} \right)^p \left(\frac{Y}{\gamma^2 Y'} \right)^q \| \delta U_x \| + \gamma \left(\frac{\frac{X}{\gamma} + \frac{Y}{\gamma^3} + \gamma}{X} \right)^p \gamma^{-2q} \| \delta T_x \|$$

$$+ \left(\frac{X}{\gamma X'} \right)^p \left(\frac{Y}{\gamma^2 Y'} \right)^q \| \delta U_y \|$$

$$\| \delta U_y' \| \leq \qquad \gamma^2 \left(\frac{X'}{\gamma X} \right)^p \left(\frac{Y'}{\gamma^2 Y} \right)^q \mid \| \delta T_y \|$$

$$\| \delta T_y' \| \leq \gamma^2 \left(\frac{X}{\gamma X'} \right)^p \left(\frac{Y}{\gamma^2 Y'} \right)^q \| \delta U_y \| + \gamma^2 \left(\frac{\frac{X}{\gamma} + \frac{Y}{\gamma^3} + \gamma}{X} \right)^p \gamma^{-2q} \| \delta T_y \| .$$

Under the conditions (4.3.4.1, 4.3.4.2) on X, Y, X', Y', we see that DN_1 decreases the norm of the perturbation by at least a factor of $C\nu^p \mu^q$, where:

$$\nu = \max \left(\frac{X'}{\gamma X}, \frac{X}{\gamma X'}, \frac{\frac{X}{\gamma} + \frac{Y}{\gamma^3} + \gamma}{X} \right) < 1 \tag{4.3.4.7}$$

$$\mu = \max \left(\frac{Y'}{\gamma^2 Y}, \frac{Y}{\gamma^2 Y'}, \gamma^{-2} \right) < 1 \tag{4.3.4.8}$$

and C is a constant depending on the relative weights assigned to δU_x etc. Hence the result. One can get ν arbitrarily close to γ^{-1}, and $\mu = \gamma^{-2}$, for the choice $X' = X$, $Y' = Y, X > Y, 1$. By choosing the weights $\underline{w}$ optimally, in:

$$\| (\delta U, \delta T) \| = w_1 \| \delta U_x \| + w_2 \| \delta T_x \| + w_3 \| \delta U_y \| + w_4 \| \delta T_y \| \tag{4.3.4.9}$$

one can get arbitrarily close to the minimum possible C of γ^3. These values are optimal because there is an eigenvector of rank p, q with eigenvalue $-\gamma^3(-\gamma)^{-p}(-\gamma^2)^{-q}$.

§4.3.5 **Significance of the simple fixed point.** If one restricts attention to class AA, table 4.3.1.1 shows that all polynomial directions from the simple fixed point are coordinate changes. Taking §4.3.4 into consideration, plus the fact that N_1 is analyticity improving, implying that N_1 is infinitely differentiable (using Cauchy estimates), and in particular C^1, this implies that, modulo coordinate changes, the simple fixed point attracts a neighbourhood, in fact faster than exponentially. This is, of course, what one should expect from KAM theory. Note also that it is attracting in the space of symmetric commuting pairs, as one expects from the reversible version of Moser's twist theorem (Moser, 1973).

I already indicated (§4.2.2) how Mather's theorem can probably be used to show that for F periodic and area preserving, convergence of $N_1^n(F,R)$ to a fixed point implies that F has a golden circle. At the simple fixed point, one could instead use Moser's twist theorem (assuming it generalises to commuting pairs). It would imply a $C^{3+\varepsilon}$-neighbourhood, for any $\varepsilon > 0$, in which all commuting pairs have a (smooth) golden curve. Convergence of $N_1^n(U,T)$ to the simple fixed point (in the $C^{3+\varepsilon}$ topology) implies that $N_1^{n_0}(U,T)$ is in this neighbourhood for some n_0, and so (U,T) has a (smooth) golden curve (cf. Escande and Doveil, 1981).

I suspect that for the simple fixed point neither Moser's twist theorem nor Mather's theorem should be necessary,[8] but I do not see how much smoothness I am going to obtain for the circle if I do not assume the maps to be analytic.[9]

Most proofs of results in KAM theory restrict one to pretty small perturbations (see §1.3.1.5). Sometime soon I hope to construct a reasonably large neighbourhood of attraction for the simple fixed point.[10] This will be along the same lines as the construction to be given in §4.7.4 for the 1-D case. Incidentally, it would be a C^4-neighbourhood, implying persistence of noble circles for C^4 perturbations, by Mather's theorem. I could also find a $C^{3+\varepsilon}$-neighbourhood for any $\varepsilon > 0$.[11] The weakest

[8] This is true. In London in October 1982 I realised that convergence of the renormalisation to the simple fixed point, plus freedom to choose the vertical $x = 0$, implies that on each vertical $x - x_0$ constant, there is a unique point (x_0, y_0) with the desired rotation number (namely, the point onto which the sequence of coordinate changes shrinks). My only problem at that time was to show that these join up to form a curve. In retrospect, that is easy: the coordinate changes depend continuously on x_0 and because they are contractions hence so does y_0. This missing step, plus a slicker proof of convergence to the simple fixed point (also generalised to a set of rotation numbers of full measure) was supplied by Khanin and Sinai(1986) (see also Sinai & Khanin, 1988). Meanwhile I learnt about Rand's approach to constructing invariant curves in the dissipative case, and in April 1984 (unpublished) I used it to construct invariant curves for any pair (U, T) which remains within some bounded set forever under renormalisation, in particular for pairs converging to the simple fixed point. Haydn extended the technique that Summer to all Diophantine rotation numbers and to deduce some smoothness (depending on the smoothness of the maps and the Liouville exponent of the rotation number) for the curves (published in 1990). Rand's technique has the advantage that one need make only one choice of vertical x 0, so that it can in principle also be applied to the critical case.

[9] This was answered by Haydn, as mentioned in the previous footnote.

[10] I never did this, but it would still be a good idea.

[11] I need at least 3 derivatives in order to be able to impose commutativity to 3rd order and hence remove all the eigenvalues of Table 4.3.1.1 not inside the unit circle. I would need Hölder continuity of the third derivative in order to be sure N_1 is a contraction in some neighbourhood of the simple fixed point.

smoothness condition known at present is $C^{3+\varepsilon}$, $\varepsilon > 0$ (Herman, 1981, Rüssmann, 1981).[12] There are counterexamples for $C^{2+\eta}$, $\eta \leq 1$ (Herman, 1981).

It is worth pointing out that even when persistence is proved only for a small neighbourhood of the linear twist, a much larger class of perturbations can be brought into this domain by coordinate changes, thereby extending the domain. For the corresponding problem for diffeomorphisms on a circle, Herman (1976) used this technique to prove global conjugacy results (see §4.7.1).

§4.3.6 **Spectrum of DN_{s1} at the simple fixed point**. In §4.4.2 it will turn out to be convenient to consider a different operator from N_1, but one which is closely related, viz.:

$$N_{s1}\colon U' = BST^{-1}SB^{-1}$$
$$T' = BTUB^{-1} \tag{4.3.6.1}$$

where S is the reflection:

$$S(x, y) = (-x, y). \tag{4.3.6.2}$$

Its action on symmetric maps is the same as that of N_1. Its value will be in improving domains.

Clearly N_1 and N_{s1} have the same symmetric fixed points. Furthermore, the behaviour of their derivatives in the space of symmetric commuting pairs is the same. Note that it is necessary to include commutation in this statement, because just as for N_1, N_{s1} preserves symmetry only for commuting pairs. It would be best, of course, to find an operator which had the same effect as N_1 on symmetric commuting pairs and preserved symmetry, but there was no obvious possibility. In §4.4.2, I obtain some non-symmetric eigenvectors for DN_{s1} at the critical fixed point, which I have not yet explained. For comparison, however, I diagonalise DN_{s1} at the simple fixed point in this section, both in maps, and in actions.

In maps, the derivative DN_{s1} is given by:

$$\delta U' = BS\,\delta(T^{-1})\,SB^{-1} \tag{4.3.6.3}$$

with the same expression as before $\delta T'$. Now:

$$T\,T^{-1} = I. \tag{4.3.6.4}$$

So

$$DT_{T^{-1}}.\delta(T^{-1}) + \delta T\,T^{-1} = 0. \tag{4.3.6.5}$$

[12] Herman (1986) has proved persistence of constant type circles in C^3 also.

Thus

$$\delta U' = -BS\left(DT_{T^{-1}SB^{-1}}\right)^{-1}.\delta T\, T^{-1}SB^{-1}. \tag{4.3.6.6}$$

In particular, at the simple fixed point:

$$\delta U' = \begin{vmatrix} -\gamma & \gamma \\ 0 & \gamma^2 \end{vmatrix} .\delta T\left(\frac{x}{\gamma}+\frac{y}{\gamma^2} - 1, -\frac{y}{\gamma^2}\right). \tag{4.3.6.7}$$

Proceeding as in §4.3.1, DN_{s1} is block upper triangular with the diagonal blocks:

$$\begin{vmatrix} 0 & -(-)^p\gamma & 0 & (-)^p\gamma \\ -\gamma & -\gamma & -\gamma & 0 \\ 0 & 0 & 0 & (-)^p\gamma^2 \\ 0 & 0 & -\gamma^2 & -\gamma^2 \end{vmatrix} \times (-\gamma)^{-p}(-\gamma^2)^{-q} \tag{4.3.6.8}$$

for rank x^py^q. Thus we obtain the following eigenvalues:

p even: $$-\gamma^2,\ 1,\ \gamma^2\omega,\ \gamma^2\bar{\omega} \times (-\gamma)^{-p}(-\gamma^2)^{-q} \tag{4.3.6.9}$$

p odd: $$\gamma\omega,\ \gamma\bar{\omega},\ -\gamma^3,\ \gamma \times (-\gamma)^{-p}(-\gamma^2)^{-q} \tag{4.3.6.10}$$

where

$$\omega, \bar{\omega} = -\frac{1}{2} \pm i\frac{\sqrt{3}}{2} \tag{4.3.6.11}$$

are the complex cube roots of unity. The eigenvectors for the real eigenvalues can be identified as B_{pq}, D_{pq}, for p even, and A_{pq}, C_{pq} for p odd.

In actions, using sum and difference variables (cf. §4.3.2), the change to the derivative is:

$$\delta\upsilon'(\sigma, \delta) = \alpha\beta\, \delta\tau\left(-\frac{\sigma}{\alpha}, \frac{\delta}{\alpha}\right). \tag{4.3.6.12}$$

So the diagonal blocks of (4.3.2.11) are changed to:

$$\begin{vmatrix} 0 & \gamma^3(-\gamma)^{-p}(-\gamma)^{-q} \\ \gamma^3(-\gamma)^{-p}(-\gamma^3)^{-q} & \gamma^3(-\gamma)^{-p}(-\gamma^2)^{-q} \end{vmatrix} \tag{4.3.6.13}$$

with eigenvalues:

p even: $$\gamma^4,\ -\gamma^2\ \times (-\gamma)^{-p}(-\gamma^2)^{-q} \tag{4.3.6.14}$$

p odd: $$-\gamma^3\omega,\ -\gamma^3\bar{\omega}\ \times (-\gamma)^{-p}(-\gamma^2)^{-q}.$$

The eigenvectors for the real eigenvalues are I_{pq}, p even, and II_{pq}, p even.

§4.3.7 **Relations between powers of golden ratio.** In §4.3 I have often used relations between powers of the golden ratio. They are collected here for convenience.

Golden ratio γ (1.3.1.5.11) is defined to be the positive solution of:

$$\gamma^2 = \gamma + 1. \tag{4.3.7.1}$$

Equivalently:

$$\gamma^{-1} = \gamma - 1. \tag{4.3.7.2}$$

One can use these relations to express higher powers of γ in terms of γ, as follows:

$$\gamma^3 = 2\gamma + 1 \tag{4.3.7.3}$$

$$\gamma^4 = 3\gamma + 2$$

$$\gamma^5 = 5\gamma + 3$$

$$\gamma^6 = 8\gamma + 5$$

and inverse powers:

$$\gamma^{-2} = 2 - \gamma \tag{4.3.7.4}$$

$$\gamma^{-3} = 2\gamma - 3$$

$$\gamma^{-4} = 5 - 3\gamma$$

$$\gamma^{-5} = 5\gamma - 8.$$

Also useful are the following two relations:

$$\begin{aligned} 1 + \gamma^{-3} &= 2\gamma^{-1} \\ 1 - \gamma^{-3} &= 2\gamma^{-2} \end{aligned} \tag{4.3.7.5}$$

and

$$1+\gamma^{-4} = 3\gamma^{-2}. \tag{4.3.7.6}$$

§4.4 CRITICAL FIXED POINT

The renormalisation also has a critical fixed point with essentially only one unstable direction. It corresponds to a map with a non-smooth noble circle. This can be found as a critical case in many one parameter families of area preserving twist maps, by using the connections between invariant circles and periodic orbits. Then I discuss how the fixed point can be found in power series in appropriate domains. Truncating at various degrees gives approximate fixed points whose properties appear to converge to the same answers as found empirically.

§4.4.1 **Critical noble circles**. The *quadratic irrationals* are those irrationals which satisfy a quadratic equation with integer coefficients. They are precisely those numbers whose continued fraction expansion is eventually periodic. For periodic maps F with a critical quadratic irrational circle, Shenker and Kadanoff (1982) found scaling behaviour in the neighbourhood of certain points for reversible area-preserving maps.[13] The behaviour appears to be the same for all quadratic irrationals with the same repeat pattern.

In particular, for nobles (repeat pattern [1]) one obtains figure 4.4.1.1 in critical cases, if one looks on a small enough scale and in appropriate coordinates (to be discussed later). Note that the picture in the smaller box repeats the whole picture on a smaller scale and turned over. Asymptotically, the self-similarity is exact, with scaling factors:

$$\begin{aligned} \alpha &= -1.4148360 \text{ in } Y \\ \beta &= -3.0668882 \text{ in } X. \end{aligned} \tag{4.4.1.1}$$

I will begin this section by describing how I find critical cases in one parameter families of reversible maps. Given a noble, the first steps are to find its convergents p_n/q_n (1.3.1.2.4), and the dominant half-line (§1.2.3.4). Then I find the parameter values P_n such that the periodic point of type (p_n, q_n) on the dominant half-line, has some given residue, e.g. 1. The P_n typically converge, and their limit is the parameter value for a critical case.

---

13 It is an open question what will be seen for non-reversible area-preserving maps, though Table 4.4.2.1 suggests there will be no new unstable directions.

This is shown in table 4.4.1.1 for rotation number $\gamma^{-2} = [0,2,1,1,...]$ around the fixed point, in the quadratic map (1.2.2.2.4). Note that the convergence is asymptotically geometric, though it takes its time to get settled down! The limiting ratio is:

$$\delta = 1.6280 \tag{4.4.1.2}$$

which I determined by superconverging the above results (cf. §3.1.2). Note that although close, this is different from golden ratio (1.3.1.5.11). This is contrary to the implications of Kadanoff (1981a) and Shenker and Kadanoff (1982). Also it indicates that the large k and small k behaviour:

$$R_{p/q} \sim a_{pq} k^q \tag{4.4.1.3}$$

derived by Greene (1979a) for the standard map, does not extend to intermediate k.[14]

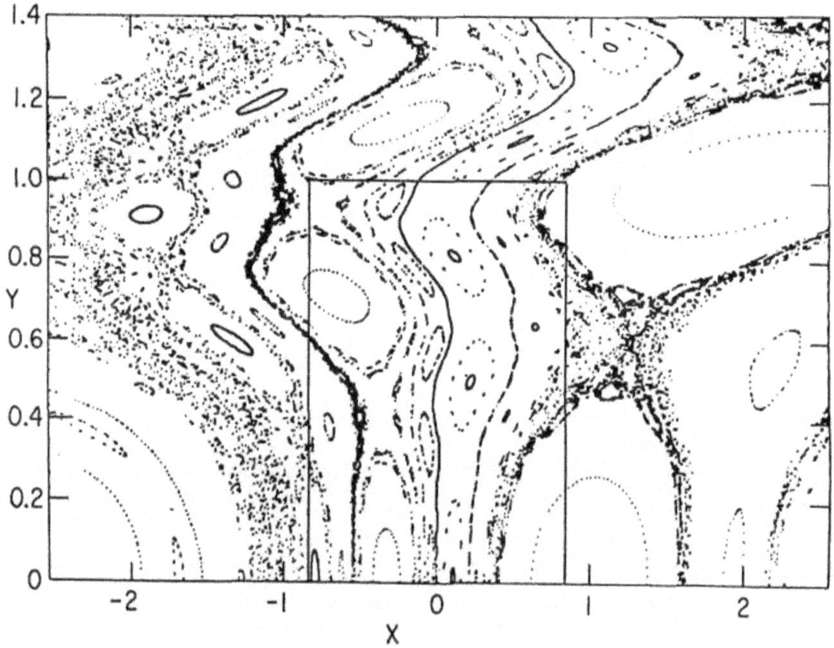

Figure 4.4.1.1: Some orbits of the universal map F^*

[14] Because if it held at k_c then we would have $a_{p_n, q_n} = R^* k_c^{-q_n}$, and so $\partial/\partial k \, R_{p_n/q_n}|_{k_c} = q_n R^*/k_c \propto \gamma^n$, whereas I observe $\partial/\partial k \, R_{p_n/q_n}|_{k_c} \propto \delta^n$.

Table 4.4.1.1: Parameter values in the quadratic map for the periodic orbits convergent to γ^{-2} to have residue 1.

p_n	q_n	parameter value P_n	ratio
1	2	4.0	
1	3	1.25	
2	5	2.82298392897775348	-1.7483
3	8	2.42465863983816842	-3.9940
5	13	2.49848604133242700	-5.3954
8	21	2.41828466488974625	-0.9205
13	34	2.41671274328787171	51.0212
21	55	2.39825342904454782	0.0852
34	89	2.39377994792488873	4.1264
55	144	2.38857137772415864	0.8589
89	233	2.38634340433683794	2.3378
144	377	2.38462836393319881	1.2991
233	610	2.38371180928814206	1.8712
377	987	2.38310001988408071	1.4981
610	1597	2.38274350586283945	1.7160
987	2584	2.3825176441328905	1.5785
1597	4181	2.38238162084892175	1.6605
2584	6765	2.38229709951522443	1.6093
4181	10946	2.38224556323407916	1.6400
6765	17711	2.38221377005868008	1.6210
10946	28657	2.38219429436231606	1.6325

Next I describe how to determine the scalings (4.4.1.1) in the critical case. The first step is to use symmetry coordinates (1.1.4.2.1) or (1.1.4.4.2). Unfortunately, I use two different conventions for symmetry coordinates in this chapter. To distinguish between them I will use (X, Y) for the first and (x, y) for the second. Then I find the positions X_n on the dominant half-line of the periodic points of types (p_n, q_n), and their convergence rate. I call these the *dominant* periodic points. This is shown in table 4.4.1.2 for the critical case for rotation number γ^{-2} in the quadratic map. Compare figure 1.3.2.1.1.

Table 4.4.1.2: Positions X_n of the convergent periodic points on the dominant half-line

p_n	q_n	position	ratio
1	3	2.17565438798993984	
2	5	1.15808946072309186	
3	8	1.34919972873801010	-5.3244911
5	13	1.30433541624393411	-4.2597391
8	21	1.31944376972476221	-2.9695038
13	34	1.31486820981790246	-3.3019682
21	55	1.31639850758256216	-2.9899801
34	89	1.31591058738277828	-3.1363689
55	144	1.31607134474930116	-3.0351343
89	233	1.31601930959112921	-3.0893990
144	377	1.31603634343353925	-3.0548104
233	610	1.31603080345748564	-3.0747141
377	987	1.31603261251445986	-3.0623558
610	1597	1.31603202319130637	-3.0697198
987	2584	1.31603221545443856	-3.0651906
1597	4181	1.31603215278571776	-3.0679281
2584	6765	1.31603217322396392	-3.0662475
4181	10946	1.31603216656061549	-3.0672636
6765	17711	1.31603216873347153	-3.0666313
10946	28657	1.31603216802500834	-3.0669992
17711	46368	1.31603216825602302	-3.0667454

To get the second scaling, evaluate the tangent maps:

$$DF^{q_n} = \begin{vmatrix} A_n & B_n \\ C_n & D_n \end{vmatrix} \tag{4.4.1.4}$$

at the dominant periodic points. Then the scaling (4.4.1.1) results if:

$$C_n \sim c\left(\frac{\beta}{\alpha}\right)^n . \tag{4.4.1.5}$$

Table 4.4.1.3 gives the elements C_n. Superconverging the results gives:

$$\frac{\beta}{\alpha} = 2.167648. \tag{4.4.1.6}$$

Hence α, as in (4.4.1.1).

Note that α can also be determined directly by finding the Y-coordinates of one of the nearest points of the periodic orbit to the dominant point. This gave the same value for α as above. They can be specified as follows. For a periodic orbit of type (p, q) of a periodic map F, find (i, j) such that:

Table 4.4.1.3: Off-diagonal elements of the tangent maps to the dominant periodic points

p_n	q_n	C_n	ratio
1	3	-3.2667522568204209d+1	
2	5	-5.4484295948402877d-1	0.01668
3	8	-1.4431072581187895d+1	26.48666
5	13	-1.1070792507950726d+1	0.76715
8	21	-4.5501683073093672d+1	4.11007
13	34	-6.7992802078947985d+1	1.49429
21	55	-1.8451626317611859d+2	2.71376
34	89	-3.4932294891324069d+2	1.89318
55	144	-8.2201003367452586d+2	2.35315
89	233	-1.6950980093940890d+3	2.06214
144	377	-3.7877987196199406d+3	2.23456
233	610	-8.0598996504873366d+3	2.12786
377	987	-1.7669831847417207d+4	2.19231
610	1597	-3.8038682194988934d+4	2.15275
987	2584	-8.2803162821299394d+4	2.17681
1597	4181	-1.7902719875319295d+5	2.16208
2584	6765	-3.8867879389525641d+5	2.17106
4181	10946	-8.4170714617582567d+5	2.16556
6765	17711	-1.8255767209324497d+6	2.16890
10946	28657	-3.9557070380450427d+6	2.16683
17711	46368	-8.5761254558232297d+6	2.16804

$$ip - jq = \pm 1. \tag{4.4.1.7}$$

Then the nearest points of the periodic orbit to the dominant point x_0 are:

$$F^i R^j x_0. \tag{4.4.1.8}$$

For a Fibonacci sequence of (p, q), the pairs (i, j) can be chosen to be another Fibonacci sequence.

In summary, defining the map R:

$$R(\theta, z) = (\theta - 1, z) \tag{4.4.1.9}$$

it looks as if, for a map with a critical noble circle there are coordinate changes B_n, such that the maps:

$$B_n\, F^{q_n}\, R^{p_n}\, B_n^{-1} \tag{4.4.1.10}$$

converge to some universal map F^*, with

$$B_{n+1} \sim B\, B_n \text{ as } n \to \infty,\ B(X, Y) = (\beta X, \alpha Y). \tag{4.4.1.11}$$

Determining the coordinate changes, however, it is not completely straightforward. One has to find scaling coordinates (§2.3.5). To see this, suppose F satisfies:

$$B^n\, F^{q_n}\, R^{p_n}\, B^{-n} \to F^* \tag{4.4.1.12}$$

with B linear diagonal. Then the effects of a quadratic shear (which preserves symmetry):

$$X \to X + cY^2 \tag{4.4.1.13}$$

grow like

$$\frac{\beta}{\alpha^2} = -1.5321. \tag{4.4.1.14}$$

So one has to determine a quadratic shear to kill the component in this direction. For example, I measured the positions (X_n, Y_n) of the nearest point of the periodic orbit to the dominant point, and evaluated:

$$c = \lim_{n \to \infty} \frac{X_n - X_\infty}{Y_n^2} = -0.7783661. \qquad (4.4.1.15)$$

Then I used the new X-coordinate:

$$\bar{X} = X - X_\infty - cY^2. \qquad (4.4.1.16)$$

This was an essential part in producing figure 4.4.1.1. Actually I also applied a quartic shear (see §4.5.1). Furthermore, I rescaled the axes so that the 89/233 dominant point is at $\bar{X} = 1$, and the next point of that orbit has $Y = 1$.

I did the same for rotation number $\gamma^{-1} = [0, 1, 1, \ldots]$ in the standard map (1.3.1.5.3). The determination of the critical parameter value is given in table 4.4.1.4.[15] Note that I have switched X and Y now, and hence also B and C. ℓ/m is the rotation number, ℓi, ℓf, ji, jf refer to initial and final lines, p is parameter value and r is the desired residue. dp is an estimate of the error in p, dr is the deviation of the residue from the desired value, yi, yf are y-coordinates of the initial and final points of the orbit, dyi is an error estimate for yi, e is a closing error, bi, bf, ci, cf are components of the tangent maps on the initial and final lines, in symmetry coordinates, w is the action, and *detm* 1 is the deviation from area preservation. yp, bp refer to the dominant point. Columns marked *ratio* mean either the ratio of successive elements of the previous column, or the ratio of differences. *Accumulation* is an estimate of the accumulation point of the first column. *Constant* is the constant in the relation:

$$p_n - p_\infty \sim C\delta^{-n} \qquad (4.4.1.17)$$

or

$$b_n \sim C\left(\frac{\beta}{\alpha}\right)^n \qquad (4.4.1.18)$$

for example.

[15] Superconverging the results give $k_c \simeq 0.9716354061062$ ($\pm$ 4 in the last place). Strangely, the ratios of differences of the accumulation estimates for p in Table 4.4.1.4 oscillate, but with a 2-step ratio of about 7.1154. I used this to estimate the above accumulation value.

Note that the y-values scale as δ. This is because $\delta < |\beta|$, so that the β-scaling is swamped by the slower movement due to change in parameter.

Table 4.4.1.4: Determination of the critical parameter value for the orbit of inverse golden rotation number in the standard map.

i0 m0	ii ji if jf	p r	dp dr	yi yf	dyi e	bi bf	ci cf	w detml
1	0 0	4.000000000000000000000000d+00	0.	9.999999999999999999999999d-01	0.	1.0000000d+00	0.	6.0132118d-01
1	1 1	1.000000000000000000000000d+00	0.	9.999999999999999999999999d-01	0.	0.	-4.0000000d+00	0.
1	0 0	2.000000000000000000000000d+00	0.	5.000000000000000000000000d-01	0.	4.0000000d+00	0.	2.5000000d-01
2	0 1	1.000000000000000000000000d+00	0.	5.000000000000000000000000d-01	0.	0.	-4.0000000d+00	0.
2	0 0	1.5175481588462763428578d+00	2.d-29	6.1385957580948204683349d-01	1.d-29	8.0941167d+00	-6.8893903d-30	6.5544110d-01
3	1 0	1.000000000000000000000000d+00	-3.d-29	7.7228084838103590633300d-01	3.d-29	2.6161729d-29	-2.0701926d+00	-3.d-29
3	0 0	1.2055842936824400261651d+00	3.d-30	5.8099659044351122998379d-01	5.d-30	2.0930160d+01	2.4770149d-29	8.8539377d-01
5	1 1	1.000000000000000000000000d+00	1.d-28	4.7663162014150383248011d-01	3.d-29	-4.6467581d-28	-1.1161359d+00	0.
5	0 0	1.1484855896130613348299d+00	2.d-29	5.9436824267502131959854d-01	5.d-31	4.7320000d+01	-8.2508024d-30	1.5423764d+00
8	0 1	1.000000000000000000000000d+00	-1.d-28	6.8940916366924441817414d-01	0.	3.0893024d-28	-1.2630185d+00	-5.d-29
8	0 0	1.0789946734614360546183d+00	2.d-30	5.9204190246843241338573d-01	3.d-32	1.0446193d+02	1.1925408d-30	2.4032315d+00
13	1 0	1.000000000000000000000000d+00	-3.d-29	6.7540626751134181237415d-01	0.	-2.0873131d-28	-3.9602045d-01	-5.d-29
13	0 0	1.0382327324943334305548d+00	2.d-29	5.9399598292136241119416d-01	7.d-30	2.3079315d+02	-7.4688018d-30	3.9815030d+00
21	1 1	1.000000000000000000000000d+00	-6.d-28	6.2578821998256420208356d-01	-2.d-28	5.1245212d-27	-3.3637256d-01	-2.d-28
21	0 0	1.0105956989556351837463d+00	1.d-28	5.9408795790635046310203d-01	2.d-30	5.0290559d+02	4.1744849d-29	6.4283243d+00
34	0 1	1.000000000000000000000000d+00	5.d-27	6.6623761516415224734655d-01	0.	-6.0047194d-26	-3.4967592d-01	5.d-29
34	0 0	9.9523263134906640514538d-01	1.d-26	5.9449955609380641250862d-01	3.d-28	1.0985669d+03	-3.0277632d-27	1.0407160d+01
55	1 0	9.999999999999999999999999d-01	-8.d-25	6.7792701746838982882888d-01	8.d-28	1.9071166d-23	-1.7440992d-01	-5.d-29
55	0 0	9.8611838652870813491946d-01	3.d-28	5.9463407283982361956166d-01	1.d-29	2.3857343d+03	-3.3264356d-29	1.6832620d+01
89	1 1	9.999999999999999999999999d-01	-2.d-26	5.2806222797867455004499d-01	-3.d-27	8.2028980d-25	-9.6746192d-02	-4.d-28
89	0 0	9.8848433448428137741811d-01	3.d-26	5.9475447181485543479399d-01	5.d-28	5.1864210d+03	-2.9989531d-27	2.7244784d+01
144	0 1	9.999999999999999999999999d-01	-4.d-24	6.6533165252602893375987d-01	-1.d-26	1.5404468d-22	-1.0044797d-01	-2.d-27
144	0 0	9.7706941345108631349925d-01	1.d-26	5.9481569222323915712936d-01	2.d-28	1.1250034d+04	7.8609848d-28	4.4062316d+01
233	1 0	9.999999999999999999999999d-01	2.d-24	6.7706672610050842229367d-01	2.d-26	-1.7641895d-22	-5.8128601d-02	3.d-27
233	0 0	9.7496661387014492855822d-01	9.d-24	5.9485733937915507822790d-01	2.d-25	2.4412927d+04	-2.1669285d-25	7.1331945d+01
377	1 1	9.999999999999999999999999d-01	-1.d-21	5.2909968553638294151164d-01	2.d-25	1.8976413d-19	-2.7877275d-02	-4.d-27
377	0 0	9.7368143224177998194049d-01	1.d-25	5.9488134198392446721721d-01	2.d-27	5.2932466d+04	-8.6714031d-27	1.1541905d+02
610	0 1	9.999999999999999999999999d-01	-1.d-22	6.6487659253233165768843d-01	3.d-26	1.5923110d-20	-2.8822328d-02	-3.d-27
610	0 0	9.7289126281167071488571d-01	9.d-26	5.9489684384637229732587d-01	2.d-27	1.1470698d+05	2.4175227d-27	1.8675575d+02
987	1 0	9.999999999999999999999999d-01	7.d-23	6.7695243409862025108219d-01	-3.d-25	-1.9243048d-20	-1.4420798d-02	2.d-27
987	0 0	9.7240688606598401563848d-01	2.d-23	5.9490689153294228712573d-01	4.d-25	2.4884348d+05	1.1615710d-26	3.0217951d+02
1597	1 1	9.999999999999999999999999d-01	7.d-22	5.2930768749368747548165d-01	1.d-23	-3.6073542d-19	-8.8127802d-03	-4.d-26
1597	0 0	9.7210911826722850654027d-01	1.d-22	5.9491181941895160584125d-01	3.d-24	5.3949240d+05	1.0309588d-24	4.8893994d+02
2584	0 1	9.999999999999999999999999d-01	1.d-19	6.6478640429355758902341d-01	-4.d-23	-6.7663321d-17	-8.2837268d-03	-1.d-25
2584	0 0	9.7192638786595275107705d-01	1.d-23	5.9491532133439185569088d-01	2.d-25	1.1694807d+06	5.5990289d-26	7.9112410d+02
4181	1 0	9.999999999999999999999999d-01	2.d-20	6.7692883559495257972468d-01	8.d-23	-1.3797901d-17	-4.1448208d-03	-6.d-25
2584	0 0	9.7192638786595275107705d-01	2.d-26	5.9491532133439185569088d-01	1.d-27	1.1694807d+06	1.2979745d-25	7.9112410d+02
4181	1 0	9.999999999999999999999999d-01	4.d-20	6.7692883559495257972468d-01	5.d-23	-3.6622962d-17	-4.1448208d-03	-4.d-25
4181	0 0	9.7181412813005934495916d-01	7.d-23	5.9491747752949088463358d-01	1.d-24	2.5351880d+06	6.7590423d-25	1.2800687d+03
6765	1 1	9.999999999999999999999999d-01	4.d-19	5.2935766193538115387943d-01	2.d-22	-7.4393880d-16	-2.3033484d-03	1.d-24
6765	0 0	9.7174518879898310176210d-01	2.d-22	5.9491850918805141690997d-01	3.d-24	5.4955107d+06	-2.9239518d-25	2.0711973d+03
10946	0 1	9.999999999999999999999999d-01	-4.d-19	6.6476463373159072083937d-01	-5.d-22	6.7495437d-16	-2.3806948d-03	7.d-25
10946	0 0	9.7170283952404638752901d-01	1.d-21	5.9491961319604826964379d-01	3.d-23	1.1912678d+07	1.3735235d-23	3.3512705d+03
17711	1 0	9.999999999999999999999999d-01	4.d-17	6.7691381238790205680390d-01	4.d-21	-1.3734187d-13	-1.1913586d-03	5.d-24
17711	0 0	9.7167682828849617729546d-01	2.d-24	5.9492011240118782245670d-01	9.d-32	2.8822812d+07	3.6558909d-23	5.4224724d+03
28657	1 1	9.999999999999999999999999d-01	4.d-16	5.2936913869169652406821d-01	1.d-20	-2.2061289d-12	-6.6202656d-04	-2.d-23

a= 0.   level= 0 b= 1.   rx= 9.999999999999999999999999999995d-01

l	m	yp	ratio	accumulation	constant
1	1	9.99999999999999999999999999995d-01			
1	2	5.00000000000000000000000000000d-01			
2	3	6.13885957580948204683349690429d-01	-4.39137416809488d+00	5.92740734441061664984061606621d-01	4.07259268558108d-01
3	5	5.80996590443511229900379719458d-01	-3.46467536474584d+00	5.88357256714000892989551230004d-01	3.06129210633523d-01
5	8	5.94368242675021319590549816600d-01	-2.45766078043493d+00	5.98508908756928734932449725893d-01	1.41088121092608d-01
8	13	5.92041902468432413385733288094d-01	-5.74793497255366d+00	5.92386630939009552074110410823d-01	2.16302515010718d+00
13	21	5.93995982921362411194166111133d-01	-1.19050380095690d+00	5.93103913958927238717842227786d-01	2.53970425912112d-03
21	34	5.94087957906350463102037055539d-01	2.12457817001421d+01	5.94092580027258644591026206693d-01	-8.87654950409875d+03
34	55	5.94499556093886412500623277785d-01	2.23455187555840d-01	5.93969516130612055703649316898d-01	3.29517727676607d-09
55	89	5.94634072839823619561661479587d-01	3.05902856084625d+00	5.94699377664402016167592494492d-01	-1.53539165092290d+00
89	144	5.94754471814855434793992031614d-01	1.11725023165572d+00	5.95701256658208148362787994044d-01	-3.11182439009849d-03
144	233	5.94815692223239157129367177765d-01	1.96664769495114d+00	5.94879024923149086917775808212d-01	-1.07806360219317d-01
233	377	5.94857339379155078227903064470d-01	1.46997813825524d+00	5.94945954472570789607501212221d-01	-9.02065725953366d-03
377	610	5.94881341983924467217211843886d-01	1.72077164960349d+00	5.94915120725650995741468612154d-01	-3.09463290833636d-02
610	987	5.94896843846372297325871019724d-01	1.58175975373718d+00	5.94923144308721157343600647824d-01	-1.61415490656483d-02
987	1597	5.94906691532942228712573088515d-01	1.65443960285137d+00	5.94920223408585557096823745944d-01	-2.69131702431166d-02
1597	2584	5.94911819410958160584125508254d-01	1.61464447383930d+00	5.94921130842098190102604373306d-01	-1.98804220212885d-02
2584	4181	5.94915321334391855690588901460d-01	1.63563778226514d+00	5.94920830641762087652900044660d-01	-2.36471014232561d-02
4181	6765	5.94917477529490884633508846379d-01	1.62412178834233d+00	5.94928932296875393423478675014d-01	-2.13876457151613d-02
6765	10946	5.94918880118005141690997385474d-01	1.63028767779532d+00	5.94920893938566613470174400850d-01	-2.26244057571066d-02
10946	17711	5.94919613196048269643797223614d-01	1.62604444709810d+00	5.94920910194036880340065991694d-01	-2.18705596811603d-02
17711	28657	5.94920112401187822456703912894d-01	1.62862104661230d+00	5.94920906528495909067033533324d-01	-2.22900493892107d-02

Table 4.4.1.4 (continued)

i	n	p	ratio	accumulation	constant
1	1	4.000000000000000000000000000000d+00			
1	2	2.000000000000000000000000000000d+00			
2	3	1.817640115054627604205786740200d+00	4.14849144816103d+00	1.36416911649002460605748377996d+00	2.63881064818097d+00
3	5	1.205584293602440026165199938 3d+00	2.07905774739925d+00	1.07077464435942672037703914695 0d+00	1.93265655660030d+00
5	8	1.148405589613861334829945 60324+00	1.69096119138050d+00	9.49872436406653750549958445394d-01	1.62318343460956d+00
8	13	1.078994673461436054616833946204+00	1.97632752415195d+00	1.00790078904720679921724560234+00	2.14351654060136d+00
13	21	1.035232732494333430554684240794+00	1.58610232129795d+00	9.60566689762767526319717439994-01	1.18880422210081d+00
21	34	1.016595690955635183746360403 4d+00	1.77626697067915d+00	9.70057808199606251038632566584-01	1.77065558227306d+00
34	55	9.952326313490664405145383214 77d-01	1.68365462151590d+00	9.69702549366774089736331400934-01	1.11320694501337d+00
55	89	9.861183865287881349194629529 2d-01	1.68560971419020d+00	9.72824752218131617058284496824-01	1.46035000617208d+00
89	144	9.804843344804281377418112716 18d-01	1.61770689164377d+00	9.71363419498997580067641340084-01	1.11953265101563d+00
144	233	9.779694134018631349925535347 2d-01	1.64983377097882d+00	9.71014344871357297008787592534-01	1.29544944085700d+00
233	377	9.749666138701449285582263137 9d-01	1.62398787887712d+00	9.71596677284994836665367618314-01	1.13400541029764d+00
377	610	9.736814322417799819404984992 3d-01	1.63618864001761d+00	9.71661308559800226522301300142d-01	1.21678108686672d+00
610	987	9.728912628116737145885710105 03d-01	1.62646336266401d+00	9.71629944948263737780431421494-01	1.14352202870353d+00
987	1597	9.724860806065984015630489958 32d-01	1.63104226979302d+00	9.71639097199431228514157981974-01	1.18079500263749d+00
1597	2584	9.721891182672285954027155678 74-01	1.62739873019647d+00	9.71634638826279859561202425794-01	1.14847346124822d+00
2584	4181	9.719263878659527510770802942 74-01	1.62918931447185d+00	9.71635920963868504117403096894-01	1.16476563226740d+00
4181	6765	9.718141281390659344959164195 594-01	1.62774658092250d+00	9.71635290425039787152113304994-01	1.15808545624357d+00
6765	10946	9.717451807989831017621005388 1d-01	1.62838446705050d+00	9.71635479958788770363190762814-01	1.15777373698151d+00
10946	17711	9.717028395240463887829017851 39d-01	1.62787512133949d+00	9.71635390960634020735373690554-01	1.15184601153030d+00
17711	28657	9.716768282584961772954664657 74-01	1.62811282115306d+00	9.71635416408552446833125194601d-01	1.15477732388688d+00

i	m	bp	ratio	constant
1	1	1.000000000000000000000000000000d+00		
1	2	4.000000000000000000000000000000d+00	4.00000000000000d+00	1.00000000000000d+00
2	3	8.894116673484249352083183083 3d+00	2.22352916837106d+00	1.79094199508394d+00
3	5	2.093816093466412083227167978 25d+01	2.35415849772507d+00	1.60403880308362 1d+00
5	8	4.732000044101451927230518468 4d+01	2.25998844490678d+00	1.81392725125103d+00
8	13	1.044619264630241934399298917 2d+02	2.20756393680170d+00	1.99246901819545d+00
13	21	2.307931407647180034419677515 8d+02	2.20935183352578d+00	1.90442012376407d+00
21	34	5.029855874972868882226252942 8d+02	2.17937833159014d+00	2.15390795703924d+00
34	55	1.098566931791772229196363413 2d+03	2.18409226645562d+00	2.12157638850458d+00
55	89	2.385734290613139596834083272 9d+03	2.17167860197829d+00	2.22855791552537d+00
89	144	5.186420998436697704780016329 8d+03	2.17393068517799d+00	2.19994001940706d+00
144	233	1.125803442625242432587281301 8d+04	2.16913251539808d+00	2.24909045268242d+00
233	377	2.441292728534473749496495975 9d+04	2.17003134038205d+00	2.23806436520930d+00
377	610	5.293246553132803230667021132 9d+04	2.16821469665682d+00	2.26147936580986d+00
610	987	1.147869756652825011161810954 0d+05	2.16855524323434d+00	2.25686508433514d+00
987	1597	2.488434803538547104767688523 15d+05	2.16787211350678d+00	2.26604272570196d+00
1597	2584	5.394924018208759156441826665 1d+05	2.16799891300752d+00	2.26405483117903d+00
2584	4181	1.169400697337910010484022712 0d+06	2.16774266586684d+00	2.26914225810254d+00
4181	6765	2.535180033058204174653883558 6d+06	2.16778954866977d+00	2.26830013151404d+00
6765	10946	5.495510650213323367874427299 4d+06	2.16769351170535d+00	2.27011771513542d+00
10946	17711	1.191267770154616036696199216 7d+07	2.16771079882881d+00	2.26977376760093d+00
17711	28657	2.582281162605861849657138065 0d+07	2.16767481672849d+00	2.27052742407331d+00

Table 4.4.1.5 gives the results for the standard map in the critical case.[16] r now refers to the actual residues, and dp, dr are meaningless. Note that I have evaluated positions of both the Birkhoff orbits, and their tangent maps. rp, rn refer to the residues of the two Birkhoff orbits, and dw is the difference in their actions. $y(i)$, $b(i)$, $i = 1, 2, 3$, refer to the points on the three subdominant half-lines, as indicated.

The periodic points on the subdominant half-planes have period 3 behaviour rather than straight geometric convergence (cf. Shenker and Kadanoff, 1982), so the ratios quoted are 3-step ratios. In the limit the subdominant scalings are:

$$\alpha_3 = -4.84581, \quad \sqrt[3]{\alpha_3} = -1.69221 \tag{4.4.1.19}$$

$$\beta_3 = -16.8597, \quad \sqrt[3]{\beta_3} = -2.56418.$$

Note that $\alpha_3 \neq \alpha^3$ and $\beta_3 \neq \beta^3$.[17] Evidence for this 3-cycle was in fact found earlier than the fixed point, by Greene (1979a).

[16] Using the estimate $k_c = 0.97163540631$ which was determined by straightforward superconverging, but can be seen to be slightly high, as the residues begin to diverge at the highest periods examined.

[17] Although $\alpha_3\beta_3 \simeq \alpha^3\beta^3$ (which one expects from scaling of the actions of the periodic orbits).

Table 4.4.1.5: Determination of scalings in the critical case for inverse golden rotation number in the standard map.

i0 m0	i1 j1 if jf	p r	dp dr	yi yf	dyi n	bi bf	ci cf	w detm1
1	0 0	9.7163540631000000000000d-01	0.	9.999999999999999999999999d-01	0.	1.0000000d+00	-7.3561657d-01	5.2461181d-01
1	1 1	2.4290885157749999999999d-01	0.	9.999999999999999999999999d-01	0.	7.5709115d-01	-9.7163541d-01	0.
1	0 1	9.7163540631000000000000d-01	0.	9.999999999999999999999999d-01	0.	1.0000000d+00	1.2076542d+00	4.7538819d-01
1	1 0	-2.4290885157749999999999d-01	0.	9.999999999999999999999999d-01	0.	1.2429089d+00	9.7163541d-01	-3.d-29
1	0 0	9.7163540631000000000000d-01	0.	5.000000000000000000000000d-01	0.	2.9716354d+00	-2.4271342d-01	2.5000000d-01
2	0 1	2.3601884069879969695402d-01	0.	5.000000000000000000000000d-01	0.	1.0283646d+00	-7.0136194d-01	0.
1	1 0	9.7163540631000000000000d-01	0.	5.7517404266948658763987d-01	0.	2.3339314d+00	5.0636429d-01	2.4413394d-01
2	1 1	-2.4026123690353439769274d-01	0.	4.2482595733051361236012d-01	0.	2.3339314d+00	5.0636429d-01	-5.d-29
2	0 0	9.7163540631000000000000d-01	0.	6.2925958820821879940653d-01	0.	6.1285401d+00	-1.2595365d-01	6.6161389d-01
3	1 0	2.6068032010133013009457d-01	0.	7.4145802342356240118692d-01	0.	1.6255635d+00	-4.7423823d-01	0.
2	1 1	9.7163540631000000000000d-01	0.	5.6576551560876404319664d-01	2.d-29	4.4260841d+00	3.0371313d-01	6.5970845d-01
3	0 1	-2.6558846671759392080022d-01	0.	7.1711724219561797840167d-01	3.d-29	2.2424380d+00	5.9957191d-01	3.d-29
3	0 0	9.7163540631000000000000d-01	0.	5.8412712869190911130486d-01	0.	1.3218940d+01	-5.5509312d-02	8.9165521d-01
5	1 1	2.4201467155025531566782d-01	0.	5.0751694087874662582436d-01	0.	3.1842235d+00	-2.3044057d-01	0.
3	0 1	9.7163540631000000000000d-01	0.	6.4682263442445343637355d-01	0.	4.1793389d+00	2.9486239d-01	8.9129023d-01
5	1 0	-2.4704917239032194884930d-01	0.	6.5924428398273446919864d-01	0.	4.5352184d+00	2.7172448d-01	-5.d-29
5	0 0	9.7163540631000000000000d-01	0.	5.9840016755377787157063d-01	0.	2.9140857d+01	-2.6104569d-02	1.5402157d+00
8	0 1	2.5552012800789443515789d-01	0.	6.7572319189212082019911d-01	0.	2.9116942d+00	-2.6133183d-01	-5.d-29
5	1 0	9.7163540631000000000000d-01	0.	6.8293845079449161981580d-01	0.	8.4960395d+00	1.5491908d-01	1.5481252d+00
8	1 1	-2.6095301594349802708088d-01	0.	5.3566520004274766243538d-01	0.	8.2532011d+00	1.5947727d-01	3.d-29
8	0 0	9.7163540631000000000000d-01	0.	5.9376383038880026193329d-01	0.	6.2221887d+01	-1.1943873d-02	2.4387535d+00
13	1 0	2.4660815330638208576983d-01	0.	6.7324160762319213492020d-01	0.	6.0296863d+00	-1.2325190d-01	-1.d-28
8	1 1	9.7163540631000000000000d-01	0.	5.2719542858771356173677d-01	1.d-29	1.5037785d+01	8.3035791d-02	2.4387334d+00
13	0 1	-2.5103199452512576298155d-01	0.	6.6167418252807785561849d-01	-1.d-28	7.9350584d+00	1.5891370d-01	-2.d-28
13	0 0	9.7163540631000000000000d-01	0.	5.9529832408438324240997d-01	0.	1.3639834d+02	-5.5321553d-03	3.9867071d+00
21	1 1	2.5229896784158527148403d-01	0.	5.3869520439431374641370d-01	0.	1.0650754d+01	-7.0847264d-02	2.d-28
13	0 1	9.7163540631000000000000d-01	0.	6.6581401826510966024359d-01	7.d-30	1.4697793d+01	8.8206198d-02	3.9867023d+00
21	1 0	-2.5769988060856006458636d-01	0.	6.7795588203812901241340d-01	-1.d-28	1.5915058d+01	8.1459737d-02	2.d-28
21	0 0	9.7163540631000000000000d-01	0.	5.9479780793573886684646d-01	1.d-29	2.9345532d+02	-2.5470121d-03	6.4254007d+00
34	0 1	2.4872940458027289209307d-01	0.	6.6411510227227466397419d-01	8.d-28	1.0425726d+01	-7.1691339d-02	-2.d-28
21	1 0	9.7163540631000000000000d-01	0.	6.7655163436627989409485d-01	8.d-30	2.9190110d+01	4.3649850d-02	6.4253996d+00
34	1 1	-2.5401324745194278082673d-01	0.	5.2899809184478161939991d-01	-2.d-28	2.8547738d+01	4.4632044d-02	-2.d-28
34	0 0	9.7163540631000000000000d-01	0.	5.9496103020029430110960d-01	2.d-29	6.3918624d+02	-1.1762948d-03	1.0412094d+01
55	1 0	2.5093749341034273737083d-01	0.	6.7713082896324359134788d-01	-2.d-27	2.0603519d+01	-3.6351235d-02	-1.d-27
34	1 1	9.7163540631000000000000d-01	0.	5.2950111315982176973048d-01	7.d-29	5.2745296d+01	2.4418631d-02	1.0412094d+01
55	0 1	-2.5630181023787271376630d-01	0.	6.6494064710629367602207d-01	1.d-27	2.7673667d+01	4.6541281d-02	-3.d-29
55	0 0	9.7163540631000000000000d-01	0.	5.9490782090140169598657d-01	2.d-29	1.3013487d+03	-5.4232282d-04	1.6837491d+01
89	1 1	2.4956827507415932853284d-01	0.	5.2929439273291434376307d-01	-2.d-27	3.7379418d+01	-2.0041403d-02	-2.d-28
55	0 1	9.7163540631000000000000d-01	0.	6.6469831187865471976961d-01	1.d-29	5.0905201d+01	2.5133497d-02	1.6837491d+01
89	1 0	-2.5480835399773489154558d-01	0.	6.7684963153521312042668d-01	1.d-27	5.5236216d+01	2.3162805d-02	2.d-27
89	0 0	9.7163540631000000000000d-01	0.	5.9492817361530933692021d-01	3.d-29	2.9999393d+03	-2.5027733d-04	2.7249584d+01
144	0 1	2.5040872330742463497436d-01	0.	6.6479633565481321520385d-01	7.d-27	3.6071207d+01	-2.0814030d-02	4.d-27
89	1 0	9.7163540631000000000000d-01	0.	6.7693298330835123394335d-01	2.d-28	1.0107736d+02	1.2609963d-02	2.7249584d+01
144	1 1	-2.5575619708654510129979d-01	0.	5.2939479775730495051027d-01	1.d-26	9.9475707d+01	1.2914406d-02	1.d-26
144	0 0	9.7163540631000000000000d-01	0.	5.9491951625282071212710d-01	2.d-30	6.4953139d+03	-1.1543490d-04	4.4087076d+01
233	1 0	2.4989322863128597316912d-01	0.	6.7689870325136731206250d-01	-2.d-27	7.2195162d+01	-1.0385405d-02	8.d-27
144	1 1	9.7163540631000000000000d-01	0.	5.2936497756389842044143d-01	7.d-29	1.8316043d+02	6.9963383d-03	4.4087076d+01
233	0 1	-2.5522305761146580437430d-01	0.	6.6474729821427060386279d-01	2.d-26	9.6201261d+01	1.3329335d-02	4.d-27
233	0 0	9.7163540631000000000000d-01	0.	5.9492136186727304968664d-01	4.d-30	1.4889699d+04	-5.3259984d-05	7.1336660d+01
377	1 1	2.5020866728484854453589d-01	0.	5.2937725120320172495841d-01	1.d-26	1.2979498d+02	-5.7815578d-03	-2.d-26
233	0 1	9.7163540631000000000000d-01	0.	6.6476185452232602451087d-01	1.d-27	1.7731450d+02	7.2381057d-03	7.1336660d+01
377	1 0	-2.5554964777648154980915d-01	0.	6.7691534934569815816155d-01	-1.d-25	1.9229045d+02	6.6743801d-03	-1.d-26
377	0 0	9.7163540631000000000000d-01	0.	5.9492075957331547401456d-01	8.d-30	3.0520377d+04	-2.4568036d-05	1.1542374d+02
610	0 1	2.5001571063644677479932d-01	0.	6.6475886616344820952677d-01	-1.d-26	1.2563452d+02	-5.9689963d-03	5.d-26
377	1 0	9.7163540631000000000000d-01	0.	6.7691040651768307678807d-01	1.d-27	3.5419962d+02	3.6200394d-03	1.1542374d+02
610	1 1	-2.5535034897815502785673d-01	0.	5.2937129267152764439930d-01	-2.d-25	3.4597750d+02	3.7060693d-03	2.d-26
610	0 0	9.7163540631000000000000d-01	0.	5.9492093570511920409283d-01	4.d-28	6.6192955d+04	-1.1334549d-05	1.8676040d+02
987	1 0	2.5013360601622122791638d-01	0.	6.7691244055273444640618d-01	1.d-24	2.5109180d+02	-2.9890913d-03	1.d-25
610	1 1	9.7163540631000000000000d-01	0.	5.2937306163582671383394d-01	4.d-27	6.3753910d+02	2.0123519d-03	1.8676040d+02
987	0 1	-2.5547220689167058866159d-01	0.	6.6475877378722758599040d-01	1.d-24	3.3477703d+02	3.8322612d-03	1.d-25

Table 4.4.1.5 (continued)

987	0	0	9.7163540631000000000000d-01	0.	5.94920891755063911197803d-01	3.d-29	1.4346037d+05	-5.2287766d-06	3.0210413d+02
1597	1	1	2.5006164318493274630003d-01	0.	5.2937233381515787283337d-01	-7.d-24	4.5179142d+02	-1.6603309d-03	4.d-25
987	0	1	9.7163540631000000000000d-01	0.	6.6475791047336772586076d-01	3.d-29	6.1675383d+02	2.0794411d-03	3.0210413d+02
1597	1	0	-2.5539779177564171912178d-01	0.	6.7691145302504664994622d-01	6.d-25	6.6893584d+02	1.9172292d-03	1.d-25
1597	0	0	9.7163540631000000000000d-01	0.	5.9492891260710573728051d-01	8.d-28	3.1100536d+05	-2.4122140d-06	4.8894453d+02
2584	0	1	2.5010576037924634790398d-01	0.	6.6475826571305870511007d-01	-8.d-24	4.3706031d+02	-1.7164942d-03	-2.d-26
1597	1	0	9.7163540631000000000000d-01	0.	6.7691174621487433711527d-01	6.d-28	1.2325466d+03	1.0407547d-03	4.8894453d+02
2584	1	1	-2.5544336299652214534731d-01	0.	5.2937268719320367311606d-01	1.d-23	1.2038337d+03	1.0655700d-03	4.d-25
2584	0	0	9.7163540631000000000000d-01	0.	5.9492090580807276766337d-01	5.d-28	6.7411382d+05	-1.1128069d-06	7.9112866d+02
4181	1	0	2.5007897126410897197440d-01	0.	6.7691162558005584849194d-01	-1.d-23	8.7344140d+02	-8.5808318d-04	5.d-25
2584	1	1	9.7163540631000000000000d-01	0.	5.2937258227329398083692d-01	7.d-28	2.2179239d+03	5.7829321d-04	7.9112866d+02
4181	0	1	-2.5541569197317311614327d-01	0.	6.6475809323978452854719d-01	1.d-23	1.1647115d+03	1.1012268d-03	-3.d-24
4181	0	0	9.7163540631000000000000d-01	0.	5.9492090802499914675357d-01	6.d-28	1.4613060d+06	-5.1337008d-07	1.2800732d+03
6765	1	1	2.5009560466313409696156d-01	0.	5.2937262544502166786262d-01	-2.d-22	1.5717234d+03	-4.7730422d-04	1.d-24
4181	0	1	9.7163540631000000000000d-01	0.	6.6475814444670558908746d-01	1.d-25	2.1459636d+03	5.9773397d-04	1.2800732d+03
6765	1	0	-2.5543287516317741741850d-01	0.	6.7691168415098523673726d-01	-9.d-23	2.3274551d+03	5.5112355d-04	2.d-24
6765	0	0	9.7163540631000000000000d-01	0.	5.9492090730214288541629d-01	6.d-29	3.1675502d+06	-2.3603026d-07	2.0712018d+03
10946	0	1	2.5008580064300933944775d-01	0.	6.6475812337695662395956d-01	-2.d-22	1.5207298d+03	-4.9329718d-04	-3.d-24
6765	1	0	9.7163540631000000000000d-01	0.	6.7691166676113270477321d-01	5.d-25	4.2881603d+03	2.9911536d-04	2.0712018d+03
10946	1	1	-2.5542283314274042112044d-01	0.	5.2937260448443899374313d-01	6.d-22	4.1083453d+03	3.0624376d-04	-7.d-24
10946	0	0	9.7163540631000000000000d-01	0.	5.9492090753784021899912d-01	7.d-27	6.8662831d+06	-1.0925635d-07	3.3512750d+03
17711	1	0	2.5009253196500664100156d-01	0.	6.7691167391652153234658d-01	1.d-21	3.0387854d+03	-2.4607002d-04	1.d-23
10946	1	1	9.7163540631000000000000d-01	0.	5.2937261070762474942865d-01	8.d-26	7.7168665d+03	1.6621982d-04	3.3512750d+03
17711	0	1	-2.5542978786608204110755d-01	0.	6.6475813360674739364005d-01	-8.d-22	4.0523521d+03	3.1653127d-04	3.d-23
17711	0	0	9.7163540631000000000000d-01	0.	5.9492090746098806634404d-01	2.d-26	1.4883681d+07	-5.0402803d-08	5.4224769d+03
28657	1	1	2.5000962714899432120572d-01	0.	5.2937260814699911486121d-01	-3.d-21	5.4685145d+03	-1.3718154d-04	-2.d-23
17711	0	1	9.7163540631000000000000d-01	0.	6.6475813056951097366927d-01	2.d-24	7.4662140d+03	1.7179766d-04	5.4224769d+03
28657	1	0	-2.5542671401856800752127d-01	0.	6.7691167044245075273603d-01	-4.d-21	8.0977143d+03	1.3840002d-04	7.d-23
28657	0	0	9.7163540631000000000000d-01	0.	5.9492090748604673902198d-01	3.d-26	3.2263037d+07	-2.3252198d-08	8.7737519d+03
46368	0	1	2.5009328691279968460243d-01	0.	6.6475813181923594164096d-01	2.d-20	5.2908746d+03	-1.4178876d-04	1.d-22
28657	1	0	9.7163540631000000000000d-01	0.	6.7691167147390185141961d-01	8.d-25	1.4919667d+04	8.5973838d-05	8.7737519d+03
46368	1	1	-2.5543050541431399609583d-01	0.	5.2937260939023338222491d-01	6.d-20	1.4572316d+04	8.8023141d-05	-1.d-22

a= 0.   level= 0 b= 1.   p= 9.7163540631000000000000000d-01

l	m	rp	ratio	accumulation	constant
1	1	9.9999999999999999999999999999995d-01			
1	2	5.0000000000000000000000000000000d-01			
2	3	6.2925590820821879949653730791d-01	-3.8682927315141 9d+00	6.0270540979660333607608870950d-01	3.9729459020339 5d-01
3	5	5.8412712069190911138486600483d-01	-2.864153365786882d+00	5.9500978754424503790867873464d-01	2.744030359027 27d-01
5	8	5.9840016755377787157063175884d-01	-3.144202425048964d+00	5.9501676557922321407607600264d-01	3.304892600769 92d-01
8	13	5.9376383038080026193329110954d-01	-3.043259702487 96d+00	5.9493029939037919163480813260d-01	3.044066699916 01d-01
13	21	5.9529832400438324240997221856d-01	-3.073546122714 71d+00	5.9492162601439864345620490345d-01	3.175637201171 84d-01
21	34	5.9479780793573086684646165651d-01	-3.065822566793 41d+00	5.9492091122920466056800399166 7d-01	3.133998530676 43d-01
34	55	5.9496103020029430110968950716d-01	-3.066469822552 03d+00	5.9492089163490189477222400123d-01	3.138133435566 48d-01
55	89	5.9490782090140169590657307997d-01	-3.067551498712 16d+00	5.9492090230095350002590507265d-01	3.146163140850 48d-01
89	144	5.9492517361530933692021628455d-01	-3.066338739623 63d+00	5.9492090621039024309924390235d-01	3.135921272402 45d-01
144	233	5.9491951625282071212710065104d-01	-3.067279839771 96d+00	5.9492090719700396707539997860d-01	3.144831305994 57d-01
233	377	5.9492136106727304968664547195d-01	-3.066629536350 59d+00	5.9492090742023295242104811305d-01	3.138006602007 33d-01
377	610	5.9492075957331547401456775021d-01	-3.067054006283 13d+00	5.9492090746757914654837821609d-01	3.142894716039 91d-01
610	987	5.9492095570511920409283195079d-01	-3.066784407915 12d+00	5.9492090747773034656676340485 2d-01	3.139513279442 99d-01
987	1597	5.9492089175506391119703768798d-01	-3.066952840490 61d+00	5.9492090747930081406559959991d-01	3.141797996076 17d-01
1597	2584	5.9492091260710573728051549167d-01	-3.066848600548 16d+00	5.9492090747978385387603135202d-01	3.140277108628 49d-01
2584	4181	5.9492090580807276766337033350d-01	-3.066912885886 13d+00	5.9492090747986490108672233319d-01	3.141280811477 69d-01
4181	6765	5.9492090802499914675357865384d-01	-3.066873592962 08d+00	5.9492090747988105338726503157d-01	3.140627048252 20d-01
6765	10946	5.9492090730214288541629287479d-01	-3.066897940385 67d+00	5.9492090747988431602992112257d-01	3.141057066339 40d-01
10946	17711	5.9492090753784021899912681300d-01	-3.066883491421 61d+00	5.9492090747988494803158144392 4d-01	3.140787068320 57d-01
17711	28657	5.9492090746098806634404827336d-01	-3.066893059497 30d+00	5.9492090747988508466587032727 8d-01	3.140975657228 61d-01
28657	46368	5.9492090748604673902198520990d-01	-3.066880404045 32d+00	5.9492090747988510629751429331d-01	3.140879128121 05d-01

l	m	rp	ratio	accumulation	constant
1	1	2.4290885157749999999999999999d-01			
1	2	2.3601884069879969695402499998d-01			
2	3	2.6068032010133013009457364466d-01	-2.793835181677 44d-01	2.4140425545507563701706497727d-01	1.504596121624 36d-03
3	5	2.4281467155025551566782832669d-01	-1.321222744286 09d+00	2.5005597159223636392250029488d-01	1.854617660092 94d-02
5	8	2.5552012088709443515789507787d-01	-1.382081938804 18d+00	2.4980052650008350837983626388d-01	2.068653550603 03d-02
8	13	2.4660815338638208857698399269 5d-01	-1.515428018914 70d+00	2.5013107942302435167348030131d-01	2.831643983072 56d-02
13	21	2.5229896704158852714840304855d-01	-1.566028332936 89d+00	2.5008121535011500064743408851d-01	3.271226445104 77d-02
21	34	2.4872040458027289209307867652d-01	-1.590251313604 26d+00	2.5010195407325291596280360092d-01	3.553260437144 61d-02
34	55	2.5093749341034273737083502409d-01	-1.614081678991 47d+00	2.5008936045111358465756832910d-01	3.907221251987 26d-02
55	89	2.4956827507410032885284590346d-01	-1.619236882435 59d+00	2.5009102975333734108961176620d-01	4.000289444840 89d-02
89	144	2.5040872330742463497436519026d-01	-1.629152495041 92d+00	2.5000905823899440451436422843d-01	4.210295430115 64d-02
144	233	2.4989322885312859731891224191 2d-01	-1.630371969178 28d+00	2.5008920643101830776345240183d-01	4.239950635434 97d-02
233	377	2.5020866728484885445335895826 11d-01	-1.634215106155 01d+00	2.5008892051366591457036677515d-01	4.344846721950 50d-02
377	610	2.5001571063644677479932382478d-01	-1.634764088721 84d+00	2.5000894550052812945980631407 4d-01	4.361509202036 96d-02
610	987	2.5013368601622122791638818089d-01	-1.635567088410 03d+00	2.5000892320965361947401678599d-01	4.380078508841 01d-02
987	1597	2.5006164318493274630035333000d-01	-1.637572783640 94d+00	2.5000895724073235560406828055d-01	4.460622759810 37d-02
1597	2584	2.5010576037924634790398015866d-01	-1.632907600625 15d+00	2.5008900481446392692637987103d-01	4.284353186000 71d-02
2584	4181	2.5007897126410897197440605076d-01	-1.646832830698 83d+00	2.5000890924605920543464048893 1d-01	4.870357618206 61d-02
4181	6765	2.5009560466313409696155837941d-01	-1.610561683568 74d+00	2.5008923300845000035597557947 5d-01	3.387190130925 82d-02
6765	10946	2.5008580064300933944775774154d-01	-1.710547573094 39d+00	2.5008946011744486704982988740d-01	9.646211951986 17d-02
10946	17711	2.5009253196500664100156014856d-01	-1.461968031122 62d+00	2.5000903033691349770569486102d-01	5.375057119199 01d-03
17711	28657	2.5000962714899432120572407727d-01	-2.289750418301 52d+00	2.5009085101332104357460996559 1d-01	3.173276845039 39d+01
28657	46368	2.5009328691279968460243244675d-01	-7.937172498194 49d-01	2.5009124658464555806448018343d-01	1.265750298193 47d-08

Table 4.4.1.5 (continued)

l	m	bp	ratio	constant
1	1	1.0000000000000000000000000000d+00		
1	2	2.9716354063099999999999999998d+00	2.9716354063100d+00	1.0000000000000d+00
2	3	6.1205401296887025765911528182d+00	2.0596537900359 4d+00	1.4427893703929d+00
3	5	1.3218940399764770076921985187d+01	2.1597669682549 9d+00	1.0121272661821 4d+00
5	8	2.9140857378137630999015417953 5d+01	2.2050827446545 1d+00	1.2320033062398 4d+00
8	13	6.2221887259282641115022951846d+01	2.1346232599922 0d+00	1.4038960689949 1d+00
13	21	1.3639833029178209574466878655 9d+02	2.1921279520725 8d+00	1.2291748918564 2d+00
21	34	2.9345531033812818835221467096d+02	2.1514581638046 8d+00	1.3733449033973 9d+00
34	55	6.3918623692081798265968827868d+02	2.1781381933733 7d+00	1.2616647577142 9d+00
55	89	1.3813486076970317422573841263d+03	2.1611951801607 4d+00	1.3434458940644 3d+00
89	144	2.9999392519011407941768011763d+03	2.1717468553654 46d+00	1.2033489397546 8d+00
144	233	6.4953139389113634972228584328d+03	2.1651484891886 1d+00	1.3250601634048 6d+00
233	377	1.4089699314472451869380528163d+04	2.1692099922443 8d+00	1.2980241887033 7d+00
377	610	3.0528376720061073322326929345d+04	2.1667159844002 8d+00	1.3160666325411 2d+00
610	987	6.6192954915213804683593964265d+04	2.1682435172426 5d+00	1.3040642124228 4d+00
987	1597	1.4346937113415577901613185333d+05	2.1673987614522 4d+00	1.3119604080069 56d+00
1597	2584	3.1109536144822292311581847232d+05	2.1678804077629 8d+00	1.3067880114716 2d+00
2584	4181	6.7411301795110696699570988635d+05	2.1675311795656 5d+00	1.3101536189523 4d+00
4181	6765	1.4613867783642942416179976764d+06	2.1677448014298 3d+00	1.3079596584475 1d+00
6765	10946	3.1675502221093115626823461962d+06	2.1676148159458 1d+00	1.3093738707631 8d+00
10946	17711	6.8662830647521161247208919287d+06	2.1676958090302 24d+00	1.3084820869309 5d+00
17711	28657	1.4883680500686863541688432919d+07	2.1676473887338 1d+00	1.3090201837658 9d+00
28657	46368	3.2263037033215151436064615222 5d+07	2.1676786088650 5d+00	1.3086312045177 6d+00

l	m	rn	ratio	accumulation	constant
1	1	-2.4298805157749999999999999999d-01			
1	2	-2.4026123690383439769274286868d-01			
2	3	-2.6550046671709392080822528233d-01	-1.0453629131180 6d-01	-2.4265827410948990345068060126d-01	-2.5057730880100 97d-04
3	5	-2.4784917239832194884930299079d-01	-1.3661377492524 8d+00	-2.5400442811203894296024739363d-01	-1.9977293524475 1d-02
5	8	-2.6095301594349802708088849604d-01	-1.3333934792228 1d+00	-2.5499437944734776982942754792d-01	-1.8835632138670 7d-02
8	13	-2.5183199450251257629815527281 3d-01	-1.5243735221550 8d+00	-2.5544517670213796588004753111 8d-01	-2.9740402535534 8d-02
13	21	-2.5769980068056806458636326644d-01	-1.5543964515931 8d+00	-2.5540270940483167533407491723d-01	-3.2401428245607 0d-02
21	34	-2.5401324745194278082673374465d-01	-1.5916652922998 8d+00	-2.5543574334274267800673169417d-01	-3.6814019997384 3d-02
34	55	-2.5630151023787271376630550982d-01	-1.6111057048577 0d+00	-2.5542510246202253033544497124d-01	-3.9780819714649 9d-02
55	89	-2.5488035399773409154058860698 6d-01	-1.6192567537370 3d+00	-2.5542787966207593087545812250 0d-01	-4.1290825192344 5d-02
89	144	-2.5575619708684510129979677617d-01	-1.6283545474587 6d+00	-2.5542601214664362761263540736d-01	-4.3275906270156 8d-02
144	233	-2.5522385761146558437430881655 3d-01	-1.6302437249552 0d+00	-2.5542624930240012866936689325d-01	-4.3749247506327 6d-02
233	377	-2.5554964777648150498091599832 5d-01	-1.6339949213852 2d+00	-2.5542596106680449297294886636 3d-01	-4.4803514103101 2d-02
377	610	-2.5535034897815302755673586530 3d-01	-1.6346820340178 2d+00	-2.5542599332375471160172134937d-01	-4.5020300853623 8d-02
610	987	-2.5547220868916708886615925876 5d-01	-1.6350014834639 5d+00	-2.5542596908030601132091425287 6d-01	-4.5300569947128 3d-02
987	1597	-2.5539779177564171912178367025d-01	-1.6375424781745 4d+00	-2.5542600558832465819113908911 1d-01	-4.6062790323437 7d-02
1597	2584	-2.5544336299652214534731180161d-01	-1.6329410226714 5d+00	-2.5542603489186591732971136998d-01	-4.4236170637107 4d-02
2584	4181	-2.5541569197317311614327629613d-01	-1.6468902249310 9d+00	-2.5542614612488094502359048706d-01	-5.0419824344012 9d-02
4181	6765	-2.5543287516317741741850398448 4d-01	-1.6103542672869 6d+00	-2.5542629248097529376897311430d-01	-3.4913231848164 9d-02
6765	10946	-2.5542283314274042112644170799d-01	-1.7111287625940 1d+00	-2.5542603714235612447740251980 0d-01	-1.0024023919929 6d-01
10946	17711	-2.5542970786688204110795993096d-01	-1.4607161291318 3d+00	-2.5542691407687098222584334033d-01	-8.4639945199403 7d-03
17711	28657	-2.5542671401856800752127370651d-01	-2.2962833852987 5d+00	-2.5542762222682431980693129539 7d-01	-3.4652968379176 9d+01
28657	46368	-2.5543050541431399609583244936d-01	-7.8964278970293 9d-01	-2.5542838689336242220845056477 3d-01	-1.1733662073428 9d-00

l	m	dw	ratio	constant
1	1	-4.9223624718066493137566868780d-02		
1	2	-5.8660589937455308654305446492d-03	1.1917161784293 6d-01	-4.9223624718066 5d-02
2	3	-1.8254395082867166410877266291d-03	3.1118669403649 4d-01	-1.8050609939180 7d-02
3	5	-3.6497696113978638129192624907d-04	1.9993922564008 9d-01	-4.5663735238510 3d-02
5	8	-9.0569611900357446722706225822d-05	2.4815158638300 2d-01	-2.3884397102002 6d-02
8	13	-2.0064767700100518162617362125d-05	2.2153973360214 6d-01	-3.7598973680436 6d-02
13	21	-4.7283871002521018252354226809d-06	2.3565621346458 0d-01	-2.7608299675043 7d-02
21	34	-1.0757895938830771162096379811d-06	2.2751723851567 5d-01	-3.4090814356477 1d-02
34	55	-2.4981613567870806947179077066d-07	2.3221653853037 2d-01	-2.9544409174761 9d-02
55	89	-5.7312235902629424498341278615d-08	2.2941767050763 9d-01	-3.2034090082591 3d-02
89	144	-1.3244036908774623617386108926d-08	2.3109091279000 3d-01	-3.0493097380140 2d-02
144	233	-3.0472441498804771487178316658d-09	2.3007982704983 4d-01	-3.1061036346290 8d-02
233	377	-7.0297536611285599171349386793d-10	2.3069217021554 7d-01	-3.0941019602157 3d-02
377	610	-1.6190063766639798171534917547d-10	2.3031907726973 5d-01	-3.1548677029269 7d-02
610	987	-3.7327430606533734796816940422d-11	2.3054630774022 8d-01	-3.1146020650304 1d-02
987	1597	-8.6005303129913566799177305009 6d-12	2.3040772777497 6d-01	-3.1410121420483 5d-02
1597	2584	-1.9823561612954491843053895049d-12	2.3049231723549 0d-01	-3.1237654690021 3d-02
2584	4181	-4.5681560146465368350187787835d-13	2.3044073027025 0d-01	-3.1349729614814 7d-02
4181	6765	-1.0528033507333792670271708634 8d-13	2.3047231836175 7d-01	-3.1276765181073 6d-02
6765	10946	-2.4262882122060398137071126007d-14	2.3045317187428 8d-01	-3.1323571022654 6d-02
10946	17711	-5.5917487469619077450528905198d-15	2.3046514914556 5d-01	-3.1292656487194 3d-02
17711	28657	-1.2806641525876364928465320948 4d-15	2.3045016458031 51d-01	-3.1311630049048 2d-02
28657	46368	-2.9701171438662766710935701401d-16	2.3048031078557 5d-01	-3.1240500924779 5d-02

l	m	y(1):TS	ratio	accumulation	constant
1	1	9.9999999999999999999999999993d-01			
1	2	5.7517404266948658763987766876d-01			
2	3	7.4148002342356240118692538415d-01			
3	5	6.5924428398273446919864270343d-01			
5	8	6.8293845079449161981508155310d-01	-1.7929558811062 0d+01	6.7724553374340694574682144267d-01	-1.0207149107392 0d-01
8	13	6.7324160762319213492020981485d-01	-1.7151353085428 2d+01	6.7700146043372192131714921575d-01	6.4486562989048 5d-02
13	21	6.7795580283812901241348238648d-01	-1.7445976607944 1d+01	6.7694140716007782145536289709d-01	3.0074355970511 34d-01
21	34	6.7655163436627959409485004732d-01	-1.6874162386333 7d+01	6.7690097555325736849416640245 8d-01	-1.0174268227259 9d-01
34	55	6.7713052896324359134780837409d-01	-1.6758619581102 3d+01	6.7691144245066985921363725520 2d-01	6.1472007120555 3d-02
55	89	6.7684963153821313042668897418d-01	-1.6702621499922 3d+01	6.7691103672261593929733479620d-01	2.9403972150416 02d-01
89	144	6.7693298330351233943059349403d-01	-1.6046294931726 6d+01	6.7691161470199615660702179407d-01	-1.0216159707994 5d-01
144	233	6.7609079325136731206250634108d-01	-1.6087214596841 6d+01	6.7691166366566008080726528153 4d-01	6.2415871517401 9d-02
233	377	6.7691534934569815816155741266d-01	-1.6874674777940 2d+01	6.7691167275793750820099309931d-01	2.9811609029337 1d-01
377	610	6.7691040651768307678887813533d-01	-1.6063173885444 21d+01	6.7691167039903212147524708139d-01	-1.0220232140653 0d-01
610	987	6.7691244805527544464061860187 9d-01	-1.6853225702713 6d+01	6.7691167109508803454166967315d-01	6.2075021562122 9d-02
987	1597	6.7691145302584664994622872669d-01	-1.6856331421818 7d+01	6.7691167122091840905261454511d-01	2.9694619198263 1d-01
1597	2584	6.7691174621487433711527349088d-01	-1.6858797776419 9d+01	6.7691167119878377878069973636d-01	-1.0216154063158 3d-01
2584	4181	6.7691162350000558484919455404 9d-01	-1.6861094473826 7d+01	6.7691167120042984271486617954d-01	6.2181073295234 1d-02
4181	6765	6.7691160415098523673726448906d-01	-1.6860372853238 30d+01	6.7691167121027338830730427333d-01	2.9727693419409 0d-01
6765	10946	6.7691166676113278477321213756d-01	-1.6859822617647 1d+01	6.7691167120987478041825774424 0d-01	-1.0217740735501 7d-01
10946	17711	6.7691167391682183234658206185d-01	-1.6859296223358 2d+01	6.7691167121000882874145628367d-01	6.2150068203826 5d-02
17711	28657	6.7691167084424507527360371810 8d-01	-1.6859451952447 4d+01	6.7691167121002942875037533906d-01	2.9718512779995 2d-01
28657	46368	6.7691167147390185141961032182d-01	-1.6859599520582 3d+01	6.7691167121002300636335929849d-01	-1.0217278310123 1d-01

Table 4.4.1.5 (continued)

i	m	γ(2):RTS	ratio	accumulation	constant
1	1	9.99999999999999999999999999995d-01			
1	2	4.24825957330513412368122331119d-01			
2	3	5.65765515608764404319664868109d-01			
3	5	5.07516940787466258243680275234d-01			
5	8	5.35665280042747662435386616864d-01	-2.04336759427670d+01	5.30494010167799392941815342794d-01	-1.05660052837286d-01
8	13	5.27195420587713561736775846204d-01	-1.66401295117692d+01	5.29381917857777008240797881004d-01	3.63835977509870d-02
13	21	5.30695284394313746413709064524d-01	-1.66430975718113d+01	5.29381550026083104223801656334d-01	3.63894824961114d-01
21	34	5.28998091844781619399911365964d-01	-1.65052361672464d+01	5.29377227395463613635804233924d-01	-1.04288830187451d-01
34	55	5.29501113159821769730489298164d-01	-1.68379732663179d+01	5.29371855618189398310850321574d-01	3.66467548590298d-02
55	89	5.29294392732914343763871510174d-01	-1.69304207569555d+01	5.29372522085739525283301988632d-01	3.79155467948030d-01
89	144	5.29394797757384950510270642254d-01	-1.69034623499470d+01	5.29372639701797630919231848754d-01	-1.07010450304882d-01
144	233	5.29364977563898420441438294804d-01	-1.68684792946379d+01	5.29372596320574032281545796074d-01	3.65609003257187d-02
233	377	5.29377251283201724950419164194d-01	-1.68426360000174d+01	5.29372607356018438538875818484d-01	3.73695672051473d-01
377	610	5.29371292671527644399307948154d-01	-1.68506319815947d+01	5.29372609436435550268588907914d-01	-1.06162959966784d-01
610	987	5.29373861635826713833946464624d-01	-1.68574308832671d+01	5.29372608935100548998134167294d-01	3.65575213235280d-02
987	1597	5.29372333815157872833370361604d-01	-1.68635487129671d+01	5.29372609090139505267804370044d-01	3.75414981487172d-01
1597	2584	5.29372687193203673116864334814d-01	-1.68616351380473d+01	5.29372609119630607108575260694d-01	-1.06414912730252d-01
2584	4181	5.29372582273293980836922748884d-01	-1.68601393601809d+01	5.29372609113091983885439878853d-01	3.65666389078061d-02
4181	6765	5.29372625445802165786262890060d-01	-1.68587338983962d+01	5.29372609115204245030093470804d-01	3.74913406401449d-01
6765	10946	5.29372604484438993743136747264d-01	-1.68591789239648d+01	5.29372609115603846328140970914d-01	-1.06342660849520d-01
10946	17711	5.29372610707624749420653419594d-01	-1.68595175865608d+01	5.29372609115510952991281080144d-01	3.63631959151966d-02
17711	28657	5.29372608146999114861210163704d-01	-1.68598357738668d+01	5.29372609115542317722713563474d-01	3.75052770184600d-01
28657	46368	5.29372609390233382224913116254d-01	-1.68597208228191d+01	5.29372609115540551629271234694d-01	-1.06362204164251d-01

i	m	γ(2):RS	ratio	accumulation	constant
1	1	9.99999999999999999999999999995d-01			
1	2	5.00000000000000000000000000000d-01			
2	3	7.17117242195617978401675699474d-01			
3	5	6.46022634424453438373557138794d-01			
5	8	6.78723191892120020199114544614d-01	-1.73007043396789d+01	6.66121201245720517825462658384d-01	-1.66121201245721d-01
8	13	6.61674152328077885616499867704d-01	-1.54542411455775d+01	6.65043604381720484687621881894d-01	3.20735578138975d-02
13	21	6.65814018265109660243598164604d-01	-1.69799245280753d+01	6.64757763339734333891690594354d-01	3.04537135382771d-01
21	34	6.64115182272274663074195277294d-01	-1.70119761948278d+01	6.64759642907509912518749662324d-01	-1.86511634320245d-01
34	55	6.64940647186203676822073820334d-01	-1.70195492122567d+01	6.64759372086210301986659830757d-01	5.25090456899564d-02
55	89	6.64695311875654719769611924624d-01	-1.68743222396902d+01	6.64757899221594519170976567274d-01	3.00722241885638d-01
89	144	6.64796335664013215203855840664d-01	-1.68161991858343d+01	6.64758103484744966196463926574d-01	-1.81808166382778d-01
144	233	6.64747298214278603562793978294d-01	-1.68333588566145d+01	6.64758140192701157311165004394d-01	5.17155186200862d-02
233	377	6.64761854522326024510879805114d-01	-1.68542208470522d+01	6.64758127524254176918471799094d-01	3.08741865582706d-01
377	610	6.64755866163448209526773039884d-01	-1.68700275350493d+01	6.64758130820919253212729402574d-01	-1.83420025060522d-01
610	987	6.64758877378722750599040117454d-01	-1.68651287029122d+01	6.64758131442295836509427959894d-01	5.19668786208881d-02
987	1597	6.64757910473367725860761226424d-01	-1.68609687919619d+01	6.64758131292798308860648661907d-01	3.00919206826948d-01
1597	2584	6.64758265713058705110072959154d-01	-1.68572347907858d+01	6.64758131338972312715157407194d-01	-1.82914107980387d-01
2584	4181	6.64758089323978452834719582185d-01	-1.68584019364582d+01	6.64758131347754402567029413044d-01	5.18916921954746d-02
4181	6765	6.64758144446705589087467383844d-01	-1.68593198321661d+01	6.64758131345795808853570379168d-01	3.00844276386753d-01
6765	10946	6.64758123376956623959860981854d-01	-1.68601767191056d+01	6.64758131346424387662026596414d-01	-1.83063621950994d-01
10946	17711	6.64758133606747393648056175324d-01	-1.68599024222091d+01	6.64758131346546784763045010274d-01	5.19131442817013d-02
17711	28657	6.64758130569510973669270151934d-01	-1.68597006231759d+01	6.64758131346322228903087422110d-01	3.00869221669421d-01
28657	46368	6.64758131819235941648968658474d-01	-1.68595086958927d+01	6.64758131346530927113436294954d-01	-1.83022554262745d-01

i	m	b(1):TS	ratio	constant
1	1	1.24290805157749999999999999990d+00		
1	2	2.38393140212460845898981870800d+00		
2	3	1.62556361113770371115154080274d+00		
3	5	4.55821837049807368021441269974d+00	3.64887447115860d+00	1.24290805157750d+00
5	8	8.49603949940174966452020739784d+00	3.60929782904183d+00	2.38393140212461d+00
8	13	6.02968634680114081180013214304d+00	3.70928992030015d+00	1.62556361113778d+00
13	21	1.59158579390256434124542792994d+01	3.50921534770642d+00	1.29287391520629d+00
21	34	2.91901899632932117561571959764d+01	3.43573143290455d+00	2.47204738790518d+00
34	55	2.06835192049369929598091331194d+01	3.43028111502217d+00	1.75770198400452d+00
55	89	5.52362156509779196262676882734d+01	3.47068894518644d+00	1.32122517120800d+00
89	144	1.01077357819417791531596390294d+02	3.49013271781194d+00	2.39635081523006d+00
144	233	7.21961617751320877464216644284d+01	3.49051634184376d+00	1.69763798017849d+00
233	377	1.92290448033885929495471019378d+02	3.48123863605879d+00	1.30924984751363d+00
377	610	3.54199621910624250982573020081d+02	3.47672563847267d+00	2.42410864334030d+00
610	987	2.51001801942622947169294689134d+02	3.47666407425388d+00	1.71001891940699d+00
987	1597	6.68935837935037966917872807114d+02	3.47877830009138d+00	1.31295760120987d+00
1597	2584	1.23254664321133026682549004174d+03	3.47980790200375d+00	2.41561108205902d+00
2584	4181	8.73441396147949700987441683794d+02	3.47982121796724d+00	1.71178450773695d+00
4181	6765	2.32745509511660600356550486264d+03	3.47933981576067d+00	1.31109848033349d+00
6765	10946	4.28816032715830269882280014054d+03	3.47910592331479d+00	2.41884905313721d+00
10946	17711	3.03878543733209473902680766404d+03	3.47909482048100d+00	1.71357234493538d+00
17711	28657	8.09771425141483109519933140024d+03	3.47921395708351d+00	1.31210324930014d+00
28657	46368	1.49196665612190869290289622734d+04	3.47926976207676d+00	2.41736593019502d+00

i	m	b(2):RTS	ratio	constant
1	1	7.57691140422499999999999999998d-01		
1	2	2.38393140212460845898981870780d+00		
2	3	4.42688405791682472564461633974d+00		
3	5	3.18422353762035445865173283024d+00	4.20506549486823d+00	7.57691140422500d-01
5	8	8.25320112356789680269724178184d+00	3.50613493499374d+00	2.35393140212461d+00
8	13	1.50377853975208262820127926354d+01	3.39692325364338d+00	4.42688405791682d+00
13	21	1.06597541560753015057079417144d+01	3.34485127386341d+00	9.51977614820067d-01
21	34	2.85477384491759264865834971064d+01	3.45898979338512d+00	2.38601488196894d+00
34	55	5.27482957293894369441126191004d+01	3.50751751984141d+00	4.28730157893283d+00
55	89	3.73794100022014085858475942694d+01	3.50935523472296d+00	8.64721379151602d-01
89	144	9.94757074155015980254105409914d+01	3.48453897994757d+00	2.35115308436192d+00
144	233	1.83160426127697167803874461384d+02	3.47254524967316d+00	4.37409191090716d+00
233	377	1.29794976423999708436284377964d+02	3.47236557282086d+00	8.92803978839936d-01
377	610	3.45977584040330760855010764684d+02	3.47000999698562d+00	2.36441964110334d+00
610	987	6.37639096130227188672521489894d+02	3.48876879710764d+00	4.34316285025030d+00
987	1597	4.51791423555391406116145741874d+02	3.48080824883386d+00	8.84173491148990d-01
1597	2584	1.20383371358468558211702404453d+03	3.47951440979580d+00	2.36833314335911d+00
2584	4181	2.21792585181810717036253831904d+03	3.47880602484241d+00	4.35257256029179d+00
4181	6765	1.57172541855428613058430169074d+03	3.47887395934870d+00	8.86634263122831d-01
6765	10946	4.18834532042344908034350449874d+03	3.47917264083900d+00	2.36149268210536d+00
10946	17711	7.71686651505074876736833436634d+03	3.47931672680806d+00	4.34987921890930d+00
17711	28657	5.46851449214339746847968283093d+03	3.47930651727544d+00	8.83973094608387d-01
28657	46368	1.45723156081061345924417313224d+04	3.47925361766323d+00	2.36116293025751d+00

Table 4.4.1.5 (continued)

l	m	b(3)1RS	ratio	ronstant
1	1	1.000000000000000000000000000000d+00		
1	2	1.020364593689999999999999999999d+00		
2	3	2.2424079583498634481758470493d+00		
3	5	4.1793389316257213953934215013d+00	4.179338931625724d+00	1.000000000000000d+00
5	8	2.9116942065780564428840467170d+00	2.831300173287074d+00	1.020364593690000d+00
8	13	7.9350584496391608344320155410d+00	3.538585502488512d+00	2.242437958349864d+00
13	21	1.4697792943968171868180217555d+01	3.516774586785004d+00	1.150481141841436d+00
21	34	1.0425726069543894999552871569d+01	3.500639081532074d+00	8.131772402294774d-01
34	55	2.7673666763601307440556291229d+01	3.487018956443194d+00	2.275273209620594d+00
55	89	5.0985200919301804747599311068d+01	3.463458804129446d+00	1.225270741374814d+00
89	144	3.6071207161016468033742792225d+01	3.459826866772314d+00	8.709587292723614d-01
144	233	9.6201261089472292845562950494d+01	3.476274469889454d+00	2.290016393896464d+00
233	377	1.7731449873991779496138308814d+02	3.463229523462804d+00	1.204525164868564d+00
377	610	1.2565452464811117560648266130d+02	3.483513154611344d+00	8.533129017711174d-01
610	987	3.0477703358162978597728069069d+02	3.479965127998824d+00	2.262730187239164d+00
987	1597	6.1675383153526861197288973555d+02	3.478304571358904d+00	1.211361644810914d+00
1597	2584	4.3706030817190653603228756509d+02	0.478269560896374d+00	8.584701271394644d-01
2584	4181	1.1647115055435774602876347884d+03	3.479066329898174d+00	2.285097954039544d+00
4181	6765	2.1459635676000123725989852437d+03	3.479449105744714d+00	1.209378628950404d+00
6765	10946	1.5207298258186952875175595071d+03	3.479458770030934d+00	8.570139416121604d-01
10946	17711	4.0520521426525556815430487605d+03	3.479275445745504d+00	2.204411327206794d+00
17711	28657	7.4662140248022357089128002400d+03	3.479189552668994d+00	1.209912047750904d+00
28657	46368	5.2980746282374362865733964190d+03	3.479168053668724d+00	8.374318718930424d-01

Also I evaluated the residues of the periodic orbits. They converge to the limits:

$$R_n^+ \to R^* = 0.2500888$$

$$R_n^- \to -0.255426 \qquad (4.4.1.20)$$

and the convergence is at rate:

$$\delta' = -0.6108. \qquad (4.4.1.21)$$

I also evaluated R^+ and δ' for the critical γ^{-2} case in the quadratic map, giving the same results. This suggests a faster way to get a critical parameter value (for a noble). Namely, find the parameter values P_n' where $R_n^+ = 0.2500888$, since they will converge at rate $\delta'/\delta = -0.3752$. I checked this out for another noble in the quadratic map, [0,4,1,1,...], actually using residue 1/4. The results are given in table 4.4.1.6.

Finally, I discuss the smoothness of the critical noble circle. Note that the bends in figure 4.4.1.1 die out towards (0, 0), as $|\alpha| < |\beta|$, and so the circle is differentiable there with $dX/dY = 0$.[18] The map on the circle is not differentiably conjugate to rotation, however, because the self-similarity implies that the function $\varphi(t)$ (semi) conjugating the map to rotation, locally satisfies (Kadanoff, 1981a):

[18] It is not smoother than $C^{1.79}$ or so as a graph, because $\log|\beta_3|/\log|\alpha_3| = 1.79006$. In fact, Sullivan has pointed out that an invariant circle for a non-trivial fixed point of renormalisation can never be C^2, otherwise the induced map on the circle is a C^2 diffeomorphism, which he proved under renormalisation would go to a rigid rotation (Sullivan, 1991).

Table 4.4.1.6: Positions and parameter values for periodic orbits with residue 1/4.

p_n	q_n	positiion	parametervalue
1	5	0.43399468788519488	0.19547971041296584
1	6	0.47369251468655915	0.34639257824780374
2	11	0.45080590424883110	0.27413390237202544
3	17	0.45002992850870050	0.29859597961391083
5	28	0.45116605421251491	0.28881814368699184
8	45	0.45027497557250210	0.29239697495949389
13	73	0.45073198040992516	0.29102577489322026
21	118	0.45052008297310977	0.29153433336392662

$$\varphi\left(\frac{t}{\gamma}\right) = \frac{1}{\alpha}\varphi(t). \tag{4.4.1.22}$$

Since φ cannot be constant, and $|\alpha| < \gamma$, this implies that φ is not differentiable at 0.[19] Incidentally, it is an open question whether this φ is continuous, giving a full conjugacy, or equivalently whether the circle is transitive (i.e. has a dense orbit).[20] I would be very surprised if it were not, as the place where a gap in the limit set would be most likely is around 0, but the orbit of 0 accumulates on 0, by self-similarity.

§4.4.2 **Critical fixed point**. The results of the previous section strongly suggest that there is another fixed point of N_1. Following the pioneering work of Kadanoff (1981b) and Shenker, I worked in the action representation, using the induced renormalisation (4.2.1.9). Actually, I composed in the opposite order from (4.2.1.9), but since the fixed point is expected to be symmetric, changing the order of composition is equivalent to reflecting (x, x') to $(-x', -x)$,

The first problem is to find good domains in which to expand (§2.3.8). Let us regard (4.2.1.9) as one second order equation:

---

19 In fact, we can deduce that φ is at most $C^{0.712}$ or so, since $\log|\alpha|/\log\gamma = 0.71208347$. It is likely that φ is always singular continuous i.e. $\varphi'(t) = 0$ for almost all t, and the same for φ^{-1}.

20 Subject to some conditions on the critical fixed point which are almost certainly true, transitivity can be proved by Rand's method.

$$\tau''(x,x') = \alpha'\beta'\,\tau'\left(\frac{x}{\alpha'}, \bar{x}\left(\frac{x}{\alpha'}, \frac{x'}{\alpha'}\right)\right) + \alpha'\beta'\,\alpha\beta\;\tau\left(\frac{1}{\alpha'}\,\bar{x}\left(\frac{x}{\alpha}, \frac{x'}{\alpha}\right), \frac{x'}{\alpha'\alpha}\right) \tag{4.4.2.1}$$

where α', β' are the values of α, β for τ'. Then we see that if we want τ'' to be defined at (x, x'), we need:

$$\tau' \text{ at } \left(\frac{x}{\alpha'}, \bar{x}\left(\frac{x}{\alpha'}, \frac{x'}{\alpha'}\right)\right) \tag{4.4.2.2}$$

and

$$\tau \text{ at } \left(\frac{1}{\alpha'}\,\bar{x}\left(\frac{x}{\alpha}, \frac{x'}{\alpha}\right), \frac{x'}{\alpha'\alpha}\right). \tag{4.4.2.3}$$

I evaluated α and $\bar{x}(x, x')$ for an approximate fixed point. Note that this was not easy to find, not yet knowing good domains. John Greene had done some exploring, however, and gave me a reasonable starting point, which I was then able to improve on. As I found better fixed points, I could evaluate better domains, giving better fixed points, and so on.

I chose an arbitrary starting point e.g. (0, 0), and asked which points are required to evaluate τ'' there. I call these the preimages. I iterated this process several times. Note that each time the number of points required doubles.[21] This is shown in figure 4.4.2.1. It appears that this process has a certain arc as a limit set. Thus any domain must contain at least this arc. It corresponds, in fact, to a segment of the golden curve.[22]

---

[21] If I had not reduced to a second order equation in τ, I could have done the same, determining simultaneous domains for υ and τ. Then the number of points would have grown like the Fibonacci numbers.

[22] This procedure has turned out to be very useful in all renormalisation problems of dynamics. A non-hyperbolic invariant set for the original dynamics is turned into a hyperbolic set for an associated group of maps (in this case generated by a pair of contractions) (some people prefer to consider the inverse of this group, which is usually generated by a single expanding map called a Markov map). This allows one, for example, to estimate the Hausdorff dimension of the Feigenbaum attractor (Falconer, 1985), and smoothness estimates on the conjugacy between the non-hyperbolic sets for different dynamical systems whose orbits converge together under renormalisation (Rand, 1991). The analogous set of contractions is used by Davie and Dutta (1992) to deduce the non-area-preserving spectrum of the area-preserving period doubling fixed point from the area-preserving spectrum, and exactly the above contractions are used by MacKay (1992d) to do the same for the critical golden circle fixed point described here.

Then I looked for a suitable parallelogram domain. The idea is to write:

$$\tau(x, x') = t(\xi, \xi') \tag{4.4.2.4}$$

with

$$\begin{aligned} x &= c + a_{11}\xi + a_{12}\xi' \\ x' &= c + a_{21}\xi + a_{22}\xi' \end{aligned} \tag{4.4.2.5}$$

which I write as:

$$\underline{x} = \underline{c} + A.\underline{\xi} \tag{4.4.2.6}$$

and look for $t(\xi, \xi')$ analytic on the product of unit disks:

$$|\xi|, |\xi'| \le 1. \tag{4.4.2.7}$$

This corresponds to a parallelogram domain in x, x'. The reason for restricting to parallelograms is that the corresponding coordinate change is only affine, which is relatively straightforward to implement.

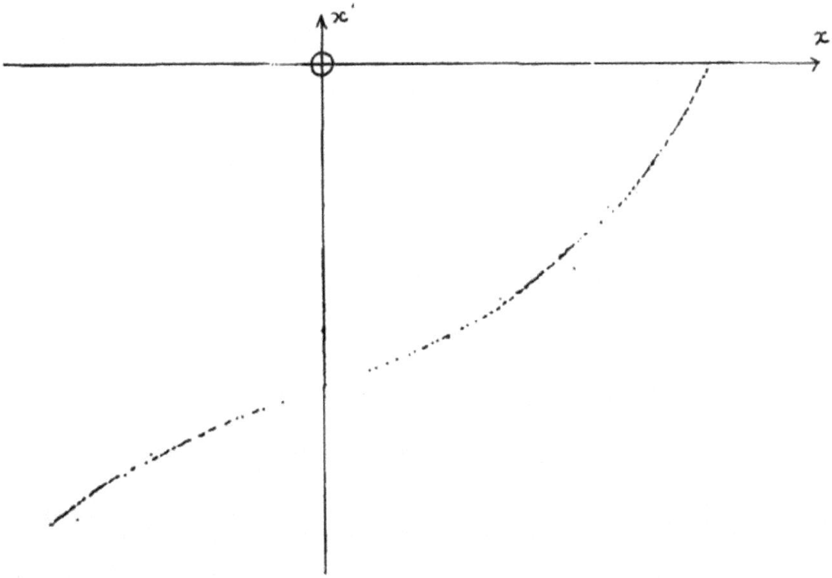

Figure 4.4.2.1: Points which it is essential to include in the domain for N_1

So I tried some parallelogram domains roughly centred on the arc, plotting the two preimages required by (4.4.2.2), (4.4.2.3), until I found some for which the preimages lay strictly inside. This is not quite sufficient, however, as I want complex domains. So I evaluated a quantity which I call the *analyticity safety factor*. This is an upper bound on the amount by which the preimages remain inside the domain, calculated by estimates of the form:

$$|\Sigma a_n z^n| \leq \Sigma |a_n| \text{ for } |z| \leq 1. \tag{4.4.2.8}$$

For our specific case, this works as follows. Given $\underline{c}$ and A as in (4.4.2.5), then evaluate the coefficients X_{ij} of:

$$\bar{x}\left(\frac{\underline{c} + A.\underline{\xi}}{\alpha}\right) = \sum_{i,j} X_{ij}\, \xi^i\, \xi'^j. \tag{4.4.2.9}$$

Then set

$$X^+ = \sum_{i+j \geq 2} |X_{ij}| \tag{4.4.2.10}$$

and

$$a = A^{-1}. \tag{4.4.2.11}$$

Set

$$f_i = \left| a_{i1}\, c\left(\frac{1}{\alpha} - 1\right) + a_{i2}(X_{00} - c') \right| + \left| \frac{a_{i1}\, A_{11}}{\alpha} + a_{i2}\, X_{10} \right| + \left| \frac{a_{i1}\, A_{12}}{\alpha} + a_{i2} X_{01} \right| + |a_{i2} X^+| \tag{4.4.2.12}$$

$$g_i = \left| \frac{a_{i1}}{\alpha}(X_{00} - c) + a_{i2} c'\left(\frac{1}{\alpha^2} - 1\right) \right| + \left| \frac{a_{i1} X_{10}}{\alpha} + \frac{a_{i2} A_{21}}{\alpha^2} \right| + \left| \frac{a_{i1} X_{01}}{\alpha} + \frac{a_{i2} A_{22}}{\alpha^2} \right| + \left| \frac{a_{i1} X^+}{\alpha} \right|. \tag{4.4.2.13}$$

Then the analyticity safety factor is:

$$\max\,(f_1, f_2, g_1, g_2). \tag{4.4.2.14}$$

I ran a minimization routine based on the Monte Carlo method of Physics Today (May 82) to minimize the safety factor over all choices of parallelogram domains. Wiggly domains with more complicated coordinate changes could be found with arbitrarily good safety factor, but they would be awful to compute with. The reason for using a Monte Carlo minimization method is that there are lots of local minima, and the function is non-differentiable, especially at the minima. One might have tried using a root finder to look for solutions of

$$f_1 = f_2 = g_1 = g_2 \tag{4.4.2.15}$$

but this is 3 equations in 6 unknowns.

The best parallelogram domain I found was in fact a rectangle with sides parallel to the axes (figure 4.4.2.2), which was comforting as that makes the coordinate changes simpler. Using the normalisation to be discussed below, it was given by:

$$|x - c| \leq r, \ |x' - c'| \leq r' \tag{4.4.2.16}$$

with

$$\begin{array}{ll} c = 0.18319019 & r = 0.86715977 \\ c' = -0.638297136 & r' = 0.725794827 \end{array} \tag{4.4.2.17}$$

and a safety factor of 0.93253. So I represented $\tau(x, x')$ by:

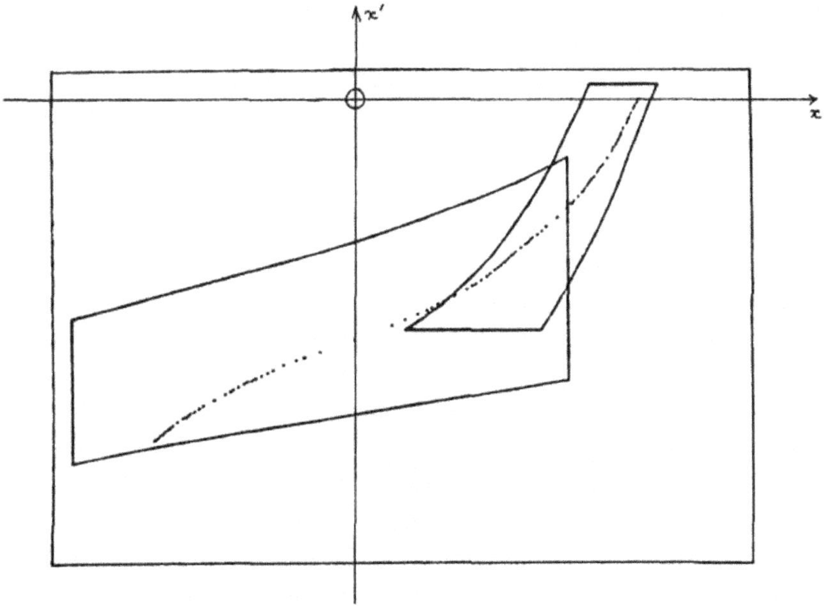

Figure 4.4.2.2: Optimal domain for N_1, and its preimages

$$t(\xi, \xi') = \tau(c + r\xi, c' + r'\xi') \quad (4.4.2.18)$$

expanding $t(\xi, \xi')$ in the product of the unit discs.

Given a domain for τ, I just used its rescaled version for υ. This is OK algorithmically, because I can do the rescaling exactly, but for a proof I would need to use slightly smaller domains, so that τ analytic on its domain implies υ' analytic on slightly more than the domain for υ.

Next I have to discuss the choice for α and β. At first I used:

$$\alpha\beta = \frac{1}{\tau(0, 0)}, \; \beta = \frac{1}{\tau_1(0,0)} \quad (4.4.2.19)$$

which forces the normalisation

$$\upsilon'(0, 0) = 1, \; \upsilon_1'(0, 0) = 1. \quad (4.4.2.20)$$

But then I realised that it was better to evaluate things closer to the minimal arc, in order to reduce truncation errors. So I switched to the following prescription. α is determined by the solution of:

$$\tau_1\left(0, \frac{1}{\alpha}\right) = 0 \quad (4.4.2.21)$$

which is unique, under the twist condition, and β by:

$$\frac{\alpha}{\beta} = \tau_{12}\left(0, \frac{1}{\alpha}\right). \quad (4.4.2.22)$$

This choice forces the normalisation:

$$\upsilon_1'(0, 1) = 0, \; \upsilon_{12}'(0, 1) = 1. \quad (4.4.2.23)$$

Note that the first condition is equivalent to forcing the image of the origin under U to have x-coordinate equal to 1. The second normalisation adjusts the y-scale to give unit twist. One might ask why I did not set the y-scale by:

$$\frac{1}{\alpha\beta} = \tau\left(0, \frac{1}{\alpha}\right) \quad (4.4.2.24)$$

forcing the normalisation:

$$\upsilon'(0, 1) = 1. \quad (4.4.2.25)$$

The answer is that $\upsilon(0, 1)$ happens to be zero for the fixed point.

Having found a domain I implemented Newton's method on a truncated version of N_1, where I truncated all operations at some degree in the expansions in the domains. Note that evaluating $N_1(\upsilon, \tau)$ involves two Newton steps itself, one for determining α, and the other for determining $\bar{x}(x, x')$. For simplicity, I give the details of the procedure for a domain centred on $(0, 0)$. For other domains, include the appropriate affine changes of coordinate. The only place the effect of the domain really enters is in the truncations.

Given $\upsilon(x, x')$, $\tau(x, x')$ as polynomials of degree d, first evaluate α as the root of:

$$\tau_1\left(0, \frac{1}{\alpha}\right) = 0 \tag{4.4.2.26}$$

by iterating the Newton step:

$$\frac{1}{\alpha} \to \frac{1}{\alpha} - \frac{\tau_1\,(0, \frac{1}{\alpha})}{\tau_{12}\,(0, \frac{1}{\alpha})}. \tag{4.4.2.27}$$

Then determine β by:

$$\frac{\alpha}{\beta} = \tau_{12}\,(0, \frac{1}{\alpha}). \tag{4.4.2.28}$$

Set

$$\upsilon'(x, x') = \alpha\beta\,\tau\left(\frac{x}{\alpha}, \frac{x'}{\alpha}\right). \tag{4.4.2.29}$$

Next find $z(x, x')$ of degree d such that:

$$\tau_2\left(\frac{x}{\alpha}, z(x, x')\right) + \upsilon_1\left(z(x, x'), \frac{x'}{\alpha}\right) = 0 \tag{4.4.2.30}$$

to degree d, by iterating the Newton step:

$$z(x, x') \to z(x, x') - \frac{\tau_2\left(\frac{x}{\alpha}, z\,(x, x')\right) + \upsilon_1\left(z(x, x'), \frac{x'}{\alpha}\right)}{\tau_{22}\left(\frac{x}{\alpha}, z\,(x, x')\right) + \upsilon_{11}\left(z(x, x'), \frac{x'}{\alpha}\right)} \tag{4.4.2.31}$$

where all quantities are truncated at degree d. Note that dividing power series is just as easy as multiplying them. If you want ideas, see Knuth (1968), but I wrote my routines before I discovered Knuth. Note also that for finding the fixed point it would suffice to find $z(x, x')$ to degree $d/2$, evaluating the quantities in (4.4.2.31) to degree $d/2$ too, as stationarity to degree $d/2$ implies that the composition is correct to degree d. The spectrum of DN_1, however, would not be determined so well.

Then evaluate:

$$\tau'(x, x') = \alpha\beta \left(\tau\left(\frac{x}{\alpha}, z(x, x')\right) + \upsilon\left(z(x, x'), \frac{x'}{\alpha}\right)\right) \tag{4.4.2.32}$$

truncating to degree d.

Next I need the derivative of the renormalisation.

$$\delta\upsilon'(x, x') = \alpha\beta\, \delta\tau\left(\frac{x}{\alpha}, \frac{x'}{\alpha}\right) + \frac{\delta(\alpha\beta)}{\alpha\beta}\, \upsilon'(x, x') + \alpha\, \delta\left(\frac{1}{\alpha}\right) (x\upsilon_1'(x, x) + x'\upsilon_2'(x, x')) \tag{4.4.2.33}$$

$$\delta\tau'(x, x') = \alpha\beta \left(\delta\tau\left(\frac{x}{\alpha}, z(x, x')\right) + \delta\upsilon\left(z(x, x'), \frac{x'}{\alpha}\right)\right)$$

$$+ \frac{\delta(\alpha\beta)}{\alpha\beta}\, \tau'(x, x') + \alpha\, \delta\left(\frac{1}{\alpha}\right) (x\tau_1'(x, x') + x'\tau_2'(x, x')). \tag{4.4.2.34}$$

Note that $\delta z(x, x')$ does not contribute, by stationarity, so I do not need to evaluate it.

The correction $\delta\left(\frac{1}{\alpha}\right)$ to the scaling α, due to the change $\delta\tau$, is given by:

$$\delta\tau_1\left(0, \frac{1}{\alpha}\right) + \tau_{12}\left(0, \frac{1}{\alpha}\right) \delta\left(\frac{1}{\alpha}\right) = 0. \tag{4.4.2.35}$$

So

$$\delta\left(\frac{1}{\alpha}\right) = - \frac{\delta\tau_1\left(0, \frac{1}{\alpha}\right)}{\tau_{12}\left(0, \frac{1}{\alpha}\right)}. \tag{4.4.2.36}$$

Similarly, $\delta(\alpha\beta)$ is given by:

$$\frac{\delta(\alpha\beta)}{\alpha\beta} = - \frac{\beta}{\alpha} \delta\left(\frac{\alpha}{\beta}\right) - 2\alpha\, \delta\left(\frac{1}{\alpha}\right) \tag{4.4.2.37}$$

with

$$\delta\left(\frac{\alpha}{\beta}\right) = \delta\tau_{12}\left(0, \frac{1}{\alpha}\right) + \tau_{122}\left(0, \frac{1}{\alpha}\right) \delta\left(\frac{1}{\alpha}\right). \tag{4.4.2.38}$$

Finally, I perform a Newton step:

$$(\upsilon, \tau) \to (\upsilon', \tau') - \left(I + (DN_1 - I)^{-1}\right).\left((\upsilon', \tau') - (\upsilon, \tau)\right). \qquad (4.4.2.39)$$

Note that it is not necessary to update DN_1 every step, nor to evaluate it to the full degree d, to get a convergent process. In practice, however, I did.

This gave me reasonable results. For example, at degree 14, I obtained a fixed point with

$$\alpha = -1.418, \quad \beta = -3.055. \qquad (4.4.2.40)$$

Compare the values from (4.4.1.1):

$$\alpha = -1.4148360, \quad \beta = -3.0668862. \qquad (4.4.2.41)$$

One can also evaluate the spectrum of DN_1. The eigenvalues larger than 0.8 in modulus are shown, and interpreted, in table 4.4.2.1, for degree 16. There is only one relevant eigenvalue not contained inside the unit circle, namely, δ, as we expected. Unfortunately, however, beyond degree 14 the results started to get worse. This indicates that the approximate fixed point used to produce the domain was not at all close to the fixed point in the ℓ_1 norm for the domain. In fact, the fixed point probably has a singularity in the

Table 4.4.2.1: Spectrum of DN_1 at the critical fixed point, showing eigenvalues greater than 0.8 in modulus.

eigenvalue	compare with	value	interpretation
7.013	$\gamma\alpha\beta$	7.021	constant terms in action
-4.334	$-\alpha\beta$	-4.339	non-zero Calabi invariant
3.055	$-\beta$	3.066	non-commuting
-3.055	β	-3.066	coordinatechange
-2.678	$-\alpha\beta/\gamma$	-2.682	constant terms in action
2.153	β/α	2.167	coordinatechange
-2.155	$-\beta/\alpha$	-2.167	non-commuting
1.633	δ	1.628	relevant direction
-1.518	β/α^2	-1.532	coordinate change
1.496	$-\beta/\alpha^2$	1.532	non-commuting
-1.327	α ?	-1.414	coordinate change
1.418	$-\alpha$	1.414	non-commuting
-1.197	$-\beta/\alpha^3$?	-1.083	non-commuting
1.070	β/α^3	1.083	coordinate change
-0.999	$-\alpha/\alpha$	-1.000	non-commuting

domain.[23]

So I had to try something else. We expect the fixed point to be symmetric and would be content with obtaining its spectrum restricted to the space of symmetric commuting pairs. One way to reduce the domain is to shorten the length of the segment of the golden curve required, by evaluating the second term in (4.4.2.1) at its reflection instead. Then I require only the segment of figure 4.4.2.3. This is equivalent to considering the modified renormalisation:

$$N_{s1}\colon \upsilon'(x, x') = \alpha\beta\, \tau\!\left(-\frac{x'}{\alpha}, -\frac{x}{\alpha}\right)$$

$$\tau'(x, x') = \alpha\beta\, \tau \oplus \upsilon\!\left(\frac{x}{\alpha}, \frac{x'}{\alpha}\right). \tag{4.4.2.42}$$

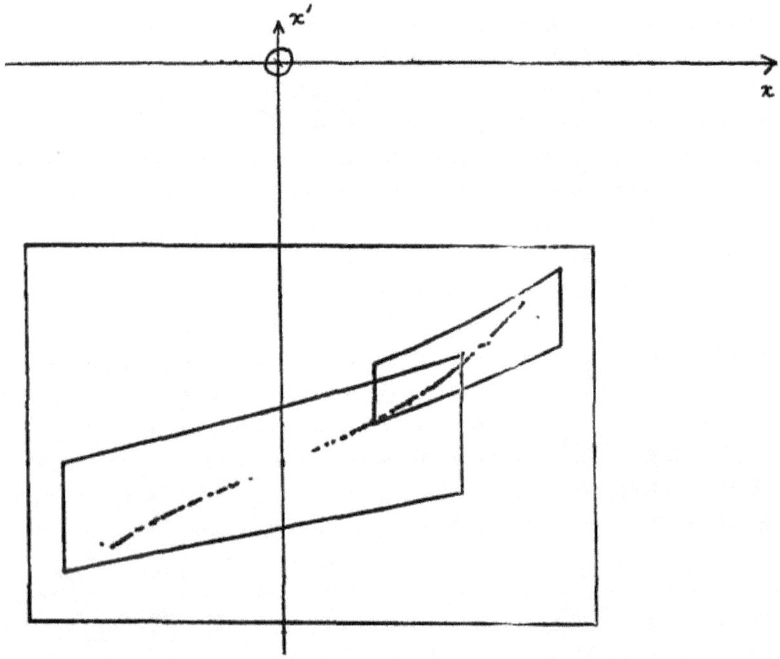

Figure 4.4.2.3: Optimal domain for N_{s1}

[23] Compare the 1-D period-doubling fixed point, which has a natural boundary in the complex plane (Epstein & Lascoux, 1981) though not close enough to the origin to make finding a suitable circular domain difficult.

It is the same as N_1, when restricted to symmetric commuting pairs, but permits good domains.[24] In the same way as before I found an optimal domain. It was a rectangle again (figure 4.4.2.3), with:

$$c = 0.050707985, \quad r = 0.502060282$$
$$c' = -0.655406307, \quad r' = 0.329680205. \tag{4.4.2.43}$$

For υ, I used its rescaled and reflected version. It has an analyticity safety factor of 0.87918.

The scheme above can be modified trivially for N_{s1}. This gave me much better results, which moreover appear to converge as the degree of truncation is increased. For example, the values of $\alpha, \beta, \delta, \delta'$ for several degrees are shown in table 4.4.2.2. The coefficients of $t(\xi, \xi')$ for the fixed point are given in table 4.4.2.3. To show that the fixed point is close to symmetric, I also expanded it in the largest symmetric square which fits inside the domain. This has:

$$c = 0.44, \quad r = 0.115$$
$$c' = -0.44, \quad r' = 0.115. \tag{4.4.2.44}$$

The expansion of $t(\xi, \xi')$ in this symmetric domain is given in table 4.4.2.4.

Tables 4.4.2.3, 4.4.2.4 do not give the fixed point to many significant figures, so I also include table 4.4.2.5, in which υ, τ and $\bar{x}$ are expanded about $(0, 0)$. The coefficients

Table 4.4.2.2: Values of $\alpha, \beta, \delta, \delta'$ for the fixed point of N_{s1}, when truncating at various degrees.

degree	α	β	δ	δ'
14	-1.414836085	-3.066888192	1.6279496	-0.61083048
15	-1.414836021	-3.066888344	1.6279506	-0.61083021
16	-1.414836072	-3.066888224	1.6279499	-0.61083040
17	-1.414836052	-3.066888269	1.6279502	-0.61083026
18	-1.414836062	-3.066888246	1.6279500	-0.61083028

[24] An alternative resolution of the problem of domains for (τ, υ), using N_1, would have been to use Tchebyshev projection rather than straight truncation of the power series. This is equivalent to using elliptical rather than circular domains (Nehari, 1952). Since ellipses can be chosen to be thin in the imaginary direction and still contain the required real domains, this procedure should allow one to avoid all complex singularities.

Table 4.4.2.3: Coefficients of t expanded in domain (4.4.2.43)

coefficients of t :

-1.2472e-03	-9.1282e-03	-4.1229e-02	1.0425e-02	-1.1061e-03	-1.0196e-04	1.6678e-05	-2.6050e-06	4.9198e-07	-9.1079e-08
6.7534e-03	7.4440e-02	-4.9490e-03	1.1141e-03	-1.0031e-04	3.9634e-05	-8.0243e-06	1.6044e-06	-3.4763e-07	7.0733e-08
-2.5112e-02	-9.2126e-04	9.0060e-04	-1.8915e-04	5.1599e-05	-1.2034e-05	2.8927e-06	-6.5918e-07	1.4503e-07	-2.9722e-08
-3.0205e-03	1.2898e-03	-1.3722e-04	5.2401e-05	-1.2020e-05	3.5427e-06	-8.7971e-07	2.0995e-07	-4.5103e-08	8.2052e-09
-8.3650e-03	-1.4029e-05	4.3600e-05	-9.9604e-06	3.4270e-06	-9.2207e-07	2.4474e-07	-5.6273e-08	1.0644e-08	-1.0054e-09
4.2331e-05	4.0835e-05	-5.1160e-06	2.0042e-06	-7.7753e-07	2.4000e-07	-6.0674e-08	1.2045e-08	-1.4172e-09	-4.9171e-10
4.5071e-05	7.4867e-08	2.0556e-06	-5.0010e-07	2.0210e-07	-5.5931e-08	1.4116e-08	-2.1649e-09	-2.0035e-10	4.7126e-10
1.4375e-06	1.5538e-06	-1.9496e-07	1.4756e-07	-4.2255e-08	1.3467e-08	-2.7768e-09	1.1314e-10	3.3224e-10	-2.5353e-10
9.5730e-07	3.0485e-08	9.0911e-08	-2.3859e-08	1.0000e-08	-2.7625e-09	4.7009e-10	1.3107e-10	-1.7341e-10	1.1109e-10
5.9529e-08	6.4543e-08	-7.1100e-09	7.5575e-09	-2.0710e-09	6.0929e-10	-3.0487e-11	-8.1938e-11	7.2731e-11	-4.3257e-11
2.9600e-08	2.4871e-09	4.7073e-09	-1.0762e-09	5.3707e-10	-9.9813e-11	-1.3995e-11	3.5825e-11	-2.7010e-11	
2.6096e-09	2.0123e-09	-2.3583e-10	3.7566e-10	-8.9500e-11	1.0463e-11	1.0775e-11	-1.3150e-11		
1.0740e-09	1.5968e-10	2.3124e-10	-4.5110e-11	2.4110e-11	-7.5774e-13	-4.3149e-12			
1.1701e-10	1.2504e-10	-6.1435e-12	1.0246e-11	-3.0749e-12	-2.0300e-13				
4.2162e-11	9.2914e-12	1.1195e-11	-1.6209e-12	9.4010e-13					
5.3057e-12	5.7445e-12	-1.9412e-15	8.7575e-13						
1.7407e-12	5.2403e-13	5.5037e-13							
2.4406e-13	2.6050e-13								
7.4615e-14									

1.7119e-08	-3.1786e-09	5.7411e-10	-9.7775e-11	1.4770e-11	-1.6242e-12	-3.9839e-14	1.0596e-13	-4.7005e-14
-1.3846e-08	2.5432e-09	-4.1103e-10	4.0193e-11	1.3372e-12	-3.5835e-12	1.6758e-12	-6.0319e-13	
5.4395e-09	-7.6724e-10	2.1372e-11	4.2413e-11	-2.3634e-11	9.3741e-12	-3.2405e-12		
-9.1417e-10	-1.5301e-10	1.5984e-10	-7.6702e-11	2.9863e-11	-1.0462e-11			
-3.9970e-10	3.3297e-10	-1.6095e-10	6.4000e-11	-2.3617e-11				
4.6007e-10	-2.4422e-10	1.0404e-10	-4.0441e-11					
-2.0247e-10	1.3200e-10	-5.5070e-11						
1.3510e-10	-6.1325e-11							
-5.6560e-11								

Table 4.4.2.4: Coefficients of t expanded in the symmetric domain (4.2.2.44)

coefficients of t abt syi

1.4793e-04	1.1754e-03	-3.1522e-03	3.2766e-04	-2.0955e-05	-2.1779e-07	1.2040e-08	-4.5306e-10	2.4391e-11	-1.5010e-12
-1.1754e-03	5.7301e-03	-5.0402e-05	6.4270e-06	-2.1793e-07	1.6168e-08	-8.1555e-10	5.1414e-11	-3.9192e-12	-8.0251e-14
-3.1522e-03	5.0402e-05	4.4169e-06	-6.5987e-08	1.4193e-08	-6.5000e-10	5.5564e-11	-4.0788e-12	-2.4784e-13	-1.6410e-13
-3.2766e-04	6.4270e-06	6.5977e-08	1.3057e-08	-2.5200e-10	5.0377e-11	-3.9525e-12	-3.1941e-13	-1.9956e-13	-4.5764e-14
-2.0955e-05	2.1792e-07	1.4199e-08	2.4751e-10	4.7235e-11	-1.0295e-12	-2.1933e-13	-1.6083e-13	-4.3662e-14	-9.0260e-15
2.1779e-07	1.6175e-08	6.4457e-10	5.2507e-11	5.0726e-13	-3.4500e-14	-1.0214e-13	-3.0474e-14	-7.6203e-15	-1.3775e-15
1.2043e-08	8.1151e-10	5.0511e-11	2.4563e-12	1.2050e-13	-4.0007e-14	-1.5604e-14	-4.5304e-15	-9.0864e-16	-1.3343e-16
4.5167e-10	5.3470e-11	3.4009e-12	2.2989e-13	-3.5598e-15	-5.5417e-15	-2.0477e-15	-4.6930e-16	-7.6013e-17	-8.4754e-18
2.5025e-11	3.0048e-12	2.5603e-13	1.2060e-14	-6.0303e-16	-6.5987e-16	-1.0636e-16	-3.5152e-17	-4.3273e-18	-3.2469e-19
1.2706e-12	1.9560e-13	1.6357e-14	9.3084e-16	-9.7651e-17	-5.2359e-17	-1.2404e-17	-1.7654e-18	-1.4079e-19	-5.7409e-21
7.1910e-14	1.1792e-14	1.1152e-15	5.0456e-17	-6.0264e-18	-3.0229e-18	-5.4035e-19	-5.4360e-20	-2.3539e-21	
3.9683e-15	7.2571e-16	7.0651e-17	4.3412e-18	-2.1445e-19	-1.0076e-19	-1.4692e-20	-7.5296e-22		
2.2336e-16	4.2249e-17	4.3163e-18	2.0061e-19	2.5229e-21	-2.1252e-21	-1.6215e-22			
1.1975e-17	2.2920e-18	2.2461e-19	1.4975e-20	4.6344e-22	-5.0200e-24				
5.9500e-19	1.0598e-19	9.3326e-21	5.1250e-22	1.5363e-23					
2.5250e-20	3.0702e-21	2.6058e-22	9.3181e-24						
8.4481e-22	9.5030e-23	3.0454e-24							
1.9215e-23	1.2322e-24								
2.2479e-25									

2.0743e-15	-2.0003e-14	-3.0020e-15	-5.3253e-16	-6.2720e-17	-5.0603e-18	-3.7367e-19	-1.5329e-20	-2.7481e-22
-8.5043e-14	-1.4958e-14	-2.7099e-15	-3.4991e-16	-3.5191e-17	-2.4595e-18	-1.1060e-19	-2.3157e-21	
-3.3301e-14	-6.3986e-15	-8.9751e-16	-9.6673e-17	-7.2929e-18	-3.5295e-19	-8.1692e-21		
-9.4473e-15	-1.4353e-15	-1.6557e-16	-1.3396e-17	-6.9247e-19	-1.7310e-20			
-1.6211e-15	-2.0898e-16	-1.7420e-17	-9.5985e-19	-2.5673e-20				
-1.8481e-16	-1.7214e-17	-1.0127e-18	-2.0067e-20					
-1.3449e-17	-8.4910e-19	-2.5016e-20						
-5.7931e-19	-1.0075e-20							
-1.1433e-20								

are given in the order:

$$1, x, x', x^2, xx', x'^2, \ldots . \tag{4.4.2.45}$$

Note, however, that one should transform this to an expansion in the domain (4.4.2.43) if one wants to use it for anything.

The eigenvalues of DN_{s1} larger than 0.4 in modulus are shown and interpreted in table 4.4.2.6. For this table I left out the contributions (4.4.2.36) and (4.4.2.38) of the changes in α and β to the derivative. The triples of eigenvalues with the same cube correspond to non-symmetric eigenvectors, which are therefore irrelevant. The beginnings

Table 4.4.2.5: The critical fixed point to degree 18, expanded about (0, 0) in the order (4.4.2.45)

1/alpha = -7.06795660190850e-01

coefficients of u expanded about 0,0:

-5.04313415062478e-01	-8.96341843330159e-01	8.96341825292403e-01	-2.08893068032901e-01	8.35030007040076e-01	-2.08891311088232e-01
-3.19327626228859e-02	-2.63671592019450e-02	2.63411879943726e-02	3.19250477313762e-02	-1.41010843337370e-01	2.03967253635506e-02
2.10107227178240e-02	2.04784280163298e-02	-1.41754700749500e-01	-2.29465226964451e-03	-3.54068103904150e-03	-1.38983904770680e-03
9.15530628453823e-04	3.10184278307104e-03	2.10507674538755e-03	1.16900763329476e-03	6.57366965622200e-04	7.03018810784434e-04
1.47483579107878e-03	1.79722349709474e-03	1.90069922292651e-03	1.64010478887484e-03	-1.11063708269253e-03	-3.31575337311830e-03
-3.90122690840528e-03	-2.45350969013422e-03	-7.79330800363496e-04	9.31608372364507e-05	3.23950516750889e-04	1.66174810086466e-04
-1.43345706501471e-03	-5.26213685007157e-03	-7.46752653118551e-03	-5.73668326013141e-03	-2.64834839938452e-03	-6.68247484363175e-04
3.15211130823725e-05	1.43057237172771e-04	7.73131482007545e-05	-1.97808292160713e-03	-7.96558911800697e-03	-1.28792560792518e-02
-1.17021298013874e-02	-6.88308105709523e-03	-2.66220769571816e-03	-6.84561444234261e-04	-8.14874957231386e-05	2.47770426362778e-05
1.38519638435950e-05	-2.06449273526380e-03	-9.26974928013174e-03	-1.68933599966258e-02	-1.75606117799307e-02	-1.19264992961347e-02
-5.63470945823571e-03	-1.88683177708672e-03	-4.31287975334363e-04	-5.09568640588421e-05	8.09004203330234e-06	5.20004077069207e-06
-1.76921767399946e-03	-8.73191986732308e-03	-1.76896643536022e-02	-2.97149957798998e-02	-1.61176028460458e-02	-8.92617515789712e-03
-3.63185740487022e-03	-1.88938669118282e-03	-2.32473362458937e-04	-2.95500729669497e-05	8.59713450722145e-07	1.17328265436190e-06
-1.21391104864113e-03	-6.54566313966110e-03	-1.46048096692719e-02	-1.98272265017326e-02	-1.66782063432361e-02	-1.05723348307927e-02
-5.02837833215228e-03	-1.81798862618184e-03	-4.95027874436966e-04	-9.69082166020472e-05	-1.12550318410546e-05	3.55042608524360e-07
4.10048338379602e-07	-6.62123281820796e-04	-3.87415680052888e-03	-9.44303003716787e-03	-1.35528425998158e-02	-1.32218828240624e-02
-9.44717468858500e-03	-5.14685569851778e-03	-2.17800193384672e-03	-7.16714108545440e-04	-1.80106840321887e-04	-3.27258396894301e-05
-3.49512147893404e-06	1.25117152720200e-07	1.12293201150664e-07	-2.80163933217989e-04	-1.77054878163409e-03	-4.68452836933614e-03
-7.34760121613623e-03	-7.89903727771252e-03	-6.28239076545903e-03	-3.85733686479983e-03	-1.06859786318672e-03	-7.18573958009266e-04
-2.17404076102465e-04	-5.02238578406020e-05	-8.17155553065739e-06	-6.46307983812503e-07	1.03214293034959e-07	3.99630564475452e-08
-8.88948136275195e-05	-6.94357695607582e-04	-1.72640331160210e-03	-2.93975559156846e-03	-3.45477138499890e-03	-3.02867407772682e-03
-2.07037779838502e-03	-1.13047872281822e-03	-4.97715086556237e-04	-1.76032720251411e-04	-4.98493231225528e-05	-1.02589221737874e-05
-1.39286950445959e-06	-2.32881146227907e-08	4.00891801119800e-08	1.01701561169510e-08	-1.60900044100069e-05	-1.45288798314197e-04
-4.46023645526514e-04	-8.20132351713625e-04	-1.04661143357910e-03	-1.00314492392329e-03	-7.55828024033557e-04	-4.59288679382377e-04
-2.27688096588443e-04	-9.19966371472863e-05	-2.98338507746533e-05	-7.46012256604245e-06	-1.28672428710047e-06	-7.90762127688521e-08
4.00632303411045e-08	1.90441198436135e-08	3.86248943033605e-09	-2.80723918971202e-06	-2.19660911570205e-05	-7.23254811690061e-08
-1.42905900928735e-04	-1.97076149657665e-04	-2.05158012415113e-04	-1.69000565397926e-04	-1.13171037105382e-04	-6.23946306382199e-08
-2.83450893368659e-05	-1.04708317876993e-05	-3.03037742592808e-06	-6.18760942449862e-07	-5.03704352079637e-08	2.35369087580673e-08
1.46059902860019e-08	4.98623973793861e-09	9.33951580430718e-10	-1.86709871117739e-07	-1.57329378848657e-06	-8.55022169644817e-06
-1.17659947588670e-05	-1.74422448401470e-05	-1.96123344230433e-05	-1.76398359354219e-05	-1.20238834853118e-05	-7.76759171095437e-06
-3.90041624141244e-06	-1.59927308804409e-06	-8.11570065672801e-07	-1.10163697715009e-07	-3.41665563106401e-09	1.04374832695487e-08
6.33079741892729e-09	2.61257753211252e-09	8.07192234380476e-10	1.52726504657575e-10		

coefficients of t expanded about 0,0:

-1.34660984534269e-01	2.92264162715622e-01	-2.92264266416105e-01	-9.63670457910960e-02	3.85603611711078e-01	-9.63678566856698e-02
2.08375111846013e-02	1.71933718005791e-02	-1.72098717303961e-02	-2.00467981612069e-02	-1.30905496150433e-01	2.62988298040231e-02
1.94100535025513e-02	2.62178394800078e-02	-1.30987342046641e-01	2.75039117639557e-03	4.05167209550520e-03	1.19619836704934e-03
-1.81549230489682e-03	-4.63653714278714e-03	-2.99880136171591e-03	3.03182583911010e-03	3.51354928990995e-03	3.32226859778920e-03
2.72631681228877e-03	1.29956421986216e-03	1.21517976514163e-03	2.16245512693950e-03	4.34614253516860e-04	8.47261458826569e-04
2.43683221035381e-04	-2.03826484812192e-03	-6.41691891760909e-03	-1.02932042305320e-02	-8.67283444586095e-03	-2.90476669756873e-03
2.86087085342017e-04	5.29364396254662e-04	1.16639712360957e-04	-2.47276148414211e-03	-9.79986318216763e-03	-2.12285781789230e-02
-2.76325948398024e-02	-1.94718418424613e-02	-5.38431030874446e-03	7.25207630714285e-05	1.29718071670185e-04	-4.26620762034001e-04
-3.58396245223402e-03	-1.39377589869600e-02	-3.56169447013577e-02	-6.12654921954734e-02	-6.74282353899986e-02	-4.17027365948113e-02
-1.03560826835279e-02	3.05180138895468e-05	5.99309081523568e-05	-3.77450347821974e-04	-3.19465084151382e-03	-1.39762390013354e-02
-4.17377145193032e-02	-8.83425891833998e-02	-1.30075081762423e-01	-1.28133379464742e-01	-6.86633715522218e-02	-1.52922183341209e-02
1.22960581333106e-05	9.00983793527101e-06	-3.09686172973188e-04	-2.43633191764489e-03	-1.14168181529014e-02	-3.80620386879098e-02
-9.35467410530340e-02	-1.68913245949545e-01	-2.17094142996123e-01	-1.85380525469100e-01	-9.15109363581170e-02	-1.85414855398227e-02
6.08001197722999e-06	5.26441180272934e-06	-1.66884540169427e-04	-1.43691136503768e-03	-7.34085024269300e-03	-2.69563160982773e-02
-7.45385279197671e-02	-1.56761816551972e-01	-2.47296927787366e-01	-2.82127140134651e-01	-2.16553836261696e-01	-9.70561432959069e-02
-1.79995378541794e-02	2.35574870895613e-06	2.62477663801421e-06	-7.35225869531762e-05	-6.86540714911539e-04	-3.77838069696809e-03
-1.50356253025260e-02	-4.36913258431930e-02	-1.07973577587015e-01	-1.98108045330390e-01	-2.77375950400447e-01	-2.84319007116750e-01
-1.90101892414191e-01	-8.12741981491021e-02	-1.38983073995215e-02	1.18615192123270e-06	3.06352523718462e-06	-1.91832037863412e-05
-2.42541666072797e-04	-1.49070493513283e-03	-6.45281631276723e-03	-2.13281454573202e-02	-5.54622479474132e-02	-1.14490430406129e-01
-1.06468967561696e-01	-2.34452994233445e-01	-2.18088703990416e-01	-1.39042476209257e-01	-5.25820324434656e-02	-8.31560489079997e-03
5.11452082238076e-07	2.01735282959219e-06	-1.06161434867701e-06	-5.84922289250719e-05	-4.30813670909167e-04	-2.03977963299388e-03
-7.39232651607469e-03	-2.09810712698486e-02	-4.74733778277816e-02	-8.69435447877027e-02	-1.27186284807461e-01	-1.45079834351585e-01
-1.23452237625443e-01	-7.24986636550797e-02	-2.53794261286268e-02	-3.73305307661942e-03	2.29488626071176e-07	1.13150054336018e-06
2.38034504290633e-06	-4.69028895341659e-06	-7.64583292772522e-05	-4.43240896546529e-04	-1.77256910295075e-03	-5.46595200915833e-03
-1.35200184967280e-02	-2.72884961635569e-02	-4.49068151514522e-02	-5.96015403002619e-02	-6.21840976537111e-02	-4.87279123958420e-02
-2.65003582412906e-02	-8.62753626115975e-03	-1.18228730867879e-03	7.05098835216689e-08	4.19153539874749e-07	1.22780950245383e-06
1.97856082789694e-06	-4.23424210067730e-06	-5.28143139911106e-05	-2.54739742160790e-04	-8.80199397274381e-04	-2.38274634956175e-03
-5.24502063191981e-03	-9.51338950935143e-03	-1.42072066685228e-02	-1.72460033325655e-02	-1.65666253720970e-02	-1.20196948419326e-02
-6.07982829722503e-03	-1.84651467830994e-03	-2.35982204343451e-04	1.81643773775155e-08	9.95786391420785e-08	3.10724352185689e-07
7.52947193876867e-07	1.24137185883135e-06	-4.06356601409801e-07	-1.31022115046879e-05	-6.00430781220100e-05	-1.90207979722109e-04
-4.63892188832924e-04	-9.23880918991921e-04	-1.52519603322784e-03	-2.08608322298101e-03	-2.33257380231659e-03	-2.07447622641724e-03
-1.39937787276876e-03	-6.60110186173592e-04	-1.87118156795053e-04	-2.22061558122713e-05		

coefficients of n expanded about 0,0:

-9.03267844338744e-01	1.83203698050406e-01	3.47264444938618e-01	-6.73306299923640e-03	4.98832824676391e-02	8.92666300816076e-02
2.42554548943780e-02	3.36190789339796e-02	5.19244507118470e-02	5.08554154069516e-02	-4.16682138843380e-03	1.66116711587309e-03
1.28049975751593e-02	2.53977335604614e-02	1.83625061267938e-02	6.42943625040771e-03	1.31323228024689e-02	2.27118649551653e-02
3.63004179172555e-02	4.31784297149220e-02	2.42696170589776e-02	-2.15272680577516e-03	-4.46552485852716e-03	-1.38736613677890e-02
-4.34857875326893e-02	-8.92574486372280e-02	-9.41168595207014e-02	-3.90957388864609e-02	2.36632961512173e-03	9.39533735814546e-03
3.98750262887448e-02	1.47668247387289e-01	3.74181261508276e-01	5.91858182726313e-01	8.22157155665937e-01	1.92104665028136e-01
-1.27137120049486e-03	-8.47732426882060e-03	-5.65329376801813e-02	-2.72736098763708e-01	-8.73628027054238e-01	-1.84189406337441e+00
-2.45614966307053e+00	-1.85822198823672e+00	-5.94829209343096e-01	1.22205204986794e-03	1.08906467794250e-02	8.22133520719399e-02
4.39355920039034e-01	1.61611373335032e+00	4.12911988416334e+00	7.23064176777780e+00	8.24880708513342e+00	5.45823275015704e+00
1.55443341629659e+00	-8.46993901398085e-04	-1.04676830789161e-02	-9.31128993504116e-02	-5.55920561591144e-01	-2.30125202685932e+00
-6.81542421556082e+00	-1.45239496468048e+01	-2.17893581172036e+01	-2.17324058004700e+01	-1.27715577229280e+01	-3.27082033249215e+00
6.93170964022139e-04	9.61691762519368e-03	9.13707776618153e-02	5.84715005908393e-01	2.63868859115428e+00	8.72754422040839e+00
2.14940339098394e+01	3.91052230945565e+01	5.13558928263974e+01	4.54731198207981e+01	2.40293047986911e+01	5.59145653953500e+00
-4.45160364549172e-04	-7.06262359320531e-03	-7.28421788396325e-02	-4.99705616262773e-01	-2.43015648215035e+00	-8.78573188674943e+00
-2.41911072165620e+01	-5.09855568675028e+01	-8.11979782259896e+01	-9.44491880066525e+01	-7.51671894452547e+01	-3.60693045270837e+01
-7.68639423960773e+00	2.76184600853132e-04	4.43223409702681e-03	4.78942782998455e-02	3.47622988830888e-01	1.79985396996942e+00
6.99644023324197e+00	2.10369398999851e+01	4.95440373655479e+01	9.11791458409625e+01	1.28827705112320e+02	1.34635394872314e+02
9.72631825879798e+01	4.27279447117535e+01	8.39521281172631e+00	-1.32230591401289e-04	-2.11051720732387e-03	-2.44217243608236e-02
-1.89766037443329e-01	-1.04683360361290e+00	-4.34365079122668e+00	-1.40576315355224e+01	-3.61674231710706e+01	-7.43811916242662e+01
-1.21428892627720e+02	-1.54195024364213e+02	-1.46141647972039e+02	-9.66132372830430e+01	-3.91174774938718e+01	-7.12555395433139e+00
5.88769579181123e-05	7.63273581058809e-04	9.25807938645273e-03	7.80046128336633e-02	4.61286991058884e-01	2.03972972408894e+00
7.04705882523945e+00	1.95149042692992e+01	4.38400867453975e+01	7.99634817603561e+01	1.17223656467973e+02	1.34949913870457e+02
1.17071582431666e+02	7.13060411511428e+01	2.67579463284035e+01	4.53990849927320e+00	-2.10634013879640e-05	-1.61745389037975e-04
-2.14023780860240e-03	-2.14195799052633e-02	-1.41477296857722e-01	-6.76872428754564e-01	-2.50135410013793e+00	-7.40691813626142e+00
-1.79234105116475e+01	-3.57202324399698e+01	-5.85047641603646e+01	-7.77839350645001e+01	-8.18897776490412e+01	-6.54249922727586e+01
-3.69188306296530e+01	-1.28997996242075e+01	-2.04610381140789e+00	1.00291544702298e-05	2.22667653988318e-05	1.73521484549025e-04
3.04270526849029e-03	2.55999236127018e-02	1.38927740155976e-01	5.58491386571509e-01	1.77316224894815e+00	4.59332090577612e+00
9.86541293984840e+00	1.76570275844342e+01	2.62250913072213e+01	3.18768343301667e+01	3.08946042985640e+01	2.28380876657156e+01
1.20055123416889e+01	3.92069590729913e+00	5.83047730232003e-01	-2.08951302234933e-06	-6.33177551734341e-07	5.48100422436517e-05
8.83117704536937e-05	-1.33128691133379e-03	-1.16627682221547e-02	-5.62167243014571e-02	-1.90138561009550e-01	-5.55269870419686e-01
-1.28217338813796e+00	-2.47818902891642e+00	-4.02565983584290e+00	-5.46738804651899e+00	-6.11660787336060e+00	-5.48724156096409e+00
-3.77636212971663e+00	-1.88249766426772e+00	-5.66829210360225e-01	-7.91176274246777e-02		

Table 4.4.2.6 Spectrum of DN_{s1} at the critical fixed point, showing eigenvalues greater than 0.4 in modulus.

eigenvalue	compare with	value	interpretation
7.0208826	$\gamma\alpha\beta$	7.0208826	constant terms in the actions
-3.0668882	β	-3.0668882	coordinate change
-2.6817385	$-\alpha\beta/\gamma$	-2.6817385	constant term in the actions
2.5641849 $\times 1, \omega, \bar{\omega}$			non-symmetric
1.6279500	δ	1.6280	relevant direction
-1.5320950	β/α^2	-1.5320951	coordinate change
1.0000001	1	1	scale change
1.0000000	1	1	scale change
-0.99999996 $\times 1, \omega, \bar{\omega}$			non-symmetric
0.89544684 $\times 1, \omega, \bar{\omega}$			non-symemtric
-0.7653736	β/α^4	-0.7653736	coordinate change
-0.6108303	δ'	-0.6108	essential convergence rate
0.4995593	α^{-2}	0.4995601	coordinate change

of the eigenvectors $\delta t(\xi, \xi')$ are given in table 4.4.2.7. The coefficients are given in the order (4.4.2.45), and the expansion is with respect to (4.4.2.44).

This procedure could in principle be carried to arbitrary precision, and a proof of existence of a fixed point and bounds on its spectrum produced in the same way as Lanford (1981) and Eckmann et al. (1982) did for period doubling in 1-D and area preserving maps, respectively.[25]

Kadanoff (1981b) and Shenker's method, while it appeared to have an eigenvalue of +1, making the fixed point problem ill-posed, did also locate approximately the 3-cycle under N_1 mentioned in §4.4.1. I think its three elements are all coordinate transforms of the fixed point. You can see "subdominant" lines in figure 4.4.1.1.

[25] As far as I know, no one has done this yet. However, Davie (private communication) is developing a proof of existence of the analogous fixed points for all quadratic irrationals with large enough partial quotients.

Table 4.4.2.7 Eigenvectors of DN_{s1} for eigenvalues larger than 0.4 in absolute value. The top row on each page gives the eigenvalues (real and imaginary parts), and the columns give the eigenvectors, expanded in the symmetric domain (4.4.2.44) and in the order (4.4.2.45).

-2.6817385e+00	0.	7.0200002e+00	0.	-1.2820924e+00	2.22e+00	-1.2820924e+00	-2.22e+00	2.5641849e+00	0.
5.2573111e-01	0.	8.5065081e-01	0.	-5.6781946e-02	3.68e-01	-5.6781946e-02	-3.68e-01	-1.4571761e-01	0.
0.	0.	0.	0.	-1.3017810e-02	5.64e-02	-1.3017810e-02	-5.64e-02	-1.9356698e-01	0.
0.	0.	0.	0.	1.3644940e-01	1.11e-01	1.3644940e-01	-1.11e-01	8.9770940e-03	0.
0.	0.	0.	0.	1.3115637e-04	1.49e-03	1.3115637e-04	-1.49e-03	-2.1307411e-02	0.
0.	0.	0.	0.	-6.8309661e-04	9.56e-04	-6.8309661e-04	-9.56e-04	-8.6439440e-03	0.
0.	0.	0.	0.	-1.7807272e-03	-1.10e-03	-1.7807272e-03	1.10e-03	-1.8448221e-02	0.
0.	0.	0.	0.	1.0459070e-06	1.19e-04	1.0459070e-06	-1.19e-04	-3.1134577e-04	0.
0.	0.	0.	0.	5.2665850e-05	1.41e-04	5.2665850e-05	-1.41e-04	-2.7860173e-04	0.
0.	0.	0.	0.	1.1793383e-04	9.11e-05	1.1793383e-04	-9.11e-05	2.3472527e-05	0.
0.	0.	0.	0.	1.4909460e-04	8.49e-05	1.4909460e-04	-8.49e-05	8.6813845e-05	0.
0.	0.	0.	0.	9.1572312e-07	5.46e-06	9.1572312e-07	-5.46e-06	-2.2276057e-05	0.
0.	0.	0.	0.	9.8114314e-07	7.00e-06	9.8114314e-07	-7.00e-06	-2.9769508e-05	0.
0.	0.	0.	0.	-1.2216376e-06	2.24e-06	-1.2216376e-06	-2.24e-06	-2.2376256e-05	0.
0.	0.	0.	0.	-5.4616686e-06	-1.74e-06	-5.4616686e-06	1.74e-06	-2.0133635e-05	0.
0.	0.	0.	0.	-5.1879565e-06	-2.83e-06	-5.1879565e-06	2.83e-06	-1.3149867e-05	0.
0.	0.	0.	0.	5.5801674e-08	3.64e-07	5.5801674e-08	-3.64e-07	-1.0641238e-06	0.
0.	0.	0.	0.	1.6774648e-07	6.02e-07	1.6774648e-07	-6.02e-07	-1.7178386e-06	0.
0.	0.	0.	0.	2.6229512e-07	4.82e-07	2.6229512e-07	-4.82e-07	-9.7243475e-07	0.
0.	0.	0.	0.	3.6263331e-07	3.64e-07	3.6263331e-07	-3.64e-07	3.6678646e-08	0.
0.	0.	0.	0.	4.9998766e-07	3.02e-07	4.9998766e-07	-3.02e-07	7.1950489e-07	0.
0.	0.	0.	0.	3.4941478e-07	1.86e-07	3.4941478e-07	-1.86e-07	4.5837267e-07	0.
0.	0.	0.	0.	3.9839787e-09	2.02e-08	3.9839787e-09	-2.02e-08	-7.0264738e-08	0.
0.	0.	0.	0.	8.1419570e-09	3.77e-08	8.1419570e-09	-3.77e-08	-1.4054605e-07	0.
0.	0.	0.	0.	5.5494518e-09	2.87e-08	5.5494518e-09	-2.87e-08	-1.3625986e-07	0.
0.	0.	0.	0.	-5.3040413e-09	1.02e-08	-5.3040413e-09	-1.02e-08	-1.1249191e-07	0.
0.	0.	0.	0.	-1.9437960e-08	-5.98e-09	-1.9437960e-08	5.98e-09	-9.8363201e-08	0.
0.	0.	0.	0.	-2.9167909e-08	-1.38e-08	-2.9167909e-08	1.38e-08	-8.4319422e-08	0.
0.	0.	0.	0.	-1.7432146e-08	-9.18e-09	-1.7432146e-08	9.18e-09	-3.8234307e-08	0.
0.	0.	0.	0.	2.4855547e-10	1.28e-09	2.4855547e-10	-1.28e-09	-4.1534289e-09	0.
0.	0.	0.	0.	6.9040177e-10	2.79e-09	6.9040177e-10	-2.79e-09	-9.4100030e-09	0.
0.	0.	0.	0.	1.0376222e-09	2.87e-09	1.0376222e-09	-2.87e-09	-9.3295136e-09	0.
0.	0.	0.	0.	1.3960338e-09	2.38e-09	1.3960338e-09	-2.38e-09	-5.3108408e-09	0.
0.	0.	0.	0.	2.0822493e-09	1.93e-09	2.0822493e-09	-1.93e-09	-2.4457160e-10	0.
0.	0.	0.	0.	2.8147070e-09	1.83e-09	2.8147070e-09	-1.83e-09	3.5564056e-09	0.
0.	0.	0.	0.	2.7755368e-09	1.53e-09	2.7755368e-09	-1.53e-09	4.3302016e-09	0.
0.	0.	0.	0.	1.2932441e-09	6.91e-10	1.2932441e-09	-6.91e-10	1.8364371e-09	0.
0.	0.	0.	0.	1.5397642e-11	7.68e-11	1.5397642e-11	-7.68e-11	-2.6726550e-10	0.
0.	0.	0.	0.	4.1394798e-11	1.83e-10	4.1394798e-11	-1.83e-10	-6.9945946e-10	0.
0.	0.	0.	0.	5.4997718e-11	2.00e-10	5.4997718e-11	-2.00e-10	-8.8836746e-10	0.
0.	0.	0.	0.	5.6622371e-11	1.47e-10	5.6622371e-11	-1.47e-10	-8.6355624e-10	0.
0.	0.	0.	0.	7.1871015e-11	9.57e-11	7.1871015e-11	-9.57e-11	-8.1913278e-10	0.
0.	0.	0.	0.	9.6414487e-11	7.45e-11	9.6414487e-11	-7.45e-11	-8.3276488e-10	0.
0.	0.	0.	0.	7.7681896e-11	5.58e-11	7.7681896e-11	-5.58e-11	-8.1801501e-10	0.
0.	0.	0.	0.	4.0727769e-12	1.53e-11	4.0727769e-12	-1.53e-11	-6.1958592e-10	0.
0.	0.	0.	0.	-2.3216818e-11	-7.56e-12	-2.3216818e-11	7.56e-12	-2.3527392e-10	0.
0.	0.	0.	0.	9.1147522e-13	4.57e-12	9.1147522e-13	-4.57e-12	-1.6168[illegible]e-11	0.
0.	0.	0.	0.	2.8103891e-12	1.22e-11	2.8103891e-12	-1.22e-11	-4.739[illegible]e-11	0.
0.	0.	0.	0.	4.8867160e-12	1.55e-11	4.8867160e-12	-1.55e-11	-6.8341512e-11	0.
0.	0.	0.	0.	9.1263201e-12	1.49e-11	9.1263201e-12	-1.49e-11	-7.4508089e-11	0.
0.	0.	0.	0.	2.2019195e-11	1.78e-11	2.2019195e-11	-1.78e-11	-7.3991311e-11	0.
0.	0.	0.	0.	4.7670053e-11	2.91e-11	4.7670053e-11	-2.91e-11	-6.8302[illegible]e-11	0.
0.	0.	0.	0.	7.4921153e-11	4.33e-11	7.4921153e-11	-4.33e-11	-5.6526[illegible]e-11	0.
0.	0.	0.	0.	7.8387753e-11	4.51e-11	7.8387753e-11	-4.51e-11	-4.401[illegible]e-11	0.
0.	0.	0.	0.	4.8853719e-11	2.84e-11	4.8853719e-11	-2.84e-11	-3.1592[illegible]e-11	0.
0.	0.	0.	0.	1.3378703e-11	8.03e-12	1.3378703e-11	-8.03e-12	-1.39[illegible]e-11	0.

Table 4.4.2.7 (continued)

-3.0660882e+00	0.	1.6279496e+00	0.	-1.5320950e+00	0.	5.0000001e-01	8.66e-01	5.0000001e-01	-[illegible].66e-01
1.2521951e+00	0.	3.3424273e-02	0.	2.4200652e-01	0.	3.8559014e-03	-8.41e-03	3.8559014e-0[illegible]	[illegible].41e-03
1.6363914e-01	0.	1.2649247e-01	0.	9.4093464e-02	0.	2.0193216e-03	-4.96e-02	2.0193216e-0[illegible]	[illegible].96e-02
-1.6363914e-01	0.	-1.2649247e-01	0.	-9.4093464e-02	0.	4.6779619e-03	4.51e-02	4.6779619e-0[illegible]	-[illegible].51e-02
-6.8927939e-15	0.	1.4170249e-02	0.	2.4592610e-02	0.	1.4585813e-03	-8.46e-03	1.458581[illegible]	[illegible].48e-03
-1.6779928e-15	0.	-1.8962623e-02	0.	-9.6724885e-16	0.	1.9298482e-02	-6.37e-03	1.92984[illegible]	[illegible].37e-03
-7.5072010e-15	0.	1.4170254e-02	0.	2.4592610e-02	0.	-2.6700233e-02	1.05e-02	-2.6700233e-0[illegible]	-1.05e-02
-3.4913212e-16	0.	-4.2209992e-04	0.	2.1425380e-03	0.	2.7219494e-04	-5.63e-04	2.7219494e-04	[illegible].63e-04
-2.0218098e-16	0.	-1.9579594e-04	0.	-1.6238066e-16	0.	9.7554251e-04	-5.56e-04	9.7554251e-0[illegible]	[illegible].56e-04
-9.2937673e-17	0.	1.9581887e-04	0.	-1.3350707e-16	0.	2.0661066e-04	-2.38e-04	2.0661066e-04	[illegible].38e-04
-3.4119128e-16	0.	4.2211128e-04	0.	-2.1425380e-03	0.	7.9979300e-04	-1.57e-04	7.9979300e-04	1.57e-04
1.0160391e-17	0.	-3.0530539e-05	0.	-1.2664326e-17	0.	2.4288021e-05	-5.11e-05	2.4288021e-05	5.11e-05
-1.7122439e-17	0.	-2.5733073e-05	0.	-9.7140894e-18	0.	1.0969155e-04	-4.93e-05	1.0969155e-0[illegible]	4.93e-05
-3.5413856e-18	0.	-1.9842204e-05	0.	4.3176236e-20	0.	8.5626318e-05	2.98e-06	8.5626318e-05	-2.98e-[illegible]
2.8085351e-17	0.	-2.5721656e-05	0.	-3.9915137e-18	0.	3.1502101e-05	3.25e-06	3.1502101e-05	-3.25e-[illegible]
-7.6003908e-18	0.	-3.0522364e-05	0.	3.4462782e-19	0.	-5.4034307e-05	1.49e-06	-5.4034307e-05	-1.49e-[illegible]
1.4983850e-18	0.	-6.5136374e-07	0.	-8.1774229e-19	0.	2.7077794e-06	-2.88e-06	2.7077794e-06	2.88e-0[illegible]
-1.1823135e-18	0.	-7.6944066e-07	0.	-9.8960376e-19	0.	7.2067229e-06	-2.97e-06	7.2067229e-06	2.97e-[illegible]
-8.4411815e-19	0.	-2.2158393e-07	0.	-6.8314821e-19	0.	4.7211973e-06	-1.42e-06	4.7211973e-06	1.42e-[illegible]
2.2726405e-18	0.	2.3558485e-07	0.	1.0141792e-19	0.	3.9525488e-07	-2.01e-06	3.9525488e-07	2.01e-[illegible]
-8.1404514e-19	0.	7.8115546e-07	0.	-1.0801616e-18	0.	-4.9911182e-07	-1.53e-06	-4.9911182e-07	1.53e-[illegible]
-1.1547107e-18	0.	6.5652881e-07	0.	-1.8364113e-19	0.	1.5696147e-06	-4.77e-08	1.5696147e-06	4.77e[illegible]
1.0574156e-19	0.	-3.7983195e-08	0.	-3.5278555e-20	0.	2.1232778e-07	-1.98e-07	2.1232778e-07	1.98e[illegible]
-1.2130177e-19	0.	-5.9478307e-08	0.	-7.4607518e-20	0.	5.9767509e-07	-2.40e-07	5.9767509e-07	2.40e[illegible]
-3.8406826e-20	0.	-5.4735912e-08	0.	-5.3286336e-20	0.	5.9162984e-07	-9.86e-08	5.9162984e-07	9.86e[illegible]
4.4101904e-20	0.	-5.4234963e-08	0.	3.7614870e-20	0.	4.0270892e-07	3.28e-09	4.0270892e-07	-3.28e[illegible]
2.1431350e-19	0.	-4.9145645e-08	0.	5.0691666e-20	0.	2.7599769e-07	1.13e-07	2.7599769e-07	-1.13e[illegible]
7.3055233e-20	0.	-5.3159996e-08	0.	-2.4961961e-20	0.	1.0809783e-07	8.95e-08	1.0809783e-07	-8.95e[illegible]
1.6364291e-19	0.	-3.5299337e-08	0.	2.4438212e-20	0.	-7.3675802e-08	-5.81e-09	-7.3675802e-08	5.81e[illegible]
-1.8184681e-21	0.	-1.4329396e-09	0.	-2.3601242e-21	0.	1.5597793e-08	-1.05e-08	1.5597793e-08	1.05e[illegible]
-1.0008090e-20	0.	-2.2886254e-09	0.	-5.2444734e-21	0.	4.1909545e-08	-1.54e-08	4.1909545e-08	1.54e[illegible]
-2.9445037e-21	0.	-1.2120277e-09	0.	-5.9201730e-21	0.	4.3198248e-08	-9.76e-09	4.3198248e-08	9.7[illegible]
6.3003060e-21	0.	6.5467611e-10	0.	-1.6363579e-21	0.	2.3613689e-08	-7.42e-09	2.3613689e-08	[illegible]
1.3749707e-20	0.	3.2107556e-09	0.	5.6189231e-21	0.	2.9789263e-09	-9.76e-09	2.9789263e-09	[illegible]
2.3221625e-20	0.	5.0852154e-09	0.	-5.6455379e-22	0.	-9.3527380e-09	-1.21e-08	-9.3527380e-09	1.21e[illegible]
2.6136062e-20	0.	5.1194713e-09	0.	-7.0757803e-23	0.	-6.2972586e-09	-8.12e-09	-6.2972586e-09	8.12[illegible]
-1.7212543e-20	0.	2.4351459e-09	0.	-1.4628197e-21	0.	1.7156496e-09	-8.37e-10	1.7156496e-09	8.37[illegible]
-4.3778629e-22	0.	-7.4521613e-11	0.	-1.4240660e-22	0.	1.0611934e-09	-6.64e-10	1.0611934e-09	[illegible]
-1.1253730e-21	0.	-1.3032932e-10	0.	-3.2774246e-22	0.	3.1265537e-09	-1.12e-09	3.1265537e-09	1.1[illegible]
-7.1551098e-22	0.	-7.2533313e-11	0.	-4.3625975e-22	0.	3.9386450e-09	-7.76e-10	3.9386450e-09	7.7[illegible]
4.0041954e-22	0.	1.2069411e-10	0.	-4.1034611e-22	0.	3.2785570e-09	-2.02e-10	3.2785570e-09	2.[illegible]
1.7949091e-21	0.	5.4023469e-10	0.	-9.4728450e-23	0.	2.2268307e-09	4.96e-10	2.2268307e-09	-[illegible]
3.0521055e-21	0.	1.0521017e-09	0.	1.7231267e-22	0.	1.3081910e-09	1.13e-09	1.3081910e-09	-1.1[illegible]
6.1257104e-21	0.	1.2381144e-09	0.	-3.6258494e-22	0.	7.1750792e-10	1.46e-09	7.1750792e-10	-1.4[illegible]
4.4435502e-21	0.	8.2012162e-10	0.	2.8436585e-22	0.	1.9386559e-10	1.03e-09	1.9386559e-10	-1.0[illegible]
-1.4302647e-21	0.	2.3384082e-10	0.	-2.4779884e-22	0.	-1.2892394e-10	2.62e-10	-1.2892394e-10	-2.6[illegible]
-3.0984478e-23	0.	-3.1511997e-12	0.	-8.9163318e-24	0.	6.7958495e-11	-3.71e-11	6.7958495e-11	3.71e-11
-1.0209937e-22	0.	-4.4982989e-12	0.	-2.2132687e-23	0.	2.1394679e-10	-7.20e-11	2.1394679e-10	7.20e-11
-7.3708852e-23	0.	5.2094721e-12	0.	-3.3071046e-23	0.	2.9926393e-10	-5.66e-11	2.9926393e-10	5.[illegible]e-11
-7.2156244e-23	0.	3.9554217e-11	0.	-3.2666713e-23	0.	2.6757894e-10	-1.56e-11	2.6757894e-10	1.5[illegible]e-11
1.4952055e-22	0.	1.2480385e-10	0.	-7.0369285e-23	0.	1.5120955e-10	3.07e-11	1.5120955e-10	-3.07e-11
3.4279857e-22	0.	2.6058359e-10	0.	-9.3623996e-23	0.	-2.0475590e-11	5.87e-11	-2.0475590e-11	-5.8[illegible]e-11
8.5047943e-22	0.	3.8470936e-10	0.	-1.3443498e-22	0.	-1.7398365e-10	6.22e-11	-1.7398365e-10	-6.22e-11
1.2271313e-21	0.	3.8999606e-10	0.	-1.1125678e-22	0.	-2.0317913e-10	5.96e-11	-2.0317913e-10	-5.9[illegible]e-11
4.9248786e-22	0.	2.4733443e-10	0.	-5.2118476e-23	0.	-9.9553438e-11	6.16e-11	-9.9553438e-11	-6.16e-11
-1.3354241e-23	0.	7.4026549e-11	0.	-2.4324589e-23	0.	-9.8398681e-12	3.42e-11	-9.8398681e-12	-3.42e-11

Table 4.4.2.7 (continued)

-9.9999996e-01	0.	-4.4772181e-01	7.75e-01	-4.4772181e-01	-7.75e-01	8.9545185e-01	0.	1.0000000e+00	0.
-1.5422089e-01	0.	-2.9578535e-02	-2.44e-02	-2.9578535e-02	2.44e-02	2.0320553e-02	0.	9.8921310e-04	0.
-5.6244036e-02	0.	-4.5033726e-03	-3.05e-02	-4.5033726e-03	3.05e-02	1.0402095e-01	0.	-7.8598875e-03	0.
6.9657246e-02	0.	4.5033697e-03	3.05e-02	4.5033697e-03	-3.05e-02	-1.0402095e-01	0.	7.8598075e-03	0.
-6.9048876e-03	0.	4.9598654e-03	-3.08e-04	4.9598654e-03	3.08e-04	2.5730250e-02	0.	-2.1079510e-02	0.
1.7360915e-03	0.	-4.4533354e-03	-3.55e-02	-4.4533354e-03	3.55e-02	-3.9222986e-02	0.	3.8318268e-02	0.
-6.3461127e-02	0.	4.9598741e-03	-3.08e-04	4.9598741e-03	3.08e-04	2.5730234e-02	0.	-2.1079510e-02	0.
3.2314082e-04	0.	9.3251780e-04	2.13e-04	9.3251780e-04	-2.13e-04	4.0941987e-03	0.	-2.1911695e-03	0.
2.4957065e-03	0.	7.1729617e-04	-3.21e-03	7.1729617e-04	3.21e-03	-3.6455800e-03	0.	3.9055057e-04	0.
-1.2011259e-04	0.	-3.6384047e-03	-2.62e-03	-3.6384847e-03	2.62e-03	-3.5268193e-04	0.	-3.9055261e-04	0.
1.0238226e-03	0.	-1.9062407e-03	-2.16e-03	-1.9062407e-03	2.16e-03	-5.4269566e-03	0.	2.1911684e-03	0.
2.7383084e-04	0.	7.9915065e-05	-2.45e-05	7.9915065e-05	2.45e-05	1.6331655e-04	0.	-1.4013350e-04	0.
1.1774158e-04	0.	1.2936705e-05	-2.06e-04	1.2936705e-05	2.06e-04	-3.7242376e-04	0.	4.2979224e-05	0.
-1.2212084e-05	0.	-7.1607174e-05	-1.57e-04	-7.1607174e-05	1.57e-04	-2.0493865e-04	0.	2.9536381e-05	0.
-3.0837405e-05	0.	1.0346360e-04	-2.54e-05	1.0346360e-04	2.54e-05	-2.4860767e-04	0.	4.2978500e-05	0.
1.1697963e-04	0.	1.2518925e-04	6.60e-05	1.2518925e-04	-6.60e-05	2.2521264e-04	0.	-1.4013394e-04	0.
1.7300399e-05	0.	6.6318781e-06	-2.69e-06	6.6318781e-06	2.69e-06	-2.2141494e-05	0.	1.4563065e-06	0.
1.2118253e-05	0.	4.1139143e-06	-1.79e-05	4.1139143e-06	1.79e-05	-2.4064854e-05	0.	1.4573745e-06	0.
6.1148460e-06	0.	-5.7603913e-06	-1.83e-05	-5.7603913e-06	1.83e-05	-1.4417179e-05	0.	4.4125572e-07	0.
4.1297892e-06	0.	-8.4643966e-06	-1.01e-05	-8.4643966e-06	1.01e-05	-5.1654357e-06	0.	-4.4158945e-07	0.
1.3212947e-06	0.	-1.5927614e-05	-5.75e-06	-1.5927614e-05	5.75e-06	7.8130141e-06	0.	-1.4579418e-06	0.
-0.5776450e-06	0.	-1.0871860e-05	-5.80e-06	-1.0871860e-05	5.80e-06	1.6303490e-05	0.	-1.4567100e-06	0.
5.6990377e-07	0.	3.8275882e-07	-3.43e-07	3.8275882e-07	3.43e-07	-1.4270626e-06	0.	8.0523625e-08	0.
7.5750148e-07	0.	1.8727591e-07	-1.28e-06	1.8727591e-07	1.28e-06	-2.0367063e-06	0.	1.0810969e-07	0.
3.7564395e-07	0.	-3.3959082e-07	-1.39e-06	-3.3959082e-07	1.39e-06	-1.7073218e-06	0.	9.5013236e-08	0.
2.3357752e-09	0.	-2.1515804e-07	-7.47e-07	-2.1515804e-07	7.47e-07	-1.3808710e-06	0.	9.2603915e-08	0.
-2.4827733e-07	0.	2.5373744e-07	-2.25e-07	2.5373744e-07	2.25e-07	-1.0137326e-06	0.	9.4605659e-08	0.
-2.1462629e-07	0.	8.8184582e-07	7.36e-08	8.8184592e-07	-7.36e-08	-1.1329222e-06	0.	1.0776875e-07	0.
-1.8270926e-10	0.	6.5027861e-07	1.75e-07	6.5027861e-07	-1.75e-07	-1.0854326e-06	0.	8.0385099e-08	0.
3.5945058e-08	0.	2.5290554e-08	-2.69e-08	2.5290554e-08	2.69e-08	-7.1295043e-08	0.	3.0169758e-09	0.
6.0298380e-08	0.	2.0200036e-08	-1.00e-07	2.0200036e-08	1.00e-07	-1.4253273e-07	0.	5.4318518e-09	0.
4.8389267e-08	0.	-1.6083034e-08	-1.28e-07	-1.6083034e-08	1.28e-07	-1.4031660e-07	0.	4.3372083e-09	0.
3.1073216e-08	0.	-3.0249522e-08	-9.55e-08	-3.0249522e-08	9.55e-08	-9.4206875e-08	0.	1.6672298e-09	0.
2.1676066e-08	0.	-3.9537472e-08	-5.42e-08	-3.9537472e-08	5.42e-08	-3.2576987e-08	0.	-1.7702170e-09	0.
2.1393381e-08	0.	-5.4252958e-08	-2.43e-08	-5.4252958e-08	2.43e-08	1.5844477e-08	0.	-4.5212903e-09	0.
1.3393613e-08	0.	-7.1412384e-08	-1.27e-08	-7.1412384e-08	1.27e-08	5.2234958e-08	0.	-5.6000108e-09	0.
-6.4118951e-10	0.	-4.1060286e-08	-9.90e-09	-4.1060286e-08	9.90e-09	4.1392099e-08	0.	-3.0697506e-09	0.
1.9863954e-09	0.	1.3704542e-09	-2.00e-09	1.3704542e-09	2.00e-09	-4.6195338e-09	0.	1.6624868e-10	0.
4.0013096e-09	0.	1.1769146e-09	-7.14e-09	1.1769146e-09	7.14e-09	-1.0897312e-08	0.	3.5775864e-10	0.
3.4100183e-09	0.	-1.0982668e-09	-9.69e-09	-1.0582668e-09	9.69e-09	-1.3263885e-08	0.	3.9819733e-10	0.
1.5026258e-09	0.	-1.5931490e-09	-7.14e-09	-1.5531490e-09	7.14e-09	-1.2460014e-08	0.	3.6157217e-10	0.
-2.4601095e-10	0.	4.9279900e-10	-2.67e-09	4.9279900e-10	2.67e-09	-1.0947922e-08	0.	3.0966284e-10	0.
-1.0917693e-09	0.	4.3743096e-09	8.75e-10	4.3743090e-09	-8.75e-10	-1.0556669e-08	0.	2.9396800e-10	0.
-1.3096072e-09	0.	7.7761506e-09	2.11e-09	7.7761500e-09	-2.11e-09	-1.0310329e-08	0.	3.0737447e-10	0.
-9.1790125e-10	0.	8.2292930e-09	1.71e-09	8.2292930e-09	-1.71e-09	-8.0790051e-09	0.	3.0231502e-10	0.
-2.3779249e-10	0.	3.9236227e-09	7.58e-10	3.9236227e-09	-7.58e-10	-4.3042124e-09	0.	1.5538647e-10	0.
1.1960052e-10	0.	8.1068010e-11	-1.33e-10	8.1068810e-11	1.33e-10	-2.7390521e-10	0.	8.2340515e-12	0.
2.7477874e-10	0.	8.7900085e-11	-4.97e-10	8.7900885e-11	4.97e-10	-7.5286340e-10	0.	2.0400053e-11	0.
2.7692268e-10	0.	-4.4118265e-11	-7.34e-10	-4.4118265e-11	7.34e-10	-1.0530012e-09	0.	2.4410234e-11	0.
1.6766764e-10	0.	-7.0767595e-11	-5.83e-10	-7.0767599e-11	5.83e-10	-1.1405602e-09	0.	1.9529277e-11	0.
0.3084974e-11	0.	1.2107525e-10	-1.80e-10	1.2107525e-10	1.80e-10	-1.1667875e-09	0.	5.8094365e-12	0.
1.2039768e-10	0.	4.5052396e-10	2.08e-10	4.5052396e-10	-2.08e-10	-1.2166740e-09	0.	-1.6259295e-11	0.
2.2474259e-10	0.	6.8425307e-10	4.21e-10	6.8425307e-10	-4.21e-10	-1.1047386e-09	0.	-3.9452557e-11	0.
2.4214438e-10	0.	6.6341993e-10	4.01e-10	6.6341993e-10	-4.01e-10	-9.6357927e-10	0.	-4.7561622e-11	0.
1.1834333e-10	0.	4.1807171e-10	2.19e-10	4.1807171e-10	-2.19e-10	-5.5130939e-10	0.	-3.3327013e-11	0.
9.3499134e-12	0.	1.5068676e-10	4.89e-11	1.5068670e-10	-4.89e-11	-1.7015406e-10	0.	-1.0380657e-11	0.

Table 4.4.2.7 (continued)

1.0000014e+00	0.	-7.6537352e-01	0.	-6.1283048e-01	0.	4.9956003e-01	0.
-1.2119637e-01	0.	-4.7450502e-02	0.	6.2645316e-03	0.	4.5777010e-03	0.
-6.3066037e-01	0.	-3.1009815e-02	0.	1.5509336e-02	0.	2.5230571e-02	0.
6.3066038e-01	0.	3.1009815e-02	0.	-1.5509335e-02	0.	-2.5230571e-02	0.
-1.4265059e-01	0.	-1.6209676e-02	0.	3.0232621e-02	0.	1.2123505e-02	0.
1.7364185e-01	0.	1.0793503e-14	0.	-5.4716406e-02	0.	-1.7953013e-02	0.
-1.4265059e-01	0.	-1.6209676e-02	0.	3.0232613e-02	0.	1.2123500e-02	0.
-1.8321338e-02	0.	-4.2366195e-03	0.	5.3623568e-03	0.	4.2335869e-03	0.
2.9742429e-03	0.	-1.1263607e-15	0.	-7.5995003e-03	0.	-2.5746669e-03	0.
-2.9745735e-03	0.	6.6331908e-16	0.	7.5995652e-03	0.	2.5746597e-03	0.
1.8321208e-02	0.	4.2366195e-03	0.	-5.3623540e-03	0.	-4.2335789e-03	0.
-1.1115332e-03	0.	-5.5364919e-04	0.	3.0502515e-04	0.	7.3779244e-04	0.
3.9206660e-04	0.	-5.1233116e-18	0.	-1.7204368e-04	0.	-2.0653857e-04	0.
2.6416110e-04	0.	1.0481024e-17	0.	1.0125249e-04	0.	1.0863047e-05	0.
3.9203077e-04	0.	7.0771121e-17	0.	-1.7202068e-04	0.	-2.0653938e-04	0.
-1.1115530e-03	0.	-5.5364919e-04	0.	3.0504609e-04	0.	7.3770995e-04	0.
1.7852160e-05	0.	-2.0940753e-05	0.	1.3176149e-05	0.	6.5859004e-05	0.
1.7674198e-05	0.	-1.0227476e-18	0.	-1.7402950e-05	0.	-1.3521902e-05	0.
5.3070975e-06	0.	-1.5003045e-18	0.	-1.2285500e-06	0.	5.3351560e-07	0.
-5.3450417e-06	0.	-2.1270983e-18	0.	1.2294967e-06	0.	-5.3910092e-07	0.
-1.7714580e-05	0.	-1.6642033e-18	0.	1.7409301e-05	0.	1.3522867e-05	0.
-1.7871304e-05	0.	2.8940753e-05	0.	-1.3165448e-05	0.	-6.5857163e-05	0.
1.1106422e-06	0.	-1.6830756e-19	0.	-1.1217005e-06	0.	2.3654434e-06	0.
1.5139812e-06	0.	-1.6925336e-20	0.	-6.9494461e-07	0.	-1.1733054e-06	0.
1.2930014e-06	0.	-2.4481766e-20	0.	1.5528425e-08	0.	-4.3571568e-07	0.
1.2327017e-06	0.	-7.4053417e-20	0.	1.3572082e-07	0.	-4.7461084e-07	0.
1.2730694e-06	0.	3.8530139e-19	0.	2.1629039e-08	0.	-4.3439987e-07	0.
1.4915639e-06	0.	-9.3025502e-19	0.	-6.8323511e-07	0.	-1.1725013e-06	0.
1.1024905e-06	0.	-1.1523116e-18	0.	-1.0540325e-06	0.	2.3640166e-06	0.
5.0687821e-08	0.	-1.0844616e-20	0.	-5.6664605e-08	0.	-7.1637452e-08	0.
9.0976760e-08	0.	-4.9551027e-21	0.	-5.1569535e-08	0.	-7.1675951e-08	0.
7.1099992e-08	0.	-2.3712133e-20	0.	-2.1509462e-08	0.	-3.0919278e-08	0.
2.5850758e-08	0.	-3.1933504e-20	0.	-8.9513984e-09	0.	-1.2132778e-08	0.
-3.4525502e-08	0.	6.3566471e-20	0.	6.7929745e-09	0.	1.0364121e-08	0.
-8.3641055e-08	0.	6.3787124e-20	0.	2.4254012e-08	0.	3.0663761e-08	0.
-1.0067299e-07	0.	-1.1677165e-19	0.	5.7262892e-08	0.	7.2150127e-08	0.
-5.3738519e-08	0.	-1.0597053e-20	0.	6.0211941e-08	0.	7.1599297e-08	0.
3.1327490e-09	0.	-3.2923951e-22	0.	-1.9646450e-09	0.	-4.0512917e-09	0.
6.7516984e-09	0.	1.3490722e-22	0.	-2.5345799e-09	0.	-5.4056394e-09	0.
7.3985181e-09	0.	-2.0276552e-21	0.	-1.0186643e-09	0.	-3.0149544e-09	0.
6.3297163e-09	0.	-6.2080147e-21	0.	-9.0396796e-11	0.	-2.9070295e-09	0.
4.4112974e-09	0.	4.0946107e-21	0.	3.4606855e-10	0.	-2.3633532e-09	0.
2.7730087e-09	0.	1.9319596e-20	0.	6.6433773e-10	0.	-2.6430395e-09	0.
2.5252430e-09	0.	1.2296205e-20	0.	1.1231669e-09	0.	-3.4771444e-09	0.
3.6109792e-09	0.	-4.1316596e-21	0.	1.3411953e-11	0.	-5.3713987e-09	0.
2.4011614e-09	0.	-4.1277073e-21	0.	-8.0662903e-10	0.	-4.2407450e-09	0.
1.7551160e-10	0.	1.9766038e-25	0.	-1.0103247e-10	0.	-1.9803619e-10	0.
4.3660959e-10	0.	3.6635225e-23	0.	-1.5807228e-10	0.	-3.5179500e-10	0.
5.2549046e-10	0.	-1.0353717e-22	0.	-1.1585189e-10	0.	-3.0111619e-10	0.
4.0142055e-10	0.	-7.7386247e-22	0.	-1.2822250e-10	0.	-2.3393645e-10	0.
-1.0709692e-11	0.	-6.6393116e-22	0.	-1.6455641e-10	0.	-1.5825389e-10	0.
-7.9460420e-10	0.	1.5253512e-21	0.	-8.0726365e-11	0.	-3.0067441e-11	0.
-1.6325734e-09	0.	2.6580487e-21	0.	3.3827312e-10	0.	1.7982996e-10	0.
-1.0466381e-09	0.	-2.2321775e-22	0.	8.5116010e-10	0.	3.4575205e-10	0.
-1.1030668e-09	0.	-1.7172495e-21	0.	9.2164076e-10	0.	3.4751152e-10	0.
-3.2166079e-10	0.	-8.1501084e-22	0.	4.0504248e-10	0.	1.4755281e-10	0.

§4.5 THE UNIVERSAL ONE PARAMETER FAMILY

The unstable manifold of the critical fixed point gives a universal one parameter family, governing the behaviour of any one parameter family as a noble circle breaks up. I describe some of its properties. In particular, I show that noble circles are robust in an important sense. In one direction the unstable manifold converges to the simple fixed point. In the other direction, it exhibits asymptotic behaviour which corresponds in some sense to a fixed point at infinity.

§4.5.1 **Properties**. The unstable manifold of the critical fixed point forms a universal one parameter family to which all one parameter families passing through the stable manifold converge, on a small enough scale and at high enough period. This statement is supported by finding self-similarity in one parameter families, to be discussed in this section. The self-similarity can be summarised by saying that it looks as if there is a reparametrisation μ, and (parameter dependent) coordinate changes B_n such that the one parameter families:

$$B_n F^{q_n}_{\mu\delta^{-n}} R^{p_n} B_n^{-1} \tag{4.5.1.1}$$

converge to a universal one parameter family F^*_μ, with:

$$B_{n+1} \sim B\, B_n \text{ as } n \to \infty, \; B(X, Y) = (\beta X, \alpha Y) \text{ independent of } \mu. \tag{4.5.1.2}$$

The universal one parameter family is the really important object for applications.

The first problem is to determine scaling-parameters/coordinates (cf. §2.3.5). The beginnings of this process were described in §4.4.1. Suppose we have a one parameter family satisfying:

$$B^n F^{q_n}_{\mu\delta^{-n}} R^{p_n} B^{-n} \to F^*_\mu \text{ as } n \to \infty. \tag{4.5.1.3}$$

Then the effect of the following parameter changes and coordinate changes (some of them parameter dependent) grow, or at least decay slower than $\delta' = -0.6108$.[26] Note that I have included only symmetric coordinate changes.

$$\begin{array}{lllll} \bar{P} = P + P_0 & + & P_1 P & + & P_2 P^2 \\ \delta = 1.628 & & 1.0 & & 1/\delta = .6143 \end{array} \tag{4.5.1.4}$$

[26] The associated rates are indicated below each term.

$$\bar{X} = X \quad + \quad x_0 \quad + \quad x_1 P \quad + \quad x_2 Y^2 \quad + \quad x_3 P^2 \quad + \; x_4 X$$

$$\beta = -3.067 \quad \beta/\delta = 1.884 \quad \beta/\alpha^2 = 1.532 \quad \beta/\delta^2 = -1.157 \quad 1.0$$

$$+ \quad x_5 Y^2 P \quad + \quad x_6 Y^4 \quad + \quad x_7 P^3 \quad + \quad x_8 XP \tag{4.5.1.5}$$

$$\beta/(\delta\alpha^2) = -.941 \quad \beta/\alpha^4 = -.765 \quad \beta/\delta^3 = -.711 \quad 1/\delta = .614$$

$$\bar{Y} = Y \; + \; y_1 Y \; + \; y_2 YP$$

$$1.0 \qquad 1/\delta = .6143. \tag{4.5.1.6}$$

Thus it is necessary to determine coordinate changes to kill components in the growing directions, and desirable to kill components in the directions that decay slower than δ'. The neutral parameter/coordinate changes with eigenvalue $+1$ correspond to scale changes, so they are arbitrary.

I did this for rotation number γ^{-2} in the quadratic map. Specifically, I determined scaling-parameters/coordinates $\Delta P, \Delta X, Y$ by determining the coefficients in the following:

$$P = P_0 + \Delta P \tag{4.5.1.7}$$

$$X = x_0 + x_1 \Delta P + x_2\, \Delta P^2 + x_3 \Delta P^3 + x_4 Y^2 + x_5 \Delta P Y^2 + x_6 Y^4 + \Delta X. \tag{4.5.1.8}$$

They have the values:

$$\begin{aligned} P_0 &= 2.38216325159 \\ x_0 &= 1.31603217783688 \\ x_1 &= 0.8315498514 \\ x_2 &= -0.46108 \\ x_3 &= 0.9 \\ x_4 &= -0.7783661 \\ x_5 &= 0.37 \\ x_6 &= -0.4. \end{aligned} \tag{4.5.1.9}$$

Note that I did not kill the components with decay rate $1/\delta$ because it is so close to δ' that it did not seem worth it.

I also scaled ΔP, ΔX, Y. The scale changes I use depend on what "level" I am working at. The level means the integer n in (4.5.1.3). I set:

$$\Delta P = \frac{P_s}{\delta^n}\mu \tag{4.5.1.10}$$

$$\Delta X = \frac{x_s}{\beta^n}\xi \tag{4.5.1.11}$$

$$Y = \frac{y_s}{\alpha^n}\eta \tag{4.5.1.12}$$

and chose P_s, x_s, y_s as follows. P_s was chosen so that the parameter values for which the periodic orbits of type (p_n, q_n) have residue 1, give:

$$\lim_{n\to\infty} \mu_n = 1. \tag{4.5.1.13}$$

x_s was chosen so that at $\Delta P = 0$, the dominant periodic point of type (p_n, q_n) has $\lim \xi_n = 1$. I chose y_s so that at $\Delta P = 0$, the nearest point of the orbit to the dominant point has $\lim \eta_n = 1$. The values this gave were:

$$\begin{aligned} P_s &= 0.0406149 \times \delta^8 \\ x_s &= -9.382d{-4} \times \beta^8 \\ y_s &= 0.0373510 \times \alpha^8 \end{aligned} \tag{4.5.1.14}$$

with level 0 corresponding to $(p_0, q_0) = (2,5)$. Figure 4.4.1.1 was actually made at level 8.

Now that I have explained how to get scaling coordinates, I can proceed with the results. Figures 4.5.1.1 and 4.5.1.2 show some orbits of F^*_μ for $\mu = -0.3, +0.3$. They are clearly subcritical and supercritical cases, respectively.

Figure 4.5.1.3 shows how R^+_n varies with μ in the limit as $n \to \infty$. It is close to exponential, putting some meat into the slogan "the stochastic transition is exponential". It is not exactly exponential, however, as a three point estimate gave, in the limit $n \to \infty$:

$$\frac{d^2}{d\mu^2} \log R^+ \sim +0.016 \text{ at } \mu = 0. \tag{4.5.1.15}$$

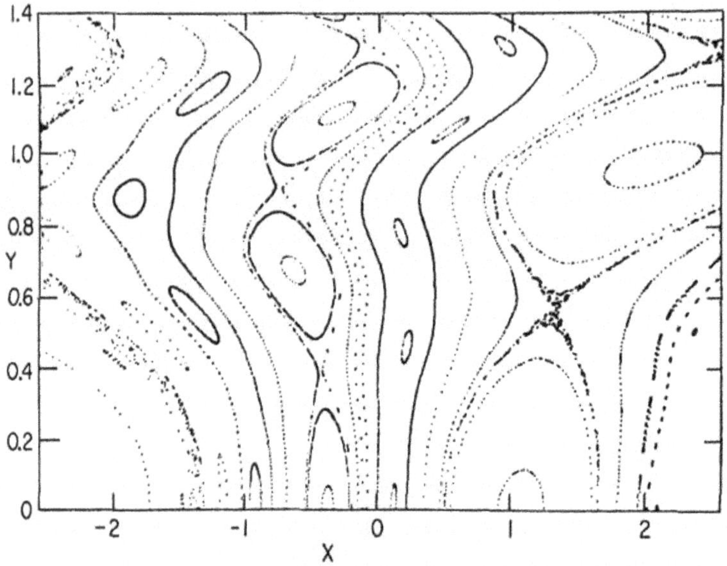

Figure 4.5.1.1: Some orbits of the universal one parameter family at $\mu = -0.3$

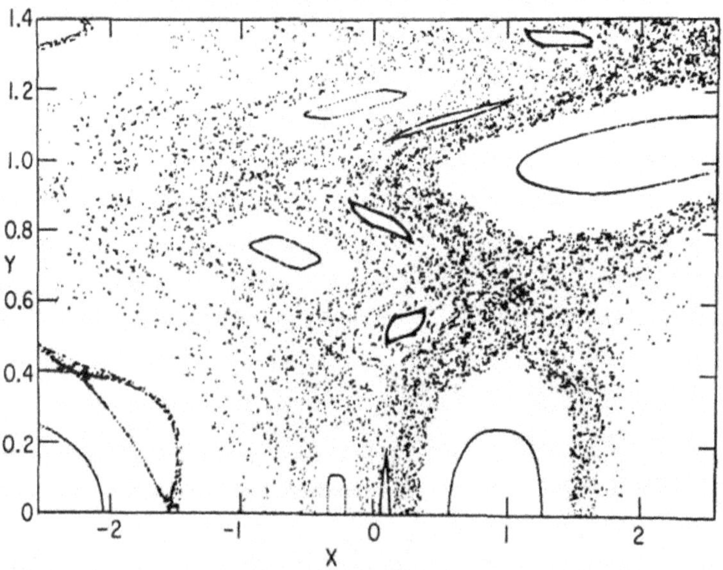

Figure 4.5.1.2: Some orbits of the universal one parameter family at $\mu = +0.3$

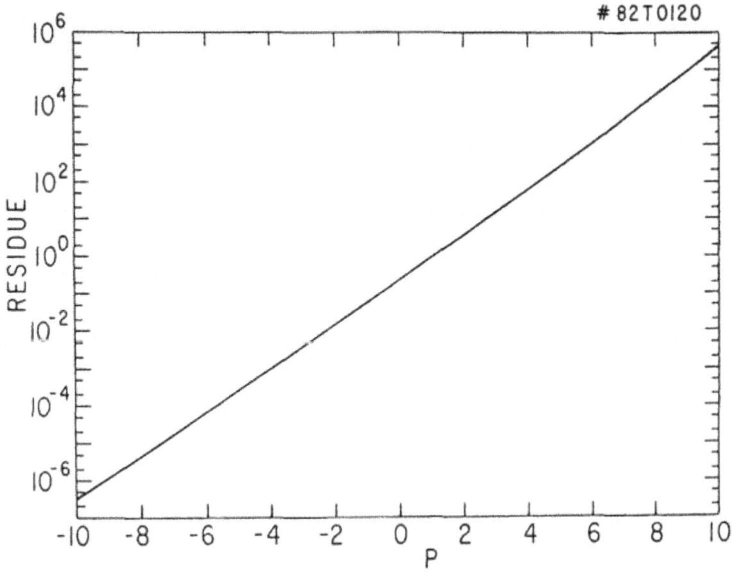

Figure 4.5.1.3: Dependence on parameter of the residue of the fixed point of the universal map

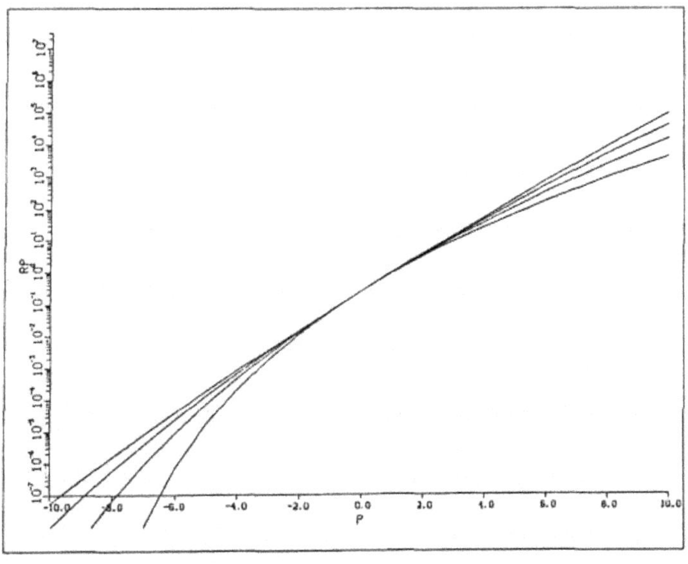

Figure 4.5.1.4: Convergence of the residue as the level is increased

In all these pictures for the universal map, I actually used some Fibonacci power of the quadratic map. I chose it large enough that going to the next level made no difference at the resolution of a picture. For some, level 8 was sufficient. Others required higher levels. I think figure 4.5.1.3 was done at level 13. To show the convergence as the level is increased, I include figure 4.5.1.4, which gives the residue of the fixed point for levels 5,6,7 and 8, I think. Note that the convergence is monotone, indicating that it is dominated by a component in the direction of the parameter change p_2 (4.5.1.4), rather than the essential convergence rate of δ'. Similarly, one can plot the residue R^- of the other Birkhoff orbit (figure 4.5.1.5).

One can also plot the residues and positions of other periodic points. Figure 4.5.1.6, for example, shows the tracks of the dominant periodic points as μ varies, also indicating their residues. They are labelled by a continued fraction expansion $[\underline{b}]$, and should be interpreted as follows. Given a noble $[\underline{a},(1,)^\infty]$:

$$R^*_{[\underline{b}]}(\mu) = \lim_{n \to \infty} R_{[\underline{a},(1,)^n \underline{b}]}(\mu/\delta^n). \tag{4.5.1.16}$$

I call figure 4.5.1.6 the *universal fractal diagram* for the neighbourhood of a noble, because Schmidt (1980) has made similar pictures for specific cases. Actually, in his plots, he uses rotation number rather than position, and plots only points of residue 1. I prefer to use position, and plot the whole track of an orbit, because it is interesting to see how they move, and to see the "island widths" associated with the periodic points. These are the intervals around each periodic point, containing no others. Note that even unstable periodic points have an island width. In fact, the island width has a precise definition here. For each (p, q), there are two points on the dominant line homoclinic to the Birkhoff orbit of non-positive residue. The interval between them is the island width about the dominant point of type (p, q).

Note that if one wants to make fractal diagrams indicating only points of certain residues, for example to reduce the clutter of figure 4.5.1.6, the optimal choice is R^* (4.4.1.15). This is because, as will be disussed in the next section, noble circles appear to be the most robust locally, and residue R^* is the fastest way to determine critical cases for nobles, as discussed in §4.4.1. As R^* is very close to $1/4$, this corresponds to plotting the points marked by circles in figure 4.5.1.6.

Finally, in the discussion of fractal diagrams, it is worth mentioning that they can be drawn in the cases of more than one parameter or no parameter, too. For two parameters, one could drop the position or rotation number coordinate, and give regions in the parameter plane where certain periodic orbits have residue less than or equal to R^*. For no parameters, one could plot the residues of the periodic points against rotation number or position on the dominant line. This will be done in figures 4.5.2.1, 4.5.2.2, for example, and implicitly in table 4.6.2.1.

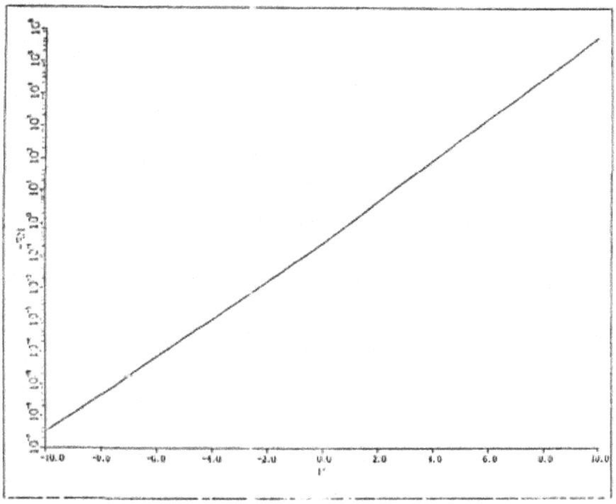

Figure 4.5.1.5: Dependence of R^- on parameter

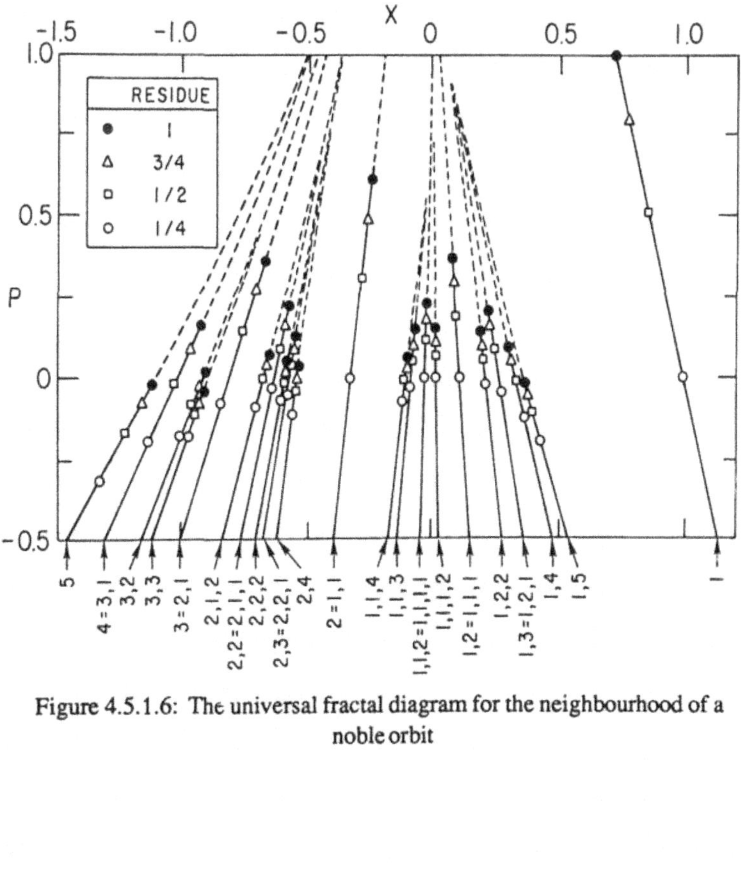

Figure 4.5.1.6: The universal fractal diagram for the neighbourhood of a noble orbit

The self-similarity leads to a related self-similarity in actions, namely, writing $\tau^{(q)}$ for the action for F^q, and using appropriate parametrisation and coordinates:

$$(\alpha\beta)^n \, \tau^{(q_n)}_{\mu\delta^{-n}} \left(\frac{\theta}{\alpha^n}, \frac{\theta'}{\alpha^n} \right) \qquad (4.5.1.17)$$

converges to a universal action. Thus, for example, for a critical noble,

$$\Delta W_{p_n/q_n} \sim \frac{K}{(\alpha\beta)^n} \qquad (4.5.1.18)$$

for some constant K, as in table 4.4.1.5. Figure 4.5.1.7 shows how ΔW for the fixed point of the universal map varies with parameter. It is given by:

$$\lim_{n \to \infty} (\alpha\beta)^n \, \Delta W_{p_n/q_n}(\mu/\delta^n). \qquad (4.5.1.19)$$

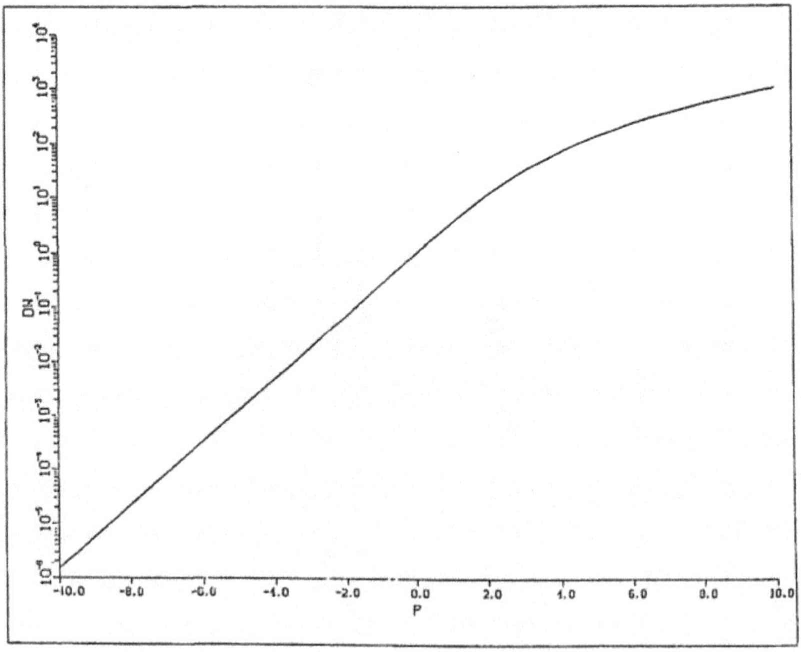

Figure 4.5.1.7: ΔW for the fixed point of the universal map

This should be distinguished from ΔW for the golden orbit, which is given by:

$$\lim_{n\to\infty} (\alpha\beta)^n \lim_{m\to\infty} \Delta W_{p_{n+m}/q_{n+m}}(\mu/\delta^n). \tag{4.5.1.20}$$

It is zero when the golden orbit is a circle, by Mather's theorem (§1.3.2.2) (i.e. for $\mu \le 0$). Figure 4.5.1.8 shows how it grows for $\mu > 0$. The graph is drawn on log-log axes, and has a slope of 3.04. This is consistent with the behaviour one would expect from the scaling relations:

$$\lim \Delta W_\mu \sim \mu^{\kappa'}\, \overline{\Delta W}\left(\frac{\ln \mu}{\ln \delta}\right),\quad \kappa' = \frac{\ln \alpha\beta}{\ln \delta} = 3.0117 \tag{4.5.1.21}$$

with $\overline{\Delta W}$ periodic, of period 1. Why ΔW appears to be 1 at $\mu = 1$ is a mystery to me, as I did not choose that normalisation. I used the same normalisation as for the previous figure.

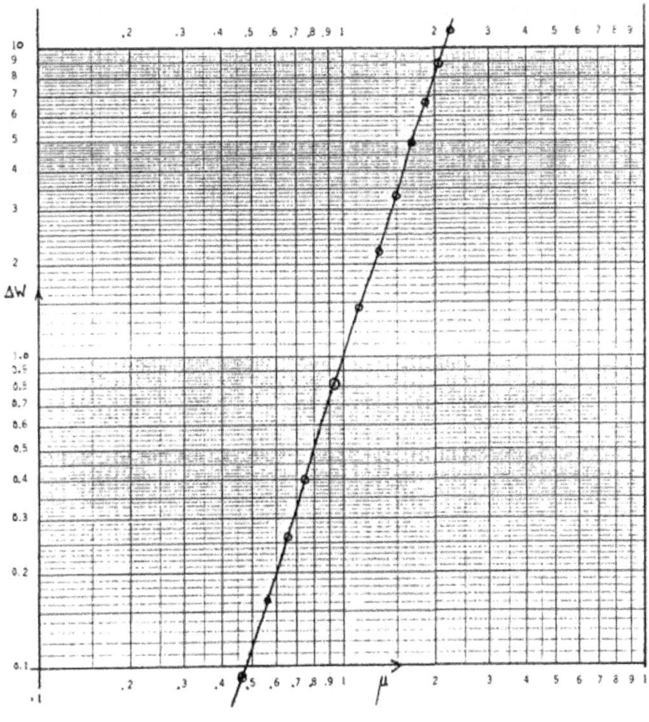

Figure 4.5.1.8: ΔW for the golden orbit

One can also measure the size of the dominant gap of the golden Cantor set, i.e. the gap through which the dominant half-line passes. The endpoints of the gap are given by the accumulation point of the Birkhoff periodic points nearest to the dominant half-line (of either sign of residue). The dependence of the gap width Δx on μ is shown in figure 4.5.1.9, on a log-log scale again, and has a slope of 0.708. This is consistent with the scaling one would expect of:

$$\Delta x = \mu^a \, \overline{\Delta x} \left(\frac{\ln \mu}{\ln \delta}\right), \quad a = \frac{\ln |\alpha|}{\ln \delta} = 0.7121. \tag{4.5.1.22}$$

The noble orbit has a Lyapunov exponent (Katok, 1982), defined by:

$$\lambda = \lim_{n \to \infty} \frac{1}{n} \log \lambda_n \tag{4.5.1.23}$$

where λ_n is the largest eigenvalue of DF^n for a typical point on the noble orbit. It is zero for a circle, but may be positive for a Cantor set. I determined an initial point on the noble Cantor set by taking the limit of the approximant periodic points on a subdominant half-line. The Lyapunov exponent was the hardest measurement of this section, because the limit is not achieved very fast, and one can not iterate too many times before one falls off

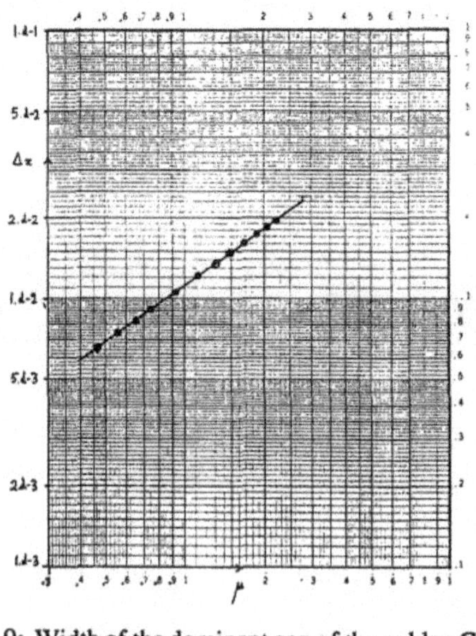

Figure 4.5.1.9: Width of the dominant gap of the golden Cantor set

the golden orbit, since it is unstable.[27] I stopped iterating when the points ceased to come in the same order as uniform rotation at the noble rotation number. To help the convergence somewhat, I took a running average of the last 20 values of (4.5.1.22). Figure 4.5.1.10 shows the results for the dependence of the Lyapunov exponent on μ. Though not as good as the rest, the results are not inconsistent with the behaviour one would expect from the scaling relations:

$$\lambda_\mu \sim \mu^\kappa \bar{\lambda}\left(\frac{\ln \mu}{\ln \delta}\right), \quad \kappa = \frac{\ln \gamma}{\ln \delta} = 0.98746 \qquad (4.5.1.24)$$

with $\bar{\lambda}$ periodic, of period 1.[28]

The data I used to make figures 4.5.1.7-10 are given in table 4.5.1.1. There are some other quantities which would be interesting to measure for the universal one parameter family. One is the fraction of the area, between two of the island chains, say, occupied by encircling invariant circles, as a function of $\mu \le 0$. Another is some sort of diffusion rate for crossing that band, for $\mu > 0$.

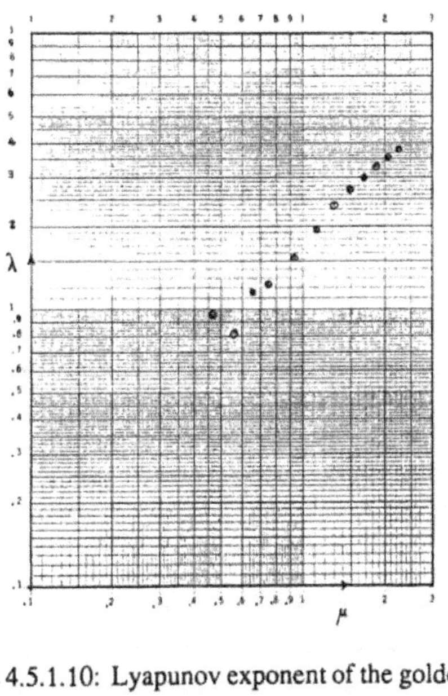

Figure 4.5.1.10: Lyapunov exponent of the golden orbit

---

27 It would have been easier to take the limit of the Lyapunov exponents for the approximating periodic orbits.

28 Several of the scaling exponents reported in this section were found independently by Peyrard and Aubry (1983).

Table 4.5.1.1: Various features of the golden orbit in the standard map at several parameter values. The position is the y-value where the golden orbit crosses $\theta = \frac{1}{2}$, and N is the number of iterations before I fell off the orbit. ΔW has been rescaled by some factor, and λ has been multiplied by 144.

parameter value	ΔW	half-gap	position	N	λ
.97163540631	0.	0.	.6647581313465286	2000	.41
.97574187231505	1.38d-5	6.500d-3	.6650001017315	609	.96
.97656316559606	2.42d-5	7.400d-3	.665048311666	594	.82
.97738445887707	3.9 d-5	8.2557d-3	.665096482224	510	1.11
.97820575215808	6.0 d-5	9.0756d-3	.66514463265	466	1.23
.97984833872011	1.23d-4	0.010629	.665240966596	422	1.53
.98149092528213	2.17d-4	0.0120928	.665337509008	404	1.95
.98313351184415	3.30d-4	0.0134845	.6654344342657	384	2.34
.98477609840617	5.0 d-4	0.014816	.6655318669061	376	2.73
.98641868496819	7.39d-4	0.0161000441	.66562987822396	369	2.96
.98806127153021	1.0 d-3	0.0173396	.6657284953140106	365	3.29
.98960385809223	1.33d-3	0.0185	.6658277122596523	361	3.53
.9913464465425	1.74d-3	0.0196	.66592749941466	356	3.80

§4.5.2 **Robustness of noble circles.** Now I will give some reasons for having concentrated so heavily on nobles. One of the most significant features of figures 4.4.1.1 and 4.5.1.1 is that in the first (the critical case), the noble circle appears to be (locally) the only circle, and in the second (supercritical case) there appear to be no circles at all. The large region of dots all belong to one orbit: Note that it took careful searching to find an orbit that had points on both sides of the picture without running off one edge or the other for ever. This is to be expected. For example, Vivaldi and I calculated that in the critical case, the probability rate (to the extent that it makes sense) for a transition closer to the critical circle is $\gamma^2 = 2.618$ less than for a transition away from it.[29]

As further evidence, I measured residues and differences in actions for other periodic orbits than the convergents of the noble. Given a noble $[\underline{a},(1,)^\infty]$, where $\underline{a}$ is a finite sequence of integers, consider the periodic orbits with rotation number $[\underline{a},(1,)^n\,\underline{b}]$, for finite

[29] For more on this idea, see MacKay, Meiss & Percival (1984).

sequences $\underline{b}$. Figures 4.5.2.1, 4.5.2.2 show the residues $R^{+}_{[\underline{b}]}$ in the limit as $n \to \infty$, in the critical case. The second figure is a detail of part of the first. The residues are plotted against position of the periodic point on the dominant line. I used the natural ambiguity

$$| b_0,\dots,b_m + 1] = [b_0,\dots,b_m, 1] \tag{4.5.2.1}$$

to group the points into a tree which branches two ways at each point. The point to notice is that they are all bounded away from 0 (assuming that one can extrapolate the trends). For a smooth Diophantine circle, however, the residues for its convergent periodic orbits must tend to zero. Thus a critical noble circle has a neighbourhood containing no smooth Diophantine circles. Note that the figures show that there are only 15 families $[(1,)^n,\underline{b}]$ of elliptic Birkhoff orbits.

The universal fractal diagram of figure 4.5.1.6 indicated that the residues of all the Birkhoff orbits increase with μ, so there are no smooth Diophantine circles in the supercritcal case either. Incidentally, it is not true that all the periodic orbits have residue increasing with μ, because when the Birkhoff orbits period double, they will typically have a period doubling sequence governed by the universal one parameter family for the period

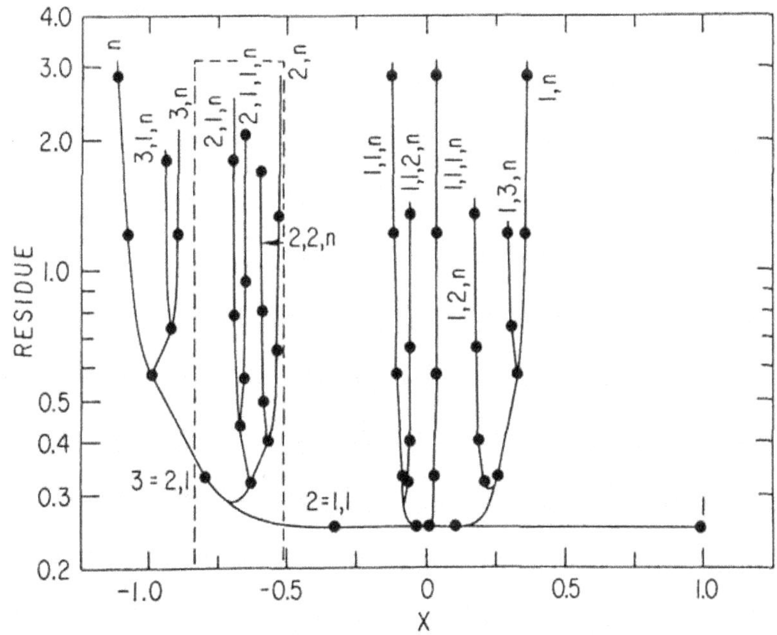

Figure 4.5.2.1: Residues of various periodic orbits of F^*

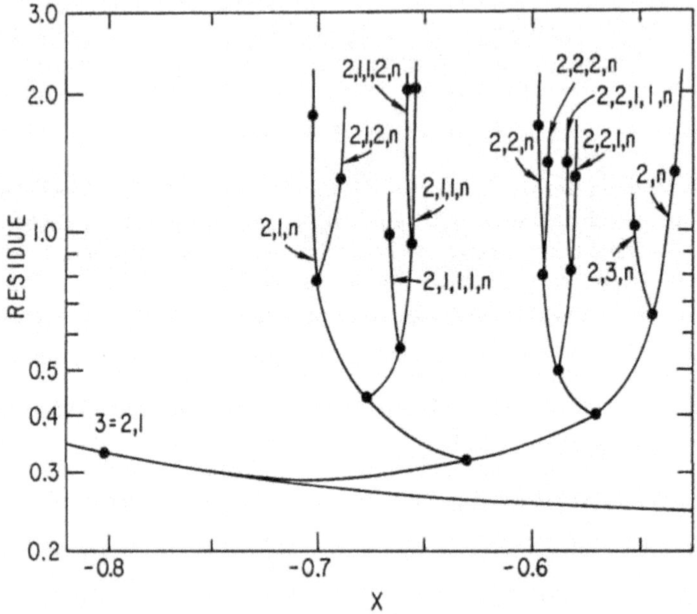

Figure 4.5.2.2: Detail of part of figure 4.5.2.1

doubling fixed point of §3.3. This has orbits whose residue first increases and then decreases (e.g. branches *j*, *k*, *l* of figure 3.3.2.3).

In the subcritical case, of course, we expect to have a smooth noble circle. Another corollary (Mather, private communication) of Moser's twist theorem is that a smooth Diophantine circle has others arbitrarily close. In fact each smooth Diophantine circle is a density point in the set of smooth Diophantine circles. So there are lots of circles in the subcritical case.

Next I consider differences in actions. Figure 4.5.2.3 shows $(\alpha\beta)^n \Delta W_{[\underline{a},(1,\,)^n \underline{b}]}$ in the limit as $n \to \infty$, for various $\underline{b}$ in the critical case. They are plotted against position of the dominant point, again. Apart from the sequence $[(1,\,)^m]$, converging to γ, I plotted points only for $\underline{b}$ with $b_0 > 1$, as the self-similarity allows one to fill in for $b_0 = 1$. They all appear to be bounded away from 0, so using Mather's result (§1.3.2.2), there are no invariant circles with irrational rotation number with continued fraction expansion $[\underline{a}, (1,\,)^n b_0, \ldots]$, $b_0 > 1$. Thus there are no irrational circles apart from the noble $[\underline{a}, (1,\,)^\infty]$. Assuming a conjectured extension of Mather's work to rational circles, there are no rational circles either. The detailed results on residues and differences in actions are given in table 4.5.2.1. Each entry is a limit extrapolated from about 10 levels.

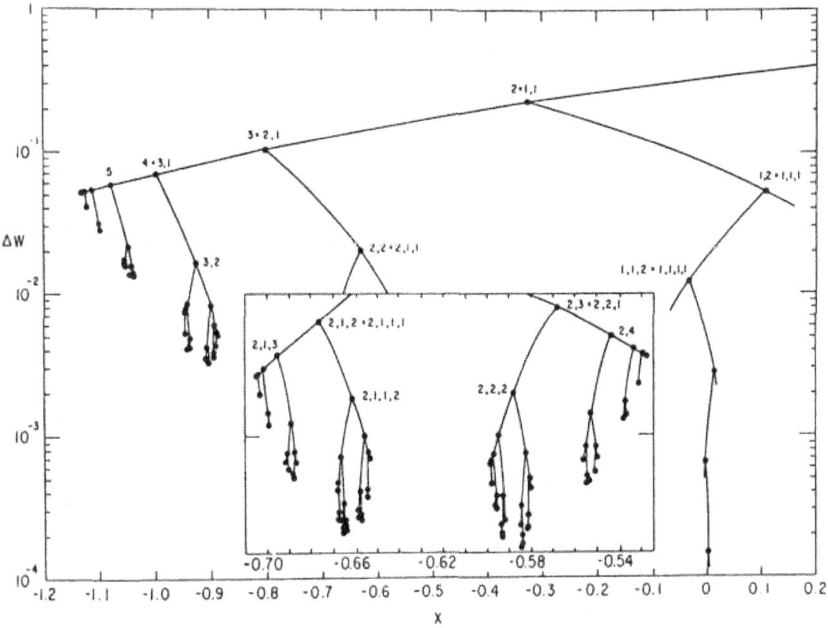

Figure 4.5.2.3: Differences in actions for various rational rotation numbers

Thus I conclude that noble circles are robust in an important sense, namely, a critical noble circle has a neighbourhood containing no other invariant circles, and all narrow enough connected neighbourhoods of a supercritical noble (Cantor set) contain no invariant circles at all. Since nobles are dense, one would like to conjecture a stronger result, namely, that isolated circles are typically nobles,[30] but deducing this from the previous statement would require estimates on the sizes of the neighbourhoods, which we do not have at present. I could look at residues near a critical [$(2,)^{\infty}$] circle, say, to see whether this conjecture has any chance. The conjecture would lead to very strong results. One would have to look for only noble circles to determine whether there are any circles at all (cf. §4.6.2). Also it would imply that typically the circles bounding the Birkhoff zones (§1.4.5) are critical nobles.[31]

30 See MacKay & Stark (1992) for progress on this question.

31 This is not correct deduction; see Greene et al (1986) for details.

Table 4.5.2.1: Position X of the dominant periodic point, residues $R^{\pm}$ and ΔW, for a selection of Birkhoff orbits for the universal map F^*.

rotation number	X	R^+	R^-	ΔW
1	1.0	0.250088	-0.25543	1.0
2=11	-.326063	0.250088	-0.25543	.230460
27	-.5280	7.745	-7.975	3.65 d-3
26=251	-.52965	3.110	-3.200	3.79 d-3
252	-.5312	16.751	-20.266	2.35 d-3
25=241	-.5337	1.339	-1.377	4.15 d-3
2412	-.5369	20.47	-25.63	1.414d-3
242=2411	-.5375	3.4802	-3.997	1.731d-3
243	-.5383	11.375	-14.065	-1.32 d-3
24=231	-.5438	.6579	-.6756	5.15 d-3
2313	-.5500	10.27	-11.935	7.06 d-4
2312=23111	-.5506	2.857	-3.221	8.334d-4
23112	-.5511	12.35	-15.78	5.69 d-4
232=2311	-.5527	1.0194	-1.093	1.46 d-3
2322=23211	-.55447	8.507	.10.54	5.24 d-4
23212	-.5546	77.99	?	4.66 d-4
2323	-.5547	45.16	59.36	4.76 d-4
233=2321	-.555	2.027	-2.225	8.42 d-4
234	-.5560	5.166	-5.755	6.78 d-4
23=221	-.5677	.3991	-.4097	8.17 d-3
2214	-.5801	4.819	-5.193	4.297d-4
2213=22121	-.5807	1.775	-1.898	5.02 d-4
221212	-.5812	45.97	-61.09	2.32 d-4
22122=221211	-.5813	5.825	-7.029	2.68 d-4
22123	-.5814	24.47	-31.7	2.29 d-4
2212=22111	-.5825	.8151	-.8618	7.61 d-4
221112	-.5838	5.629	-6.765	1.97 d-4
22112=221111	-.5843	1.642	-1.810	3.22 d-4
22113	-.5848	4.35	-4.938	2.34 d-4
222=2211	-.589	.4945	-.5145	2.02 d-3
2223	-.5922	3.976	-4.526	2.599d-4
2222=22211	-.5928	1.561	-1.718	3.78 d-4
222112	-.5932	36.50	-49.3	1.93 d-4
22212=222111	-.5933	5.322	-6.321	2.379d-4
22213	-.5934	23.06	-28.5	2.12 d-4
223=2221	-.5949	.8003	-.8368	1.02 d-3
2233	-.5964	21.72	-27.90	3.13 d-4
2232=22311	-.5965	5.372	-6.391	3.74 d-4
22312	-.5967	40.05	-52.15	3.23 d-4
224=2231	-.5974	1.684	-1.768	7.50 d-4
2242	-.59797	29.08	-36.02	4.65 d-4
225=2241	-.5983	4.233	-4.460	6.636d-4
226	-.5986	11.71	-12.36	6.35 d-4

Table 4.5.2.1 (continued)

22=211	-.631	.31915	-.3279	.020085
2115	-.6534	5.260	-5.563	6.94 d-4
2114=21131	-.6541	2.050	-2.161	7.707d-4
211312	-.6547	70.6	?	3.80 d-4
21132=211311	-.6548	7.74	-9.39	4.22 d-4
21133	-.6549	36.7	-48.2	3.69 d-4
2113=21121	-.656	.9410	-.9876	1.00 d-3
211213	-.6575	42.39	-54.13	2.58 d-4
211212=2112111	-.6576	8.38	-10.26	2.79 d-4
2112112	-.6577	77.87	?	?
21122=211211	-.6580	2.067	-2.317	4.04 d-4
21123	-.6584	5.901	-6.916	2.94 d-4
21124	-.6586	20.72	-24.84	2.67 d-4
2112=21111	-.66162	.5581	-.58043	1.844d-3
211114	-.6647	24.34	-28.89	2.40 d-4
211113=2111121	-.6648	6.750	-7.87	2.588d-4
2111122	-.66486	66.16	?	2.17 d-4
211112=2111111	-.6651	2.249	-2.53	3.32 d-4
21111112	-.6654	96.2	?	?
2111112=21111111	-.6655	9.41	-11.76	2.26 d-4
2111113	-.6655	49.5	-64.9	2.08 d-4
21112=211111	-.66653	.983	-1.05	7.06 d-4
211123	-.6674	46.7	-62.7	2.59 d-4
211122=2111211	-.6675	9.22	-11.47	2.88 d-4
2111212	-.6676	100.	-130.	2.63 d-4
21113=211121	-.6679	2.2998	-2.495	4.800d-4
21114	-.6684	6.68	-7.32	4.18 d-4
212=2111	-.677	.43481	-.44931	6.42 d-3
2124	-.6873	8.28	-9.48	6.55 d-4
2123=21221	-.6878	2.954	-3.319	7.75 d-4
212212	-.6882	226.0	?	5.11 d-4
21222=212211	-.6882	16.54	-21.3	5.46 d-4
21223	-.6883	114.5	-155.3	5.17 d-4
2122=21211	-.6894	1.335	-1.456	1.22 d-3
212112	-.6906	25.77	-34.33	5.88 d-4
21212=212111	-.6909	4.453	-5.188	7.64 d-4
21213	-.6912	18.87	-22.79	6.77 d-4
213=2121	-.6955	.7846	-.814	3.74 d-3
2133	-.6992	22.13	-28.52	1.21 d-3
2132	-.6996	5.615	-6.669	1.460d-3
214=2131	-.7019	1.799	-1.873	2.98 d-3
2142	-.7034	35.04	-43.0	1.948d-3
215=2141	-.7041	4.787	-4.995	2.73 d-3
216	-.7049	13.72	-14.33	2.65 d-3
3=21	-.802	.33142	-.33817	.10409

Table 4.5.2.1 (continued)

36	-.889	18.43	-20.05	5.21d-3
35=341	-.890	6.55	-7.11	5.44d-3
342	-.8910	70.7	?	4.34d-3
34=331	-.893	2.571	-2.771	6.14d-3
3312	-.8953	147.6	-130.	?
332=3311	-.8956	12.67	-15.85	3.89d-3
333	-.8960	77.48	-103.6	3.59d-3
33=321	-.9010	1.2069	-1.290	8.36d-3
3212	-.9065	18.93	-24.60	3.26d-3
322=3211	-.9079	3.582	-4.169	4.23d-3
323	-.9094	13.72	-16.83	3.51d-3
32=311	-.926	.73614	-.77732	.01673
3113	-.9358	21.59	-27.62	4.24d-3
3112=31111	-.9369	5.469	-6.607	4.92d-3
31112	-.9378	43.37	-58.74	4.20d-3
312=3111	-.942	1.789	-1.956	8.65d-3
3122	-.9455	42.52	-55.60	5.37d-3
313=3121	-.947	5.62	-6.29	6.96d-3
314	-.948	20.81	-23.54	6.54d-3
315	-.9485	82.5	-93.7	6.44d-3
4=31	-.99742	.5740	-.586	.0696
45	-1.0393	105.6	-133.9	.0138
44	-1.0398	27.22	-34.05	.0142
43=421	-1.0417	7.77	-9.39	.0157
422	-1.043	94.5	-128.2	.0138
42=411	-1.048	2.731	-3.086	.0219
4112	-1.0535	178.6	-183.2	.0158
412=4111	-1.0545	14.41	-17.7	.0173
413	-1.0554	90.5	-114.3	.0166
5=41	-1.079	1.208	-1.234	.0582
53	-1.099	80.8	-105.2	.0289
52=511	-1.100	13.82	-16.60	.0318
512	-1.102	173.4	-220.	.0302
6=51	-1.113	2.858	-2.92	.0539
62	-1.122	83.09	?	.0412
7=61	-1.126	7.18	-7.34	.0523
8	-1.131	18.58	-19.00	.0516

Note, however, that there are special examples where this conjecture is false. For example, de la Llave has an example with a circle of rotation number 0, for which there are arbitrarily small perturbations with no invariant circles at all (referred to by Mather, 1982a). The example is of second difference form (1.3.1.5.5). The idea is to saturate the bounds (1.3.1.5.7) of Mather's result on non-existence of invariant circles. An invariant circle can be specified by the induced map on θ:

$$\theta \to g(\theta). \tag{4.5.2.2}$$

In the proof of Mather's result, the bounds are saturated if $g^{-1\prime}$ reaches its maximum at the same point as g' has its minimum. This is the case for the example of figure 4.5.2.4(a).[32]

[32] This example has been studied in further detail by Bullett (1986).

It has rotation number 0, since all points of $(0, 1]$ converge to 1 under forward iteration. Working backwards, this choice of g requires h to be as in figure 4.5.2.4(b). This indeed saturates the bound (1.3.1.5.7). Thus any perturbation of h which breaks the bound has no invariant circles. Note that if you want the invariant circle as the graph of a function, then introduce vertical coordinate:

$$z_n = \theta_n - \theta_{n-1}. \tag{4.5.2.3}$$

Then the circle is:

$$z = \theta - g^{-1}(\theta) \tag{4.5.2.4}$$

as shown in figure 4.5.2.4(c).[33]

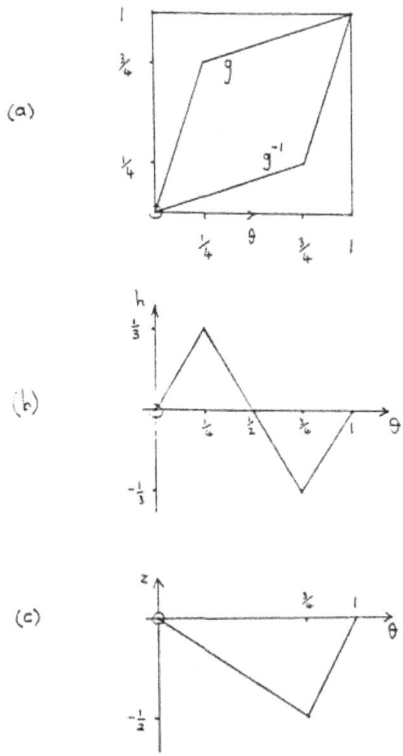

Figure 4.5.2.4: Example of a twist map with a circle of rational rotation number, such that there are nearby perturbations with no circles at all

33 Note that, given $h(\theta)$, the equation to solve for $g(\theta)$ is $g(\theta) + g^{-1}(\theta) - 2\theta = h(\theta)$.

§4.5.3 **Asymptotic behaviour for supercritical nobles.** The asymptotic behaviour under N_1 in the critical case is given by a fixed point (§4.4). Furthermore, subcritical cases appear to converge to the simple fixed point. For example, measurement of how things scale in a subcritical case give scaling along the symmetry line, of $-\gamma^{-2}$, and across it, of $-\gamma^{-1}$. The residues and differences in actions go to zero like $c_1 c_2^{\gamma^n}$, $c_2 < 1$. Put otherwise, their logarithms go to $-\infty$ asymptotically geometrically at rate γ.[34] One finds similar behaviour on subdominant half-lines.

There is also asymptotic behaviour in the supercritical case, namely,

$$X_n \to X^* \text{ like } (-)^n c_1 c_2^{g^n},\ g \sim 1.618 \tag{4.5.3.1}$$

$$(-)^n Y_n \to c_3 \tag{4.5.3.2}$$

$$R_n^{\pm} \sim c_4^{\pm} c_5^{h_{\pm}^n},\ h_{\pm} \sim 1.618 \tag{4.5.3.3}$$

$$\Delta W_n \to c_6 \text{ like } c_7 \eta^{-n},\ \eta \sim 3.30 \tag{4.5.3.4}$$

$$\log B_n \sim c_8 \kappa^n,\ \kappa \sim 1.618. \tag{4.5.3.5}$$

Here X_n is the position of the dominant periodic point of type (p_n, q_n), Y_n is the Y-coordinate of the nearest point of the orbit to the dominant point, $R_n^{\pm}$ are the residues of the Birkhoff orbits,[35] ΔW_n is the difference of their actions, and B_n is an off-diagonal component of the tangent map to the dominant point, in symmetry coordinates.[36]

So in some sense there is an attracting fixed point with infinite actions and residues, and $\beta = -\infty$, $\alpha = -1$. I have not figured out how to make sense of it.[37] The above quantities are not too easy to measure because periodic points of large residue are hard to find, mainly because there are so many tangent bifurcations that it is difficult to know which periodic orbit to take (cf. §1.2.3.4).[38]

---

34 This was observed by Greene (1979a), who called c_2 the "mean residue".

35 Meiss found that $c_4^-/c_4^+ \to -1.434$ (MacKay, Meiss & Percival, 1984).

36 The number c_5 is Greene's "mean residue" again, provided $h_{\pm}$ are really γ.

37 The answer probably lies in the "anti-integrable" limit. (Aubry and Abramovici (1991), Veerman and Tangerman (1991)). Kadanoff (1983) studied supercritical scaling in circle maps.

38 This problem can be alleviated by using continuation from the anti-integrable limit, e.g. MacKay and Baesens (1993).

§4.6 DISCUSSION

The results of §4.5 inspire a conjectured picture for the global action of the renormalisation in class AA, in which the stable manifold of the critical fixed point separates maps with a smooth golden curve from those with no golden curve. I discuss an approximate criterion for existence of noble circles, based on properties of the universal one parameter family. I also discuss some possible extensions of the work of this chapter to rotation numbers other than nobles, to dissipative systems, and to higher dimensions.

§4.6.1 **Renormalisation picture**. The results on the universal one parameter family lead me to speculate a picture like figure 4.6.1.1, for the action of N_1 in class AA (modulo coordinate changes). The critical fixed point has a codimension 1 stable manifold W^s, which, I believe, separates the space, at least locally, into pairs with a smooth golden curve and those with no golden curve. Proof of this picture would constitute a global KAM theorem.[39]

Furthermore, any one parameter family crossing W^s transversally (in which I include non-zero speed) will have asymptotically the same behaviour, on a small enough scale in space and parameter, and on a long enough timescale, as the "universal" one parameter family given by a natural parametrisation of the one dimensional unstable manifold W^u.

Note that I expect the set of maps with zero linear twist to give part of the boundary of the domains of attraction for the fixed points. There are probably other fixed points on this surface, corresponding to quadratic twist, cubic twist and so on.[40]

§4.6.2 **Criterion for existence of invariant circles**. The results on the universal one parameter family also lead to an approximate criterion for existence of invariant circles. This is a slight extension of a criterion of Greene, implicit in Greene (1979), and used, for example, by Schmidt (1980) and Schmidt and Bialek (1982).

The critical residue in the universal one parameter family is $R^* = 0.2500888$ (4.4.1.15). Thus an approximate criterion would be that a noble circle exists or not according as the residue of one of the approximant periodic orbits, is significantly less/greater than R^*. A better criterion is to take the average of the residues of two successive approximant periodic

[39] Existence of a smooth golden curve for all systems on one side of the stable manifold would follow from entry of one branch of the unstable manifold into a neighbourhood of attraction of the simple fixed point. Non-existence on the other side would follow from entry of the other branch of the unstable manifold into a region where a non-existence criterion like that of Mather (1982a) can be applied.

[40] I have not found any of these yet, though have evidence for their existence when the partial quotients of the rotation number are large enough (MacKay, 1992a). What is more likely to be at the boundary of the critical stable manifold are invariant Cantor sets under renormalisation connected with systems that have $\mathbf{Z}_n$ symmetry, $n \geq 2$ (MacKay, 1984; Ketoja & MacKay, 1989).

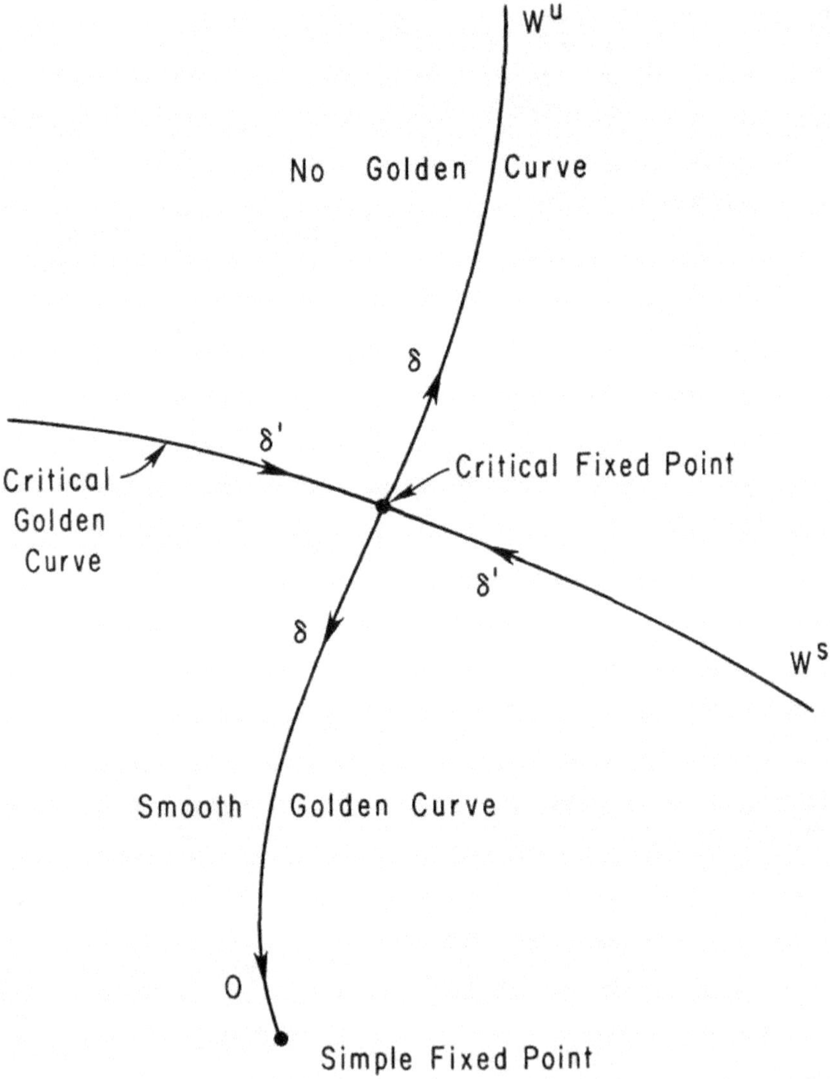

Figure 4.6.1.1: Schematic of the action of N_1 in the space of commuting pairs of area preserving twist maps with zero Calabi invariant (modulo coordinate changes)

orbits. In fact, since figure 4.5.1.3 shows that the dependence of residue on parameter is close to exponential, it is best to take logarithms. Furthermore, it is best to take a weighted average, to take into acount the essential convergence rate $\delta' = -0.6108$ (4.4.1.16). Lastly, since noble circles appear to be locally the most robust, if the noble circle between two island chains is broken, then there is unlikely to be any circle between them. Thus, I propose the following criterion:[41]

Given two island chains of rotation numbers p_0/q_0 and p_1/q_1, with:

$$|p_0q_1 - p_1q_0| = 1 \tag{4.6.2.1}$$

whose minimaximizing orbits have residues $R_1 \leq R_0$, then there is an invariant circle/no invariant circle between the island chains, according as

$$\rho = \frac{\log R_1 + 0.6108 \log R_0}{1.6108} \tag{4.6.2.2}$$

is significantly less/greater than $\log 0.2500888$.

Note that the condition (4.6.2.1) is essential to guarantee that the rotation numbers of the island chains are successive convergents of some noble. In fact, the circle given by this criterion has rotation number:

$$\frac{p_0 + \gamma p_1}{q_0 + \gamma q_1} \tag{4.6.2.3}$$

the one which one expects to be the most robust between p_0/q_0 and p_1/q_1. If R_0 and R_1 are comparable, then the rotation number given by interchanging 0 and 1 in (4.6.2.3) is probably comparably robust.

I derive the above weighting factors by approximating the behaviour of the residues near the critical fixed point by:

$$\log R_i = \log R^* + a\,\delta^i + b\,\delta'^i. \tag{4.6.2.4}$$

Apart from a change of scale, a is the value to which the map $U^{q_0}T^{p_0}$ corresponds, of the parameter μ in the universal family, and b measures how far it is from the universal family. Thus, given two successive residues, one can evaluate:

$$a = \frac{\log R_1 - \delta' \log R_0 - (1 - \delta')\log R^*}{\delta - \delta'}. \tag{4.6.2.5}$$

There is likely to be a circle of rotation number (4.6.2.3) or not according as a is negative or positive. The only other part of the criterion to explain is the choice of order. Evaluating

---

[41] See, however, the discussion of this in MacKay & Stark (1992).

(4.6.2.2) for the other order, gives a larger answer, so I chose the order to correspond to the more robust looking rotation number.

Now I give two examples of the use of this criterion. The first gives an approximate value for the "stochastic threshold" in the standard map. This means the critical value k_c of the parameter k, such that there is an invariant circle or not (restricting attention to the class of circles which encircle the cylinder) according as:

$$|k| \lessgtr k_c. \tag{4.6.2.6}$$

Of course, there need not exist such a k_c, as circles could come and go as k is varied. But it looks as if there is, and its value is (Greene, 1979a):[42]

$$k_c = 0.9716\ldots. \tag{4.6.2.7}$$

The standard map is also periodic in the vertical direction, so we need consider only one vertical period. The fixed point of positive residue of type (0, 1) has residue $|k|/4$, and so does that of type (1, 1). Thus the above criterion suggests that there is an invariant circle or not according as:

$$\log |k|/4 \lessgtr \log 0.2500888 \tag{4.6.2.8}$$

i.e.

$$|k| \lessgtr 1.0004. \tag{4.6.2.9}$$

This is very close to the empirical value (4.6.2.7).

Secondly, I chose a subcritical parameter value in the standard map, in fact $k = 0.9$, and evaluated the residues for a bunch of orbits. Then I used the criterion to indicate which island chains had invariant circles between them and which did not. This is shown in table 4.6.2.1.[43] If the predictions are compared with figure 4.6.2.1, which shows some orbits of this map, the criterion can be seen to be quite good. One should note, however, that when the residues for rotation numbers p_0/q_0 and p_1/q_1, say, are close to R^*, it is best to subdivide the interval by the rotation number:

$$\frac{p_0}{q_0} \oplus \frac{p_1}{q_1} \equiv \frac{p_0+p_1}{q_0+q_1}. \tag{4.6.2.10}$$

For example, the criterion would lead one to believe that there is a circle between rotation numbers 1/2 and 5/9, but on subdividing at 6/11, it says there are none. One might have to subdivide further to resolve this particular example.

[42] Compare the value 0.9716354061062 (± 4 in the last place) super-extrapolated from Table 4.4.1.4.

[43] Their rotation numbers are organised in what is now called the Farey tree.

Table 4.6.2.1: Residues of some periodic orbits in the standard map at $k = 0.9$.

rotation number	residue	is there a circle in between?
1/1	.225	
		no
6/7	2.15585	
		no
5/6	.958061	
		no
9/11	1.780549	
		no
4/5	.472615	
		no
11/14	1.011059	
		no
7/9	.506758	
		no
10/13	.687894	
		no
3/4	.2771156	
		no
11/15	.380245	
		no
8/11	.2514778	
		yes
13/18	.219597	
		yes
5/7	.2201389	
		yes
12/17	.190467	
		yes
7/10	.2243723	
		no
9/13	.303247	
		yes
2/3	.206959	
		yes
9/14	.220325	
		yes
7/11	.1516184	
		yes
12/19	.0793195	
		yes
5/8	.137211	
		yes
13/21	.048626	
		yes
8/13	.089359	
		yes
11/18	.0768797	
		yes
3/5	.164406	
		yes
10/17	.08758209	
		yes
7/12	.1074246	
		yes
11/19	.074387	
		yes
4/7	.170643	
		yes
9/16	.152655	
		yes
5/9	.234222	
		no
6/11	.402416	
		no
1/2	.2025	

§4.6.3 **Possible extensions.** From §4.2 up to this point, I have essentially restricted attention to noble rotation numbers, though I indicated in §4.5.2, that nobles are probably the most significant. One could do exactly parallel analysis for any quadratic irrational. What to do about other rotation numbers, however, is not yet clear to me.[44] I think the behaviour corresponding to convergence to the simple fixed point could be got hold of by including a vertical shift of origin in the rescaling B of (4.1.1.10) (cf. Escande, 1982a,b and Escande and Mehr, 1982).

[44] I made some progress in December 1982, however, and presented it at Dynamics Days, La Jolla in January 1983. The simple fixed point generalizes to a simple line, invariant under the renormalisation for arbitrary rotation number. It is attracting in class AA, for good enough rotation numbers (in fact, Haydn showed this later for all Diophantines). A generalisation of the critical fixed point is given in MacKay (1986), and is most clearly presented in MacKay & Stark (1992).

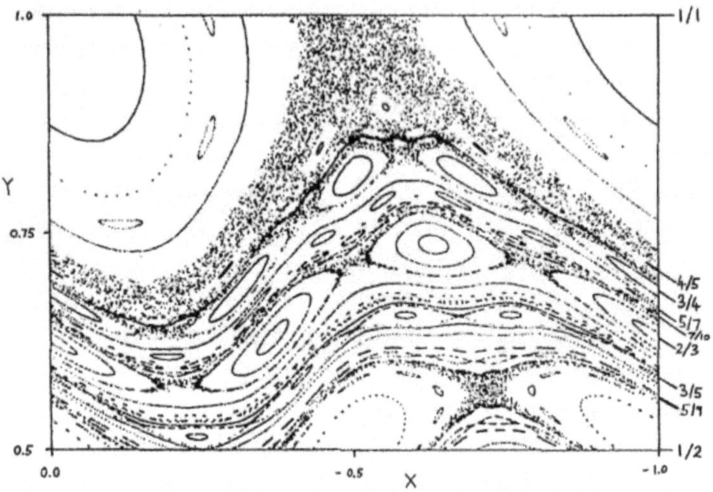

Figure 4.6.2.1: Some orbits of the standard map at $k = 0.9$

The ideas of this chapter also carry over directly to the problems of existence and breakup of invariant tori in dissipative systems.[45] The simple fixed point there is very simple, and its spectrum much simpler than in the area preserving case. Construction of a neighbourhood of attraction for the simple fixed point is easier. The critical fixed point can be easily obtained to very good accuracy, provided one takes care with the domains. This problem will be discussed in §4.7. Probably Siegel's centre problem can be approached in a similar way.[46]

Lastly, the ideas of this chapter can be extended to invariant tori of arbitrary dimension. For a $2n$-dimensional symplectic map, for example, one would consider commuting $(n+1)$-tuples of maps. The renormalisation would generalise in an obvious way. Rotation numbers would lie in $\mathbb{R}P^n$, and continued fraction expansions generalise (Khovanskii, 1963). The set corresponding to quadratic irrationals is the eigenvectors of those $n \times n$ matrices of integers whose eigenvalues are algebraics of degree n (I got this idea from Feigenbaum, who was suggesting using it for maps on the circle, but I do not see any application there). I expect there would be numbers analogous to nobles too.[47]

[45] See Rand (1992) for details.

[46] See Manton and Nauenberg (1983), Widom (1983), MacKay & Percival (1987).

[47] Kosygin (1992) and Khanin (talk given at Porto in August, 1992) have worked out this program for KAM theory for $4D$ symplectic maps. It is not as simple as I had envisaged, however, because for convergence of renormalisation are needs at least convergence of domains for the Jacobi–Perron algorithm, which requires a single expanding direction. The search for analogues of the critical fixed point, describing breakup of tori for $4D$ maps, has been fruitless, despite much searching e.g. by van Zeijts.

§4.7 RENORMALISATION FOR MAPS ON A CIRCLE

A problem closely related to that of existence of invariant circles in area preserving maps, is conjugacy of maps on a circle to rotation. The same renormalisation can be applied, and has a simple fixed point and a critical fixed point analogous to the area preserving case. The simple fixed point has one unstable direction corresponding to change in rotation number. I construct a C^3 neighbourhood of attraction for it, in the space of commuting pairs with golden rotation number. The critical fixed point has two unstable directions, one for change of rotation number, and the other corresponding to the transition from differentiable invertibility to non-invertibility. I discuss properties of the resulting two parameter family. This section contains two appendices in which I generalise Poincaré's theorem to commuting pairs, and classify irrational numbers.

§4.7.1 **Introduction**. There is a closely related problem in one dimension to that of existence of invariant circles in area preserving maps. It is the question of when a map on a circle is conjugate to rotation. As already mentioned in §1.3.1.3, there are several well-known results on this subject.

Two maps f and g are said to be *conjugate* if there is an invertible map h (coordinate change) such that:

$$hf = gh. \tag{4.7.1.1}$$

They are *topologically, differentiably*, C^r, C^∞ or C^ω conjugate according as h and its inverse are continuous, differentiable, C^r, C^∞ or analytic. f is said to be *semi-conjugate* to g if h is not required to be invertible.

Clearly, invertibility of a circle map is necessary for conjugacy to rotation, as invertibility is preserved by a coordinate change. Similarly, orientation preservation is necessary. So I will restrict attention in this section to orientation preserving invertible maps.

An orientation preserving homeomorphism of a circle has a rotation number, as discussed in §1.3.1.3. For rational rotation number p/q in lowest terms, there is always at least one periodic orbit, and all periodic orbits have type (p, q). Typically, they do not have multiplier $+1$, so they are isolated, and the map is not topologically conjugate to rotation. One can show in this case that there are an even number of periodic orbits, the periodic points are alternately attracting and repelling, and all the orbits are eventually periodic (Nitecki, 1971). Since periodic orbits with no multiplier of $+1$ persist under perturbation (§1.2.4.1), this gives rise to *frequency locking*, i.e. as one varies parameters, the set of parameter values for each rational p/q, such that the rotation number is p/q and no periodic orbit has multiplier $+1$, is open.

For irrational rotation number, Denjoy showed that provided the map is a diffeomorphism (i.e. differentiable with differentiable inverse) and its derivative has bounded variation, i.e. there exists C such that:

$$\sum_{i=1}^{n} |f'(x_i) - f'(x_{i-1})| \leq C \tag{4.7.1.2}$$

for all finite sequences $0 \leq x_0 < ... < x_n = 1 + x_0$, then it is topologically conjugate to rotation (Nitecki, 1971).[48] He also gave an example to show that C^1 is insufficient.

One might ask about smoother conjugacy. There is an analogue of the KAM theorem (e.g. Herman, 1977, 1979):

Theorem (Arnold, Moser, Rüssmann): Given $\varepsilon > 0$, $\exists\, C^{2+\varepsilon}$–neighbourhood V of uniform rotation R_α, such that for $f \in C^r, 2 + \varepsilon \leq r \leq \omega, f \in V, \rho(f) = \alpha \in$ Roth, then f is $C^{r-1-\beta}$-conjugate to R_α, $\forall \beta > 0$ (C^∞, C^ω conjugate in the cases $r = \infty, \omega$).

The Roth numbers are defined in §4.7.8. This gives conjugacy to rotation for maps close enough to rotation. Herman (1977) has proved a global result for a restricted set of rotation numbers, which nevertheless has full measure:

Theorem (Herman): For $3 \leq r \leq \omega, f$ an orientation preserving C^r diffeomorphism of the circle, with rotation number α satisfying "condition A", then the same conclusion holds as in the previous theorem.

Condition A is defined in §4.7.8.[49]

The renormalisation approach I used for invariant circles in area preserving maps can be carried over directly to this problem. As I will again restrict myself to noble rotation numbers (which satisfy condition A), the renormalisation analysis will not have anything to add to Herman's theorem, but is interesting neverthless.

§4.7.2 **Fixed point analysis.** The same motivation as in §4.1.1 leads one to consider the renormalisation:

$$\begin{aligned} N_m&: U' = BTB^{-1} \\ &T' = BTUB^{-1} \end{aligned} \tag{4.7.2.1}$$

this time on commuting pairs of maps on the line. Restricting attention to nobles, I will consider only N_1.

48 A renormalisation proof is given in MacKay (1988).

49 Both these results have subsequently been generalised to much larger sets of rotation numbers by Yoccoz.

N_1 leaves invariant several important spaces:

i) commuting pairs
ii) commuting pairs for which 0 has golden rotation number
iii) invertible pairs
iv) diffeomorphisms
v) coordinate transforms of a fixed point
vi) cubic maps
vii) all intersections of the above

The simplest behaviour under N_1 is convergence to a fixed point. There is a simple fixed point and a critical one, which will be discussed in the next sections. I do not know an analogue of Mather's theorem, however, which would allow me to deduce something from convergence to a fixed point.[50]

The stability of a fixed point can be analysed as in §4.2.1. First I discuss rotation number. If 0 has rotation number ω under (U, T), then it has rotation number:

$$\omega' = \frac{1}{\omega - 1} \tag{4.7.2.2}$$

under (U', T') (§4.1.3). This has two fixed points γ, $-1/\gamma$. I will consider only the first case. A deviation in ω from one of these values changes under the renormalisation by a factor:

$$\frac{d\,\omega'}{d\,\omega} = -\frac{1}{(\omega - 1)^2} = -\gamma^2. \tag{4.7.2.3}$$

So we expect one unstable direction with eigenvalue $-\gamma^2$, corrsponding to change of rotation number.

Non-commuting directions can be analysed exactly as in §4.2.4, giving non-commuting eigenvalues $-\alpha^{1-p}$. The only difference is that $D(B^{-1}T)_0$ (4.2.4.15) need not be invertible. In fact, it is zero at the critical fixed point. This is because the critical fixed point is a function of x^3 only (see §4.7.5). One can modify the analysis as follows. Write:

$$T(x) = t(x^3). \tag{4.7.2.4}$$

Then

[50] The deductions one would hope to make are transitivity or smooth conjugacy.

$$(T'U' - U'T')x = Bt\left((UTB^{-1}x)^3\right) - Bt\left((TUB^{-1}x)^3\right)$$

$$= 3B\, t'\left((UTB^{-1}x)^3\right).(UTB^{-1}x)^2.(UT - TU)B^{-1}x \qquad (4.7.2.5)$$

Compare

$$\frac{d}{d(x^3)} BTUTB^{-1}x = 3B\, t'\left(UTB^{-1}x)^3\right).(UTB^{-1}x)^2 \frac{d}{d(x^3)} UTB^{-1}x. \qquad (4.7.2.6)$$

The left hand side is $\frac{d}{d(x^3)} UTx$, which is non-zero, as $UT = B^{-1}TB$. Thus

$$3B\, t'\left((UTB^{-1}x)^3\right).(UTB^{-1}x)^2 |_{x=0} = (B_0')^3 = \alpha^3. \qquad (4.7.2.7)$$

So a commutator of x^p grows like $-\alpha^{3-p}$, plus terms of higher degree.

The coordinate changes are the same, giving eigenvalues α^{1-p}, but there is one new eigenvalue α^2 in the cubic case, which corresponds in a sense to the singular coordinate change:[51]

$$\delta\sigma(x) = \frac{1}{x}. \qquad (4.7.2.8)$$

This was figured out by Kadanoff (Feigenbaum et al., 1982).

§4.7.3 **Simple fixed point.** N_1 has a simple fixed point:

$$U(x) = x + 1$$
$$T(x) = x - \frac{1}{\gamma} \qquad (4.7.3.1)$$
$$B(x) = -\gamma x.$$

Its stability is given by:

$$\delta U'(x) = -\gamma\, \delta T\left(-\frac{x}{\gamma}\right)$$

$$(4.7.3.2)$$

$$\delta T'(x) = -\gamma\, \delta T\left(-\frac{x}{\gamma} + 1\right) - \gamma\, \delta U\left(-\frac{x}{\gamma}\right)$$

[51] And has the effect of unfolding the cubic critical point.

with additions for variation of B with (U, T). Taking powers of x as basis, this has block upper triangular form, with 2×2 diagonal blocks:

$$\begin{vmatrix} 0 & 1 \\ 1 & 1 \end{vmatrix} \times (-\gamma)^{1-p}. \tag{4.7.3.3}$$

Hence eigenvalues $1, -\gamma^2 \times (-\gamma)^{-p}$, with eigenvectors:

$$\underset{A_p}{\begin{vmatrix} \gamma \\ -1 \end{vmatrix}}, \underset{B_p}{\begin{vmatrix} 1 \\ \gamma \end{vmatrix}} \tag{4.7.3.4}$$

Note that all the eigenvalues are simple, so the Jordan normal form is diagonal, and the eigenvectors are uniquely determined (up to scale). The A_p are all coordinate changes, because $\delta\sigma(x) = x^p, p \neq 0$ produces:

$$\delta U(x) = (x+1)^p - x^p = px^{p-1}$$
$$\delta T(x) = \left(x - \frac{1}{\gamma}\right)^p - x^p = -\frac{p}{\gamma}x^{p-1} \tag{4.7.3.5}$$

which is A_{p-1}. The case $p = 0$ produces the zero vector, corresponding to translational invariance of the simple fixed point.

The $B_p, p > 1$ are in non-commuting directions. The commutation condition is:

$$\delta T\, U + DT.\delta U = \delta U\, T + DU.\delta T \tag{4.7.3.6}$$

i.e.

$$\delta T(x+1) + \delta U(x) = \delta U(x - \frac{1}{\gamma}) + \delta T(x). \tag{4.7.3.7}$$

Since x^j does not contribute to (4.7.3.7) at its own degree, we can evaluate the amount of x^{p-1} due to B_p from its terms of maximal degree only. For $p \neq 0$, one sees that the commutation condition fails.

Lastly, B_0 corresponds to change of rotation number. This is because the perturbed map it gives is:

$$U(x) = x + 1 + \varepsilon$$
$$T(x) = x - \frac{1}{\gamma} + \varepsilon\gamma \tag{4.7.3.8}$$

which has rotation number:

$$\frac{1+\varepsilon}{\frac{1}{\gamma} - \varepsilon\gamma} = \gamma + \varepsilon\gamma(1+\gamma^2). \tag{4.7.3.9}$$

The diagonalisation of DN_1 at the simple fixed point was also figured out by Feigenbaum et al. (1982).

§4.7.4 **A C^3-neighbourhood of attraction for the simple fixed point.** Diagonalisation of DN_1 at the simple fixed point, as in the previous section, shows that in the *subspace* C of commuting pairs with golden rotation number, the simple fixed point is attracting. One might like to know how large a neighbourhood it attracts. In this section, I derive a C^3-neighbourhood of attraction. Of course, Herman's results (1977) show that all periodic C^3-diffeomorphisms with golden rotation number are attracted to the simple fixed point, so this will not be a new result (except for the generalisation to commuting pairs). My neighbourhood will be quite small. The purpose of this section is not to get the best possible result, but just to show how this sort of proof works, and to provide a warmup for the analogous problem in the area preserving case. Unfortunately, I did not work out the area preserving case in time to include it in this thesis,[52] but I expect to obtain stronger results than any known at present.

The main complication in deriving a neighbourhood of attraction for the simple fixed point, is that it is not attracting in the whole space. There are three bad directions in which DN_1 is not contracting, corresponding to change of rotation number, and two measures of non-commutation. I will deal with these bad directions by showing that in a neighbourhood B, the points of G are restricted to a cone C, with vertex at the simple fixed point (figure 4.7.4.1). Thus, restricted to G, the bad directions can not grow.

[52] And still have not done so!

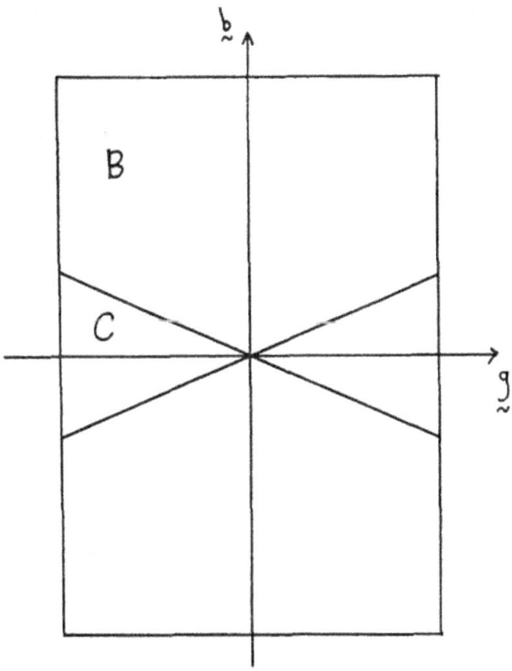

Figure 4.7.4.1: Box B and cone C

Here is the plan for this section:

(i) Define domains for the renormalisation
(ii) Choose coordinates for the space, corresponding to the adjoint eigenvectors of DN_1. Label the coordinates in the bad directions by b_0, b_1, b_2, and the good directions by g.
(iii) Take a box B around the simple fixed point, and show that points in $B \cap G$ must lie in a cone C (figure 4.7.4.1), of the form:

$$|b_i| \leq n_i(g), \quad i = 1,2,3 \tag{4.7.4.1}$$

where n_i are three norms on the good coordinates. Note that this process can be iterated, giving a succession of smaller cones.
(iv) Show that N_1 is defined on $B \cap C$.
(v) Show that on $B \cap C, N_1$ shrinks the good directions by at least some factor. Note that one may need to reduce B in some directions to show this.

(vi) Show that points in $B \cap C$ remain in B as far as the bad directions are concerned. If not, then reduce B in the good directions by the failure factor.

Then points in $B \cap G$ remain in $B \cap G$, under N_1, and get closer to the fixed point by at least some factor. So all points of $B \cap G$ converge to the fixed point, under iteration of N_1. The estimates appearing in this section were obtained by using a calculator with 14 significant figures, and adding at least 10^{-5} or so to each result. I did not check that this was guaranteed to give true estimates. Sometime I will write a calculator program to do interval arithmetic (cf. Eckmann et al., 1982).

(i) Renormalisation

$$N_1 \colon \tilde{U}(x) = \frac{1}{\lambda} T(\lambda x)$$
$$\tilde{T}(x) = \frac{1}{\lambda} TU(\lambda x) \tag{4.7.4.2}$$

with

$$\lambda = T(0). \tag{4.7.4.3}$$

Since (4.7.4.3) forces $\tilde{U}(0) = 1$, I will work in the space with the normalisation:

$$U(0) = 1 \tag{4.7.4.4}$$

If T is defined on $[0, 1]$, and:

(a)
$$\lambda = T(0) < 0 \tag{4.7.4.5}$$

and U is defined on $[\lambda, 0]$, then the following conditions guarantee that $\tilde{T}$ is defined on $[0, 1]$, $\tilde{\lambda} = \tilde{T}(0) < 0$, and $\tilde{U}$ is defined on $[\tilde{\lambda}, 0]$ (see figure 4.7.4.2):

(b)
$$T(x) \geq \lambda \text{ on } [0, 1] \tag{4.7.4.6}$$

(c)
$$T(x) \leq 1 \text{ on } [0, 1] \tag{4.7.4.7}$$

(d)
$$T(1) > 0 \tag{4.7.4.8}$$

(e)
$$U(x) \geq 0 \text{ on } [\lambda, 0] \tag{4.7.4.9}$$

(f)
$$U(x) \leq 1 \text{ on } [\lambda, 0]. \tag{4.7.4.10}$$

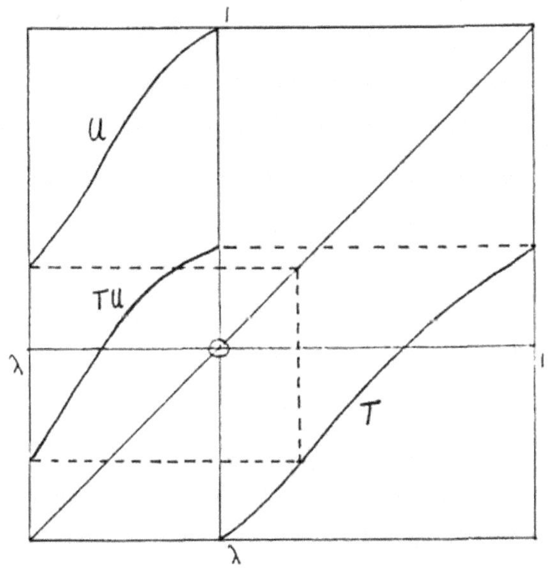

Figure 4.7.4.2: Composition of T and U

Actually, if T and U commute at 0, then

$$U(\lambda) = T(1) \tag{4.7.4.11}$$

which is positive by (d), so that (e) could be replaced by:

(e′) $$U'(x) \geq 0 \text{ on } [\lambda, 0]. \tag{4.7.4.12}$$

(ii) Choose coordinates

N_1 has the simple fixed point:

$$U(x) = 1 + x$$
$$T(x) = -\frac{1}{\gamma} + x. \tag{4.7.4.13}$$

For nearby C^3 maps with $U(0) = 1$, write:

$$T(0) = \lambda = -\frac{1}{\gamma} + \Delta T_0$$

$$U'(0) = U_1 = 1 + \Delta U_1$$

$$T'(0) = T_1 = 1 + \Delta T_1$$

$$U''(0) = U_2$$

$$T''(0) = T_2$$

$$U'''_{mx} = \max_{[\lambda,\, 0]} |U'''(x)|$$

$$T'''_{mx} = \max_{[0,\, 1]} |T'''(x)| \tag{4.7.4.14}$$

Then a point is specified by $\Delta T_0, \Delta U_1, \Delta T_1, U_2, T_2, U'''(x), T'''(x)$.

Truncated to degree 2, the adjoint eigenvectors (in my opinion, beter called eigenforms) of DN_1 at the simple fixed point are:

$$b_0 = \Delta T_0 - \frac{1}{2\gamma^2}\,\Delta U_1 + \frac{1}{2}\Delta T_1 + \frac{1}{6\gamma^3}\,U_2 + \frac{1}{6}T_2$$

$$b_1 = -\frac{1}{\gamma}\Delta U_1 - \Delta T_1 + \frac{1}{2\gamma^2}\,U_2 - \frac{1}{2}T_2$$

$$g_1 = \Delta U_1 - \frac{1}{\gamma}\,\Delta T_1 - U_2 - T_2$$

$$b_2 = -\frac{1}{\gamma}U_2 - T_2$$

$$g_2 = U_2 - \frac{1}{\gamma}T_2. \tag{4.7.4.15}$$

They measure perturbations from the simple fixed point in the following directions:

b_0: change of rotation number

b_1: $UT(0) \neq TU(0)$

g_1: coordinate changes $x \to x + \varepsilon x^2$

b_2: $(UT)'(0) \neq (TU)'(0)$

g_2: coordinate change $x \to x + \varepsilon x^3$.

I will use b_0, b_1, g_1, b_2, g_2 as coordiantes to replace $\Delta T_0, \Delta U_1, \Delta T_1, U_2, T_2$. The point is that to first order, the effect of N_1 is given by:

$$\begin{aligned}\tilde{b}_0 &= -\gamma^2 b_0\\ \tilde{b}_1 &= \gamma b_1\\ \tilde{b}_2 &= -b_2\\ \tilde{g}_1 &= -\gamma^{-1} g_1\\ \tilde{g}_2 &= \gamma^{-2} g_2\end{aligned} \tag{4.7.4.16}$$

and all higher order coefficients decay at least as fast as γ^{-1}. Near enough to the fixed point, the second order terms can be bounded in terms of first order terms, giving small corrections to (4.7.4.16). The bad directions will be dealt with in (iii), so this will give convergence to the fixed point for close enough points in G.

The inverse coordinate change to (4.7.4.15) is:

$$\begin{vmatrix}\Delta T_0\\ \Delta U_1\\ \Delta T_1\\ U_2\\ T_2\end{vmatrix} = \frac{1}{d}\begin{vmatrix} d & 1/\gamma^2 & 1/2 & -1/2 & 1/6\gamma^2\\ 0 & -1/\gamma & 1 & -1 & 1/2\\ 0 & -1 & -1/\gamma & 1 & 1/2\gamma^2\\ 0 & 0 & 0 & -1/\gamma & 1\\ 0 & 0 & 0 & -1 & -1/\gamma\end{vmatrix}\begin{vmatrix} b_0\\ b_1\\ g_1\\ b_2\\ g_2\end{vmatrix} \tag{4.7.4.17}$$

where

$$d = 1 + \gamma^{-2}. \tag{4.7.4.18}$$

This gives the following estimates:

$$\begin{vmatrix}|\Delta T_0|\\ |\Delta U_1|\\ |\Delta T_1|\\ |U_2|\\ |T_2|\end{vmatrix} \leq \begin{vmatrix} 1 & .277 & .362 & .362 & .047\\ 0 & .448 & .724 & .724 & .362\\ 0 & .724 & .448 & .724 & .139\\ 0 & 0 & 0 & .448 & .724\\ 0 & 0 & 0 & .724 & .448\end{vmatrix}\begin{vmatrix}|b_0|\\ |b_1|\\ |g_1|\\ |b_2|\\ |g_2|\end{vmatrix} \tag{4.7.4.19}$$

(iii) Good cone

Commuting pairs with golden rotation number lie in a restricted part of the space, which can be expressed as a cone, near the fixed point. First I look at the constraints imposed by commutation.

$$UT(0) - TU(0) = U(\lambda) - T(1) \tag{4.7.4.20}$$

$$= \lambda\, \Delta U_1 + \frac{\lambda^2}{2} U_2 + \frac{\lambda^3}{6} U'''(\xi_1) - \Delta T_1 - \frac{1}{2} T_2 - \frac{1}{6} T'''(\xi_2)$$

$$= b_1 + \Delta T_0 \Delta U_1 + \frac{1}{2}(\lambda^2 - \gamma^{-2}) U_2 + \frac{\lambda^3}{6} U'''(\xi_1) - \frac{1}{6} T'''(\xi_2)$$

for some $\xi_1 \in [\lambda, 0]$, $\xi_2 \in [0, 1]$, using Taylor's theorem. Thus commutation at 0 implies that:

$$|b_1| \leq |\Delta T_0||\Delta U_1| + \frac{1}{2}|\lambda^2 - \gamma^{-2}||U_2| + \frac{1}{6}|\lambda|^3 U'''_{mx} + \frac{1}{6} T'''_{mx}. \tag{4.7.4.21}$$

Similarly, one can look at the first derivative at 0:

$$(UT)'(0) - (TU)'(0) = U'(\lambda)T_1 - T'(1)U_1 \tag{4.7.4.22}$$

$$= \left(\lambda U_2 + \frac{\lambda^2}{2} U'''(\xi_1)\right)(1 + \Delta T_1) - \left(T_2 + \frac{1}{2} T'''(\xi_2)\right)(1 + \Delta U_1)$$

$$= b_2 - \left(T_2 + \frac{1}{2} T'''(\xi_2)\right)\Delta U_1 + \lambda\left(U_2 + \frac{\lambda}{2} U'''(\xi_1)\right)\Delta T_1 + \Delta T_0 U_2 + \frac{\lambda^2}{2} U'''(\xi_1) - \frac{1}{2} T'''(\xi_2)$$

for some ξ_1, ξ_2. So commutation implies that:

$$|b_2| \leq (|T_2| + \frac{1}{2} T'''_{mx})|\Delta U_1| + |\lambda|(|U_2| + \frac{|\lambda|}{2} U'''_{mx})|\Delta T_1| + |\Delta T_0||U_2| + \frac{\lambda^2}{2} U'''_{mx} + \frac{1}{2} T'''_{mx}. \tag{4.7.4.23}$$

Lastly, rotation number imposes a constraint. Note that the domains are large enough that if (U, T) commute then they can be extended to the whole line. Close enough to the fixed point, U moves right by approximately 1 and T moves left by approximately γ^{-1}. This is measured by:

$$d(x, x') \equiv \frac{1}{\gamma}(U(x) - x) + T(x') - x' \qquad (4.7.4.24)$$

$$= \frac{x}{\gamma}\Delta U_1 + \frac{x^2}{2\gamma}U_2 + \frac{x^3}{6\gamma}U'''(\xi_1) + \Delta T_0 + x'\Delta T_1 + \frac{x'^2}{2}T_2 + \frac{x'^3}{6}T'''(\xi_2)$$

$$= b_0 + \left(\frac{x}{\gamma} + \frac{1}{2\gamma^2}\right)\Delta U_1 + \left(\frac{x^2}{2\gamma} - \frac{1}{6\gamma^3}\right)U_2 + \frac{x^3}{6\gamma}U'''(\xi_1)$$

$$+ \left(x' - \frac{1}{2}\right)\Delta T_1 + \left(\frac{x'^2}{2} - \frac{1}{6}\right)T_2 + \frac{x'^3}{6}T'''(\xi_2).$$

If $d(x, x')$ has constant sign for x, x' in the domains, then (U, T) can not have rotation number γ. Thus golden rotation number implies that:

$$|b_0| \le \max\left(\frac{1}{2\gamma^2}, \left|\frac{\lambda}{\gamma} + \frac{1}{2\gamma^2}\right|\right)|\Delta U_1| + \frac{1}{2}|\Delta T_1| + \max\left(\frac{1}{6\gamma^3}, \left|\frac{\lambda^2}{2\gamma} - \frac{1}{6\gamma^3}\right|\right)|U_2|$$

$$+ \frac{1}{3}|T_2| + \frac{|\lambda|^3}{6\gamma}U'''_{mx} + \frac{1}{6}T'''_{mx}. \qquad (4.7.4.25)$$

Let us work in the *box B*:

$$|b_0| \le .2$$

$$|b_1| \le .1$$

$$|g_1| \le .05$$

$$|b_2| \le .1$$

$$|g_2| \le .02$$

$$U'''_{mx} \le .007$$

$$T'''_{mx} \le .01 \qquad (4.7.4.26)$$

Then (4.7.4.19) implies that:

$$|\Delta T_0| \le .2830 \quad \text{and} \quad |\lambda| \le .9011$$
$$|\Delta U_1| \le .1607$$
$$|\Delta T_1| \le .1700$$
$$|U_2| \le .0593$$
$$|T_2| \le .0814 \tag{4.7.4.27}$$

I could use these to get cones, but they would not be very strong cones, so instead I will use them to get overall bounds on the bad directions from (4.7.4.21, 23, 25), which will improve the bounds (4.7.4.27). This procedure can be iterated until the bounds look good enough.

(4.7.4.27) implies that in $B \cap G$:

$$|b_1| \le .068$$
$$|b_2| \le .0481$$
$$|b_0| \le .1857. \tag{4.7.4.28}$$

Using (4.7.4.19), this improves (4.7.4.27) to:

$$|\Delta T_0| \le .2340 \quad \text{and} \quad |\lambda| \le .8571$$
$$|\Delta U_1| \le .1056$$
$$|\Delta T_1| \le .1041$$
$$|U_2| \le .0361$$
$$|T_2| \le .0438 \tag{4.7.4.29}$$

This improves (4.7.4.28) to:

$$|b_1| \le .0335$$
$$|b_2| \le .0247$$
$$|b_0| \le .1114 \tag{4.7.4.30}$$

Hence (4.7.4.29) is improved to:

$$|\Delta T_0| \le .1487 \text{ and } |\lambda| \le .7667$$

$$|\Delta U_1| \le .0764$$

$$|\Delta T_1| \le .0674$$

$$|U_2| \le .0256$$

$$|T_2| \le .0269 \qquad (4.7.4.31)$$

Now I use (4.7.4.31) to evaluate a cone from (4.7.4.21, 23, 25):

$$|b_1| \le .1487\,|\Delta U_1| + .1030\,|U_2| + .0752\,U'''_{mx} + .1667\,T'''_{mx}$$

$$|b_2| \le .0319\,|\Delta U_1| + .0217\,|\Delta T_1| + .1487\,|U_2| + .2940\,U'''_{mx} + .5\,T'''_{mx}$$

$$|b_0| \le .2829\,|\Delta U_1| + .5\,|\Delta T_1| + .1424\,|U_2| + .3334\,|T_2| + .0465\,U'''_{mx} + .1667\,T'''_{mx} \qquad (4.7.4.32)$$

I want the cone expressed in terms of the new coordinates, so using (4.7.4.19):

$$\begin{vmatrix} |b_1| \\ |b_2| \\ |b_0| \end{vmatrix} \le \begin{vmatrix} .0667 & .1077 & .1539 & .1285 & .0752 & .1667 \\ .0301 & .0329 & .1055 & .1223 & .2940 & .5 \\ .4888 & .4289 & .8720 & .4244 & .0465 & .1667 \end{vmatrix} \begin{vmatrix} |b_1| \\ |g_1| \\ |b_2| \\ |g_2| \\ U'''_{mx} \\ T'''_{mx} \end{vmatrix} \qquad (4.7.4.33)$$

Next I want to eliminate bad directions from the right hand side of (4.7.4.33). The first two inequalities can be written as:

$$\begin{vmatrix} .9333 & -.1539 \\ -.0301 & .8945 \end{vmatrix} \begin{vmatrix} |b_1| \\ |b_2| \end{vmatrix} \le \begin{vmatrix} .1077 & .1285 & .0752 & .1667 \\ .0329 & .1223 & .2940 & .5 \end{vmatrix} \begin{vmatrix} |g_1| \\ |g_2| \\ U'''_{mx} \\ T'''_{mx} \end{vmatrix} \qquad (4.7.4.34)$$

The matrix on the left has inverse less than or equal to:

$$\begin{vmatrix} 1.0775 & .1854 \\ .0363 & 1.1242 \end{vmatrix}. \tag{4.7.4.35}$$

Thus (4.7.4.34) can be inverted, and the result substituted in the inequality (4.7.4.33) for $|b_0|$, to get *cone C*:

$$\begin{vmatrix} |b_1| \\ |b_2| \\ |b_0| \end{vmatrix} \leq \begin{vmatrix} .1222 & .1612 & .1356 & .2724 \\ .0409 & .1422 & .3333 & .5682 \\ .5234 & .6272 & .4035 & .7954 \end{vmatrix} \begin{vmatrix} |g_1| \\ |g_2| \\ U'''_{mx} \\ T'''_{mx} \end{vmatrix} \tag{4.7.4.36}$$

This also gives the final improvements I will make on (4.7.4.30) and (4.7.4.31):

$$|b_0| \leq .0496$$
$$|b_1| \leq .0131$$
$$|b_2| \leq .0130 \tag{4.7.4.37}$$

$$|\Delta T_0| \leq .0770 \text{ and } |\lambda| \leq .6951$$
$$|\Delta U_1| \leq .0588$$
$$|\Delta T_1| \leq .0441$$
$$|U_2| \leq .0204$$
$$|T_2| \leq .0184 \tag{4.7.4.38}$$

(iv) N_1 is defined on $B \cap G$

Check that the conditions of (i) hold on $B \cap C$.

(a) $\lambda < 0$ holds if $|\Delta T_0| < \frac{1}{\gamma}$ (4.7.4.39)

(b) $T(x) - \lambda = (1 + \Delta T_1)x + \frac{x^2}{2}T_2 + \frac{x^3}{6}T'''(\xi).$ (4.7.4.40)

So $\quad T(x) \geq \lambda$ on $[0, 1]$ if $|\Delta T_1| + \frac{1}{2}|T_2| + \frac{1}{6}T'''_{mx} \leq 1$ (4.7.4.41)

(c) $$T(x) = -\frac{1}{\gamma} + \Delta T_0 + (1 + \Delta T_1)x + \frac{x^2}{2}T_2 + \frac{x^3}{6}T'''(\xi). \qquad (4.7.4.42)$$

So $\quad T(x) \leq 1$ on $[0, 1]$ if $|\Delta T_0| + |\Delta T_1| + \frac{1}{2}|T_2| + \frac{1}{6}T'''_{mx} \leq \frac{1}{\gamma}$ (4.7.4.43)

(d) $$T(1) = -\frac{1}{\gamma} + \Delta T_0 + 1 + \Delta T_1 + \frac{1}{2}|T_2| + \frac{1}{6}T'''(\xi) > 0$$

if $$|\Delta T_0| + |\Delta T_1| + \frac{1}{2}|T_2| + \frac{1}{6}T'''_{mx} < \frac{1}{\gamma^2} \qquad (4.7.4.44)$$

(e) $$U(x) = 1 + (1 + \Delta U_1)x + \frac{x^2}{2}U_2 + \frac{x^3}{6}U'''(\xi). \qquad (4.7.4.45)$$

So $U(x) \geq 0$ on $[\lambda, 0]$ if $-1 \leq \lambda \leq 0$ and

$$|\Delta T_0| + |\Delta U_1| + \frac{1}{2}|U_2| + \frac{1}{6}U'''_{mx} \leq \frac{1}{\gamma^2} \qquad (4.7.4.46)$$

(f) $$U(x) - 1 = (1 + \Delta U_1)x + \frac{x^2}{2}U_2 + \frac{x^3}{6}U'''(\xi). \qquad (4.7.4.47)$$

So $\quad U(x) \leq 1$ on $[\lambda, 0]$ if $-1 \leq \lambda \leq 0$ and

$$|\Delta U_1| + \frac{1}{2}|U_2| + \frac{1}{6}U'''_{mx} \leq 1. \qquad (4.7.4.48)$$

Note that all the conditions are implied by (4.7.4.44) and (4.7.4.46). As mentioned in (i), I might as well replace (e) by:

(e′) $$U'(x) = 1 + \Delta U_1 + xU_2 + \frac{1}{2}x^2\, U'''(\xi). \qquad (4.7.4.49)$$

So $\quad U'(x) \geq 0$ on $[\lambda, 0]$ if $-1 \leq \lambda \leq 0$ and

$$|\Delta U_1| + |U_2| + \frac{1}{2}U'''_{mx} \leq 1. \qquad (4.7.4.50)$$

Thus I need check only (4.7.4.44) and (4.7.4.50).

(d) $\ell.\text{h.s} \le .1320 \le \gamma^{-2}$ (4.7.4.51)

(e') $\ell.\text{h.s.} \le .0827 \le 1.$ (4.7.4.52)

Now I will apply the renormalisation:

$$\tilde{U}(x) = \frac{1}{\lambda}T(\lambda x)$$
$$\tilde{T}(x) = \frac{1}{\lambda}TU(\lambda x). \tag{4.7.4.53}$$

So

$$\tilde{\lambda} \equiv \tilde{T}(0) = \frac{1}{\lambda}T(1) = \frac{1}{\lambda}\left(\lambda + 1 + \Delta T_1 + \frac{1}{2}T_2 + \frac{1}{6}T'''(\xi)\right). \tag{4.7.4.54}$$

Now

$$\frac{1}{\lambda} = \frac{1}{-\frac{1}{\gamma} + \Delta T_0} = -\gamma + \frac{\gamma}{\lambda}\Delta T_0. \tag{4.7.4.55}$$

So

$$\Delta\tilde{T}_0 = \tilde{\lambda} + \frac{1}{\gamma} = \frac{1}{\lambda}\left(\gamma\Delta T_0 + \Delta T_1 + \frac{1}{2}T_2 + \frac{1}{6}T'''(\xi)\right). \tag{4.7.4.56}$$

Next I see how the first derivatives change.

$$\tilde{U}'(x) = T'(\lambda x)$$
$$\tilde{T}'(x) = T'(U(\lambda x))U'(\lambda x). \tag{4.7.4.57}$$

So

$$\Delta\tilde{U}_1 = \Delta T_1 \tag{4.7.4.58}$$

and

$$\tilde{T}_1 = T'(1)U_1 \text{ so } \Delta\tilde{T}_1 = \Delta T_1 + T_2 + \frac{1}{2}T'''(\xi) + T'(1)\Delta U_1. \tag{4.7.4.59}$$

Next look at the second derivatives.

$$\tilde{U}''(x) = \lambda T''(\lambda x)$$
$$\tilde{T}''(x) = \lambda T''(U(\lambda x)) \left(U'(\lambda x)\right)^2 + \lambda T'(U(\lambda x))U''(\lambda x). \tag{4.7.4.60}$$

So

$$\tilde{U}_2 = \lambda T_2 \tag{4.7.4.61}$$

and

$$\tilde{T}_2 = \lambda U_1^2 (T_2 + T'''(\xi)) + \lambda T'(1)U_2. \tag{4.7.4.62}$$

Lastly, the third derivatives.

$$\tilde{U}'''(x) = \lambda^2 T'''(\lambda x)$$
$$\tilde{T}'''(x) = \lambda^2 T'''(U(\lambda x))\left(U'(\lambda x)\right)^3 + 3\lambda^2 T''(U(\lambda x))U''(\lambda x)U'(\lambda x) + \lambda^2 T'(U(\lambda x))U'''(\lambda x). \tag{4.7.4.63}$$

So

$$\tilde{U}'''_{\rm mx} \le \lambda^2 T'''_{\rm mx} \tag{4.7.4.64}$$

and

$$\tilde{T}'''_{\rm mx} \le \lambda^2 U'^{3}_{\rm mx} T'''_{\rm mx} + \lambda^2 T'_{\rm mx} U'''_{\rm mx} + 3\lambda^2 U'_{\rm mx} (|T_2| + T'''_{\rm mx})(|U_2| + |\lambda| U'''_{\rm mx}). \tag{4.7.4.65}$$

Thus

$$\tilde{T}'''_{\rm mx} \le (3\lambda^2 U'_{\rm mx} |T_2|)|U_2| + (\lambda^2 T'_{\rm mx} + 3|\lambda|^3 |T_2|)U'''_{\rm mx}$$
$$+ (\lambda^2 U'_{\rm mx} + 3\lambda^2 U'_{\rm mx} |U_2| + 3|\lambda|^3 U'_{\rm mx} U'''_{\rm mx})T'''_{\rm mx}. \tag{4.7.4.66}$$

(v) <u>Good directions shrink</u>

I will need to bound the original coordinates by the good coordinates in the cone. So using (4.7.4.19) and (4.7.4.36), I get:

$$\begin{vmatrix} |\Delta U_1| \\ |\Delta T_1| \\ |U_2| \\ |T_2| \end{vmatrix} \leq \begin{vmatrix} .8084 & .5372 & .3021 & .5335 \\ .5661 & .3587 & .3305 & .6086 \\ .0184 & .7878 & .1494 & .2546 \\ .0297 & .5510 & .2414 & .4114 \end{vmatrix} \begin{vmatrix} |g_1| \\ |g_2| \\ U'''_{mx} \\ T'''_{mx} \end{vmatrix} \qquad (4.7\ 4.67)$$

Let us start with the third derivatives. (4.7.4.64) and (4.7.4.38) imply that:

$$\tilde{U}'''_{mx} \leq .4831\, T'''_{mx}. \qquad (4.7.4.68)$$

For $\tilde{T}'''_{mx}$, I will need U'_{mx} and T'_{mx}. Now:

$$U'(x) = 1 + \Delta U_1 + xU_2 + \frac{x^2}{2} U'''(\xi). \qquad (4.7.4.69)$$

So
$$U'_{mx} \leq 1 + |\Delta U_1| + |\lambda||U_2| + \frac{\lambda^2}{2} U'''_{mx} \leq 1.0747. \qquad (4.7.4.70)$$

Also
$$T'(x) = 1 + \Delta T_1 + xT_2 + \frac{x^2}{2} T'''(\xi). \qquad (4.7.4.71)$$

So
$$T'_{mx} \leq 1 + |\Delta T_1| + |T_2| + \frac{1}{2} T'''_{mx} \leq 1.0675. \qquad (4.7.4.72)$$

Thus, using (4.7.4.67) for $|U_2|$, (4.7.4.66) gives:

$$\tilde{T}'''_{mx} \leq .0006|g_1| + .0226|g_2| + .5385\, U'''_{mx} + .6463\, T'''_{mx}. \qquad (4.7.4.73)$$

Next, let us look at $|\tilde{g}_2|$.

$$\tilde{g}_2 = \tilde{U}_2 - \frac{1}{\gamma}\tilde{T}_2 = \lambda\Big(T_2 - \frac{1}{\gamma}U_1^2 T_2 - \frac{1}{\gamma}U_1^2 T'''(\xi) - \frac{1}{\gamma}T'(1)U_2\Big)$$

$$= -\lambda\Big(\frac{1}{\gamma}g_2 + \frac{1}{\gamma}\Delta T'(1)U_2 + \frac{1}{\gamma}(U_1^2 - 1)T_2 + \frac{1}{\gamma}U_1^2 T'''(\xi)\Big). \qquad (4.7.4.74)$$

Thus

$$|\tilde{g}_2| \leq \frac{|\lambda|}{\gamma}|g_2| + |\frac{\lambda}{\gamma}\Delta T'(1)||U_2| + |\frac{\lambda}{\gamma}(U_1^2 - 1)||T_2| + |\frac{\lambda}{\gamma}U_1^2|T_{mx}'''. \quad (4.7.4.75)$$

The bounds (4.7.4.38) imply that $|U_1^2 - 1| \leq .1211.$ (4.7.4.76)

So

$$|\tilde{g}_2| \leq .4296|g_2| + .0290|U_2| + .0521|T_2| + .4816\, T_{mx}'''. \quad (4.7.4.77)$$

Using (4.7.4.36), this implies that:

$$|\tilde{g}_2| \leq .0021|g_1| + .4812|g_2| + .0170\, U_{mx}''' + .5105\, T_{mx}'''. \quad (4.7.4.78)$$

Lastly, we come to $|\tilde{g}_1|$.

$$\tilde{g}_1 = \Delta\tilde{U}_1 - \frac{1}{\gamma}\Delta\tilde{T}_1 - \tilde{U}_2 - \tilde{T}_2 \quad (4.7.4.79)$$

$$= \Delta T_1 - \frac{1}{\gamma}\Big(\Delta T_1 + T_2 + \frac{1}{2}T'''(\xi_1) + T'(1)\Delta U_1\Big) - \lambda\Big(T_2 + U_1^2\, T_2 + U_1^2 T'''(\xi_2) + T'(1)U_2\Big)$$

$$= -\frac{1}{\gamma}g_1 - \frac{1}{\gamma}\Delta T'(1)\Delta U_1 - (\lambda T'(1) + \frac{1}{\gamma})\, U_2 - (\Delta T_0 + \frac{1}{\gamma} + \lambda U_1^2)T_2 - \frac{1}{2\gamma}T'''(\xi_1) - \lambda U_1^2\, T'''(\xi_2).$$

So

$$|\tilde{g}_1| \leq \frac{1}{\gamma}|g_1| + |\frac{\Delta T'(1)}{\gamma}||\Delta U_1| + |\lambda T'(1) + \frac{1}{\gamma}||U_2| + (|\Delta T_0| + |\frac{1}{\gamma} + \lambda U_1^2|)|T_2|$$

$$+ \Big(\frac{1}{2\gamma} + |\lambda U_1^2|\Big)\, T_{mx}'''. \quad (4.7.4.80)$$

The bounds (4.7.4.38) imply that

$$|\lambda T'(1) + \frac{1}{\gamma}| \leq .1239 \quad (4.7.4.81)$$

and

$$|\lambda U_1^2 + \frac{1}{\gamma}| \leq .1612. \quad (4.7.4.82)$$

So

$$|\tilde{g}_1| \leq .6181\,|g_1| + .0418\,|\Delta U_1| + .1239\,|U_2| + .2382\,|T_2| + 1.0883\,T'''_{mx}. \quad (4.7.4.83)$$

Converting to good coordinates again by (4.7.4.36), gives:

$$|\tilde{g}_1| \leq .6613\,|g_1| + .2514\,|g_2| + .0887\,U'''_{mx} + 1.2402\,T'''_{mx}. \quad (4.7.4.84)$$

Now I collect together the estimates (4.7.4.68, 73, 78, 84):

$$\begin{vmatrix} |\tilde{g}_1| \\ |\tilde{g}_2| \\ |\tilde{U}'''_{mx}| \\ |\tilde{T}'''_{mx}| \end{vmatrix} \leq \begin{vmatrix} .6613 & .2514 & .0887 & 1.2402 \\ .0021 & .4812 & .0170 & .5105 \\ 0 & 0 & 0 & .4831 \\ .0006 & .0226 & .5385 & .6463 \end{vmatrix} \begin{vmatrix} |g_1| \\ |g_2| \\ U'''_{mx} \\ T'''_{mx} \end{vmatrix}. \quad (4.7.4.85)$$

We hope that this matrix is a contraction on $B \cap G$. Let us see if it is. For $\varepsilon \leq 1$, (4.7.4.85) shows that:

$$\begin{aligned} |g_1| &\leq .05\,\varepsilon \\ |g_2| &\leq .02\,\varepsilon \\ U'''_{mx} &\leq .007\,\varepsilon \\ T'''_{mx} &\leq .01\,\varepsilon \end{aligned} \quad (4.7.4.86)$$

implies that:

$$\begin{aligned} |\tilde{g}_1| &\leq .0512\,\varepsilon \\ |\tilde{g}_2| &\leq .0150\,\varepsilon \\ \tilde{U}'''_{mx} &\leq .0049\,\varepsilon \\ \tilde{T}'''_{mx} &\leq .0108\,\varepsilon \end{aligned} \quad (4.7.4.87)$$

Thus we need to restrict attention to a set slightly smaller than B in some directions. Try reducing the bounds on the third derivatives to:

$$U'''_{mx} \le .005\ \varepsilon$$
$$\qquad (4.7.4.88)$$
$$T'''_{mx} \le .0091\ \varepsilon.$$

Then for $\varepsilon \le 1$, (4.7.4.85) gives:

$$|\tilde{g}_1| \le .0499\ \varepsilon$$
$$|\tilde{g}_2| \le .0145\ \varepsilon$$
$$\tilde{U}'''_{mx} \le .0044\ \varepsilon$$
$$\tilde{T}'''_{mx} \le .00906\ \varepsilon \qquad (4.7.4.89)$$

which is a contraction.

Note that to get an idea of the relative lengths to take for the domain of contraction, (4.7.4.85) is a contraction on some subset of B iff its largest eigenvalue is less than 1. A coordinate box with sides of lengths proportional to the components of the eigenvector corresponding to the largest eigenvalue is guaranteed to work. Alternatively, a diamond given by the adjoint eigenvector for the largest eigenvalue will work, since the adjoint eigenvector acts as a Lyapunov function. To construct a large neighbourhood, a Lyapunov function would be the best approach, as the shape of the neighbourhood can be moulded as required, by varying the Lyapunov function.

(vi) Check bad directions

Lastly, I have to check that the growth in the bad directions is not too large, so that the cone conditions can be applied to the new point.

$$\tilde{b}_2 = -\frac{1}{\gamma}\tilde{U}_2 - \tilde{T}_2 = -\lambda\left(\frac{T_2}{\gamma} + U_1^2(T_2 + T'''(\xi)) + T'(1)U_2\right)$$
$$= \lambda\gamma b_2 - \lambda\left(\Delta T'(1)U_2 + (U_1^2 - 1)T_2 + U_1^2 T'''(\xi)\right). \qquad (4.7.4.90)$$

So

$$|\tilde{b}_2| \le |\lambda|\,|\gamma|\,|b_2| + |\lambda|\left(|\Delta T'(1)U_2| + |U_1^2 - 1|\,|T_2| + U_1^2 T'''_{mx}\right) \qquad (4.7.4.91)$$

$$\le .0250 \le .1. \qquad (4.7.4.92)$$

Next

$$\tilde{b}_1 - \frac{1}{\gamma}\Delta\tilde{U}_1 - \Delta\tilde{T}_1 + \frac{1}{2\gamma^2}\tilde{U}_2 - \frac{1}{2}\tilde{T}_2 \tag{4.7.4.93}$$

$$= -\frac{1}{\gamma}\Delta T_1 - \Delta T_1 - T_2 - \frac{1}{2}T'''(\xi_1) - T'(1)\Delta U_1 + \frac{\lambda}{2}\Big(\frac{T_2}{\gamma^2} - U_1^2(T_2 + T'''(\xi_2)) - T'(1)U_2\Big)$$

$$= \gamma b_1 - \Delta T'(1)\Delta U_1 - \frac{1}{2}(\lambda T'(1) + \frac{1}{\gamma})U_2 + \frac{1}{2}(\frac{\Delta T_0}{\gamma^2} + \lambda U_1^2 + \frac{1}{\gamma})T_2 - \frac{1}{2}T'''(\xi_1) - \frac{\lambda}{2}U_1^2 T'''(\xi_2).$$

Thus:

$$|\tilde{b}_1| \le \gamma|b_1| + |\Delta T'(1)\Delta U_1| + |\lambda T'(1) + \frac{1}{\gamma}|\frac{|U_2|}{2} + \Big(\frac{|\Delta T_0|}{\gamma^2} + |\lambda U_1^2 + \frac{1}{\gamma}|\Big)\frac{|T_2|}{2} + \frac{1+|\lambda|U_1^2}{2}T'''_{mx}$$

$$\le .03708 \le .1. \tag{4.7.4.95}$$

Lastly,

$$\tilde{b}_0 = \Delta\tilde{T}_0 - \frac{1}{2\gamma^2}\Delta\tilde{U}_1 + \frac{1}{2}\Delta\tilde{T}_1 + \frac{1}{6\gamma^3}\tilde{U}_2 + \frac{1}{6}\tilde{T}_2 \tag{4.7.4.96}$$

$$= \frac{1}{\gamma}\Big(\gamma\Delta T_0 + \Delta T_1 + \frac{1}{2}T_2 + \frac{1}{6}T'''(\xi_1)\Big) - \frac{1}{2\gamma^2}\Delta T_1 + \frac{1}{2}\Big(\Delta T_1 + \frac{1}{2}T'''(\xi_2) + T'(1)\Delta U_1\Big)$$

$$+ \frac{\lambda}{6\gamma^3}T_2 + \frac{\lambda}{6}\Big(U_1^2(T_2 + T'''(\xi_3)) + T'(1)U_2\Big)$$

$$= -\gamma^2 b_0 + \frac{\gamma}{\lambda}\Delta T_0^2 + \frac{\gamma}{\lambda}\Delta T_0(\Delta T_1 + \frac{1}{2}T_2) + \frac{1}{2}\Delta T'(1)\Delta U_1 + \frac{1}{6}(\lambda T'(1) + \frac{1}{\gamma})U_2$$

$$+ \frac{1}{6}\Big(\frac{\Delta T_0}{\gamma^3} + \lambda U_1^2 + \frac{1}{\gamma}\Big)T_2 + \frac{1}{6\lambda}T'''(\xi_1) + \frac{1}{4}T'''(\xi_2) + \frac{\lambda}{6}U_1^2 T'''(\xi_3)$$

using (4.7.4.55).

So

$$|\tilde{b}_0| \le \gamma^2 |b_0| + \frac{\gamma^2}{|\lambda|}\Delta T_0^2 + \frac{\gamma}{|\lambda|}|\Delta T_0|(|\Delta T_1| + \frac{1}{2}|T_2|) + \frac{1}{2}|\Delta T'(1)\Delta U_1|$$

$$+\frac{1}{6}|\lambda T'(1) + \frac{1}{\gamma}||U_2| + \frac{1}{6}\Big(\frac{|\Delta T_0|}{\gamma^3} + |\lambda U_1^2 + \frac{1}{\gamma}|\Big)|T_2| + \Big(\frac{1}{6|\lambda|} + \frac{1}{4} + \frac{1}{6}|\lambda U_1^2|\Big)T'''_{\mathrm{mx}}. \tag{4.7.4.97}$$

Now

$$\frac{1}{|\lambda|} \le 1.8484. \tag{4.7.4.98}$$

Thus

$$|\tilde{b}_0| \le .1807 \le .2. \tag{4.7.4.99}$$

(vii) Conclusion

In conclusion, the simple fixed point attracts all commuting pairs with golden rotation number lying in the set:

$$\begin{aligned} &|b_0| \le .2 \\ &|b_1| \le .1 \\ &|b_2| \le .1 \\ &|g_1| \le .05 \\ &|g_2| \le .02 \\ &U'''_{\mathrm{mx}} \le .005 \\ &T'''_{\mathrm{mx}} \le .0091. \end{aligned} \tag{4.7.4.100}$$

As an example, consider the periodic map:

$$F : x' = x + \Omega + \frac{a}{2\pi}\sin 2\pi x. \tag{4.7.4.101}$$

Rather than considering the pair (F, R) directly, I will consider its second iterate $(U, T) = (FR, FR^2)$ under N_1. This gives me a little head start. After normalising $U(0) = 1$, one obtains:

$$U: x' = x + 1 + \frac{a}{2\pi(\Omega - 1)} \sin 2\pi(\Omega - 1)x$$

$$T: x' = x + \frac{\Omega - 2}{\Omega - 1} - \frac{a}{2\pi(\Omega - 1)} \sin 2\pi(\Omega - 1)x. \tag{4.7.4.102}$$

Thus

$$\Delta U_1 = \Delta T_1 = a$$

$$U'''_{mx} = T'''_{mx} = a\left(2\pi(\Omega - 1)\right)^2 \tag{4.7.4.103}$$

$$\Delta T_0 = \frac{\Omega - 2}{\Omega - 1} + \frac{1}{\gamma}.$$

So

$$b_0 = \frac{\Omega - 2}{\Omega - 1} + \frac{1}{\gamma} - \frac{a}{2\gamma}$$

$$b_1 = \gamma a$$

$$g_1 = -a\gamma^{-2}$$

$$b_2 = g_2 = 0.$$

Conditions (4.7.4.100) are satisfied if

$$|a| \le .0003, \quad 1.57 \le \Omega \le 1.65. \tag{4.7.4.105}$$

Thus for all parameter values in this range with golden rotation number, (F, R) is attracted to the simple fixed point. Incidentally, since F moves right by between $\Omega \pm a/2\pi$, golden rotation number implies that:

$$|\Omega - \gamma| \le \frac{a}{2\pi} \tag{4.7.4.106}$$

so the constraint on Ω in (4.7.4.105) is automatically satisfied.

The result above is very weak, considering that Herman's results (1977) imply the same result for $|a| < 1$, but it could be improved considerably by making more careful estimates.

§4.7.5 **Critical fixed point**. Shenker (1982) found self-similar behaviour in some two parameter families (2pfs) of maps on a circle, near *critical* parameter values where the map is invertible with noble rotation number, but the inverse is not differentiable. It appears to be the same for most exmaples, and so it is called *universal*. This suggests another fixed point of the renormalisation.

The fixed point was found to a good approximation by Feigenbaum et al. (1982) and

Rand et al. (1982). The first point to note is that typically a map which is invertible, but whose inverse is not differentiable, has a cubic inflection point. Just as maps with a quadratic maximum can be made even by a coordinate change (§2.3.6), so maps with a cubic inflection can be made functions of x^3 only, which I call *cubic* maps. The renormalisation preserves the space of cubic maps, so one should look for the critical fixed point in this space. Feigenbaum et al. (1982) and Rand et al. (1982) represented their functions by polynomials, evaluated the composition on a carefully chosen grid of points, and then fitted a polynomial to the new points by Lagrange interpolation. I did not do this, but adopted the same approach as I have used throughout this thesis, composing polynomials formally with a truncation at some degree. Again one has to choose domains carefully.

Unlike the area preserving case (§4.4.2), the order of composition is crucial here. Feigenbaum (Feigenbaum et al., 1982) showed that one must use:

$$U' = BTB^{-1}$$

$$T' = BTUB^{-1}. \qquad (4.7.5.1)$$

For the other order, there are not even any real domains (let alone complex domains) which work near the critical fixed point.[53] The choice:

$$\alpha = \frac{1}{T\,(0)} \qquad (4.7.5.2)$$

forcing the normalisation:

$$U(0) = 1 \qquad (4.7.5.3)$$

makes the interval [0, 1] a good domain for T. I needed a complex domain, however. Searching around gave the following as close to optimal domains. Write:

$$U(x) = U\left(\frac{x^3 + 0.518}{0.540}\right),\ T(x) = t\left(\frac{x^3 - 0.518}{0.540}\right) \qquad (4.7.5.4)$$

and expand u, t in power series in unit discs. These domains have analyticity safety factor of 0.95901. Note that the domain for U is not crucial, but that for T is.

To find the above domains, and to iterate the renormalisation, I needed a good approximate fixed point. I constructed one as follows. Kadanoff told me about (4.7.5.10). Write:

$$H(x) = \alpha T(x). \qquad (4.7.5.5)$$

53 Note that the renormalisation can be defined in the other order (e.g. Feigenbaum, 1988), but the domain for one of the maps does not contain the critical point.

Then at the fixed point:

$$H(x) = \alpha H(H(\frac{x}{\alpha^2})). \tag{4.7.5.6}$$

The normalisation implies:

$$H(0) = 1. \tag{4.7.5.7}$$

Evaluating (4.7.5.6) at $x = 0$ gives:

$$H(1) = \frac{1}{\alpha}. \tag{4.7.5.8}$$

Evaluating at $x = 1$ gives:

$$H(H(\alpha^{-2})) = \alpha^{-2} \tag{4.7.5.9}$$

so it is consistent to suppose that α^{-2} is a fixed point of H:

$$H(\alpha^{-2}) = \alpha^{-2}. \tag{4.7.5.10}$$

Next, write $H(x) = h(x^3)$. Then:

$$h(z) = \alpha h\left((h(\frac{z}{\alpha^6}))^3\right). \tag{4.7.5.11}$$

Evaluate $h'(0)$:

$$h'(0) = \alpha h'(1)\, 3\left(h(0)\right)^2 h'(0)\, \alpha^{-6}. \tag{4.7.5.12}$$

So, provided $h'(0) \neq 0$:

$$h'(1) = \frac{\alpha^5}{3}. \tag{4.7.5.13}$$

Evaluate $h'(1)$:

$$h'(1) = \alpha h'(\alpha^{-6})\, 3\left(h(\alpha^{-6})\right)^2 h'(\alpha^{-6})\alpha^{-6}. \tag{4.7.5.14}$$

So

$$h'(\alpha^{-6}) = \frac{\alpha^7}{3} \tag{4.7.5.15}$$

choosing the sign to make T increasing. Thus I know h at three points, and the derivative at two points, in terms of α, for which Shenker gave me a value. Solving linear equations gave me a quartic approximation for h.

Newton's method on the renormalisation with truncation at various degrees gave me good approximate fixed points. The results are given in table 4.7.5.1.

Table 4.7.5.1: Fixed points of N_1 truncating at various degrees, giving also the values of α and the largest eigenvalues of DN_1.

degree	17	18	19	
α	-1.288574580	-1.288574212	-1.288574487	
coefficients	2.0921 e-4	2.0860 e-4	2.0906 e-4	
of $t(\xi)$	0.82270223	0.82270346	0.82270255	
	-0.10249655	-0.10249591	-0.10249640	
	-0.10524988	-0.10525091	-0.10525015	
	0.03364574	0.03364587	0.03364578	
	0.01178887	0.01178905	0.01178892	
	-7.92730e-3	-7.92732e-3	-7.02731e-3	
	-7.5174 e-4	-7.5180 e-4	-7.5175 e-4	
	1.59088e-3	1.59090e-3	1.59088e-3	
	-1.3734 e-4	-1.3733 e-4	-1.3733 e-4	
	-2.7568 e-4	-2.7569 e-4	-2.7568 e-4	
	7.530 e-5	7.530 e-5	7.530 e-5	
	3.909 e-5	3.909 e-5	3.909 e-5	
	-2.144 e-5	-2.144 e-5	-2.144 e-5	
	-3.68 e-6	-3.68 e-6	-3.68 e-6	
	4.80 e-6	4.80 e-6	4.80 e-6	
	-1.5 e-7	-1.5 e-7	-1.5 e-7	
	-9.1 e-7	-9.1 e-7	-9.1 e-7	
		1.9 e-7	1.9 e-7	
			1.4 e-7	
spectrum	-2.8336103	-2.8336109	-2.8336108	δ
	2.1395803	2.1395790	2.1395804	$-\alpha^3$
	-1.0000538	-0.9996956	-0.9998218	-1
	-0.5293062	-0.5284276	-0.5276189	δ'
	-0.463670	-0.467009	-0.468317	α^{-3}
	0.469945	0.468941	0.467021	$-\alpha^{-3}$

§4.7.6 **Universal two parameter family.** The fixed point has essentially only two unstable directions under the renormalisation. Thus its unstable manifold forms a *universal 2pf*, to which all 2pfs, containing a critical case attracted to the fixed point, generically converge when looked at on long enough times and small enough scales and for parameters sufficiently close to critical. Loosely speaking, the unstable directions correspond to the transition from differentiable invertibility to non-invertibility, and to changing the rotation number. Papers on this subject have not emphasised the second direction. The impression has been given that one has to keep the rotation number fixed in order to see the universality. Although this is true if you want to see it in a one parameter family, maintaining the rotation number typically requires adjustment of another parameter, so I say why not look at the whole behaviour in a 2pf?

The universal 2pf $f^*_{\mu,\nu}(x)$ can be determined from any typical 2pf $f_{a,b}(x)$ of periodic maps on the line, containing a critical case (a_∞, b_∞) for some noble number ω, by:

$$f^*_{\mu,\nu}(x) = \lim_{n\to\infty} \alpha^n R^{P_n} f^{Q_n}_{\frac{\mu}{\delta^n},\frac{\nu}{\eta^n}}\left(\frac{x}{\alpha^n}\right). \tag{4.7.6.1}$$

In this equation, R is back-rotation through one period:

$$R(x) = x - 1 \tag{4.7.6.2}$$

and P_n/Q_n are the convergents of ω. The number

$$\alpha = -1.2885746... \tag{4.7.6.3}$$

is the scaling in position. The numbers

$$\delta = -2.8336108... \tag{4.7.6.4}$$

$$\eta = \alpha^2 = 1.6604244... \tag{4.7.6.5}$$

are the eigenvalues in the unstable directions (in the opposite order from the previous paragraph). (μ, ν) is a reparametrisation of the 2pf, and x is a new position coordinate.

I will now explain why the reparametrisation is necessary. It is an example of finding scaling-parameters, as discussed in §2.3.5. Clearly one has to shift the origin of the parameters to (a_∞, b_∞). Also one has to choose the parameter directions correctly to correspond to the eigenvalues δ, η. In fact, more is required. Suppose we had a parametrisation for which the limit (4.7.6.1) existed. Then the effect of a reparametrisation:

$$\begin{aligned}\mu &\to \mu + \delta\mu(\mu, \nu)\\ \nu &\to \nu + \delta\nu(\mu, \nu)\end{aligned} \tag{4.7.6.6}$$

changes as n increases, like:

$$\delta\mu'(\mu, \nu) = \delta\, \delta\mu\left(\frac{\mu}{\delta}, \frac{\nu}{\eta}\right)$$

$$\delta\nu'(\mu, \nu) = \eta\, \delta\nu\left(\frac{\mu}{\delta}, \frac{\nu}{\eta}\right). \qquad (4.7.6.7)$$

So we see that the effect of the following reparametrisations grow with n:

parameterchange	eigenvalue	interpretation
$\delta\mu = 1$	$\delta = -2.833$	shift of origin in μ
$\delta\mu = \nu$	$\delta/\eta = -1.7066$	shear in μ with respect to ν
$\delta\mu = \nu^2$	$\delta/\eta^2 = -1.0278$	bend in μ with respect to ν
$\delta\nu = 1$	$\eta = 1.6604$	shift of origin in ν

Thus it is essential to "kill" components of the parametrisation in these directions from the universal 2pf. For example, killing the shifts in origin corresponds to choosing $(\mu, \nu) = (0, 0)$ at the critical case. Parameters chosen in this way, I call *scaling-parameters* (as discussed in §2.3.5).

In addition there are two scale changes which are neutral:

$\delta\mu = \mu$	1	scale change in μ
$\delta\nu = \nu$	1	scale change in ν.

They correspond to the arbitrariness in choosing scales. The effect of higher order parameter changes decay. But the essential attraction rate of the fixed point can be measured to be:

$$\delta' = -0.528... \qquad (4.7.6.8)$$

(see table 4.7.5.1). This would be the convergence rate for coordinate-free quantities (e.g. stability of some periodic orbit) in critical cases, though I have not checked, it.[54] Thus it may be worthwhile also killing parameter changes which decay slower than δ'. These are:

$\delta\mu = \nu^3$	$\delta/\eta^3 = -0.61899$
$\delta\mu = \mu\nu$	$1/\eta = 0.60226$
$\delta\nu = \nu^2$	$1/\eta = 0.60226$
$\delta\nu = \mu$	$\eta/\delta = -0.58597$

54 This was checked in March 1990 by Kapur, an undergraduate student of mine, using the multipliers of some complex periodic orbits.

Scaling-parameters (and the values of δ, η) can be determined in a systematic way. Provided the original parametrisation is vaguely adapted to the expected directions, define (μ, ν) by:

$$a = a_\infty + c_1\nu + c_2\nu^2 + s_\mu\mu + d_1\nu^3 + d_2\mu\nu$$
$$b = b_\infty + s_\nu\nu + d_3\nu^2 + d_4\mu \tag{4.7.6.9}$$

where the coefficients are to be determined as follows. Find parameter values (a_n, b_n) for some codimension-two property to hold for $f_{a,b}^{Q_n}$. For example, one could find the parameter values (a_n, b_n) for the simultaneous existence of a superstable Q_n-cycle and a superstable Q_{n+1}-cycle. Then choose the coefficients $a_\infty, b_\infty, c_1, c_2$ and the scaling factors δ, η so that the sequence

$$\left(\frac{\mu_n}{\delta^n}, \frac{\nu_n}{\eta^n}\right) \tag{4.7.6.10}$$

converges to something non-trivial. The limit can be scaled as desired (e.g. to (1,1)) by choosing s_μ, s_ν. Then the remaining coefficients d_1, d_2, d_3 can be determined to make the convergence of (4.7.6.10) no slower than δ'^n.

Note that certain choices of the codimension-two property will automatically force one component of the limit of (4.7.6.10) to be zero. For example, parameters for existence of a super-superstable Q_n-cycle (i.e. when both the first and second derivatives of f^{Q_n} are zero for the orbit) lie on $\nu = 0$. In fact, it is often a good idea to choose some special property like this to determine some of the coefficients independently of the others, in this case $a_\infty, b_\infty, \delta, s_\mu, d_4$. Note also that one could find parameter values for several codimension two properties, and choose the coefficients so that all the sequences (4.7.6.10) converge. This gives more information without having to go to high period.

Similar considerations apply to finding "scaling coordinates", if they are needed. Note that parameter dependent coordinate changes should be considered too.

Returning to properties of 2pfs of maps on a circle, one of their most notable features is *frequency locking bands*. As mentioned in §4.7.1, these are bands in the parameter plane in which there is an attracting periodic orbit. It is called "frequency locking" because the periodic orbit has rational rotation number. When the map is invertible, all points have the same rotation number. For rational rotation number, all orbits are eventually periodic. For irrational rotation number and differentiable invertibility, the motion is conjugate to irrational rotation. When invertibility is lost, the frequency locking bands can overlap, giving rise to several different asymptotic behaviours, and hysteresis as parameters are varied. The periodic orbits in the frequency locking bands can period double, giving all the complicated behaviour that is found in maps on an interval (e.g. Collet and Eckmann, 1980). But also there can be other sorts of aperiodic orbits, which do not have a rotation number, for example. See Curry and Yorke (1978) for a study of a typical case.[55]

Shenker is currently evaluating scaling parameters for the map:

$$\theta' = \theta + a - \frac{b}{2\pi}\sin 2\pi\theta \qquad (4.7.6.11)$$

and the noble $[0, 2, (1,)^{\infty}]$. Note that for this map, it is clear that the critical cases are all at $b = 1$, where the map loses invertibility, so $\nu = 0$ should correspond to $b = 1$. Thus $b_{\infty} = 1, d_4 = 0$ straight off. I am looking forward to seeing the picture for frequency locking in the universal 2pf.[56]

The fixed point is believed to apply also to continuous time and higher dimensional dissipative systems. As an illustration, I constructed the electronic circuit of figure 4.7.6.1. It is an autonomous and roughly three-dimensional system as it has three integrators. The values of the components have not been included, because the ones I used were not particularly well-chosen. You would do best to choose your own if you want to build it. Note that the multiplier requires biasing which is not shown either. I ran it at audio frequencies (~ 5kHz), so that I could listen to the frequency locking. The parameters R_1, R_2 were adjustable. Roughly, R_1 controls the frequency ratio of an invariant torus, and R_2 its fatness. I obtained figure 4.7.6.2 for the behaviour in this parameter plane, using an oscilloscope to see what was going on.

You can see a convergent sequence of bands with rotation numbers $1/2, 1/3, 2/5, 3/8, 5/13, 8/21, \ldots$ converging to the noble $[0, 2,(1,)^{\infty}]$. The picture does not look very self-similar, but determining scaling parameters as above, gave figure 4.7.6.3. It is reasonably self-similar, but does not give δ, η very precisely. To get a good estimate of δ, η would require going to high periods, as the essential convergence rate δ' (4.7.6.8) is not small (unlike the case of the period doubling fixed point for maps on the interval, when it is -0.1237). The best estimates I could get from figure 4.7.6.3 were $\delta \sim -2.0$ to -3.0, $\eta \sim 1.5$ to 2.3.

The flow on a critical noble torus deserves some discussion. The critical torus is a differentiable surface, even though the motion on it is only topologically conjugate to a linear quasiperiodic flow. What happens is that there are streaks on the torus which the orbits visit only very infrequently. Above critical, they develop into gaps which some orbits do not visit at all. These orbits form an attracting "cantorus" (cf. Percival, 1980).

---

[55] I saw behaviour which I believe fits in this scheme in a plasma physics experiment I performed in November, 1978 (unpublished). As the magnetic field in a plasma column was increased, drift waves destabilised, giving first a periodic oscillation, and then a second frequency destabilised. The two frequencies seemed to evolve independently at first, but then passed through a sequence of mode-locked intervals. Then a broad band spectrum appeared.

[56] I produced this in MacKay & Tresser (1986).

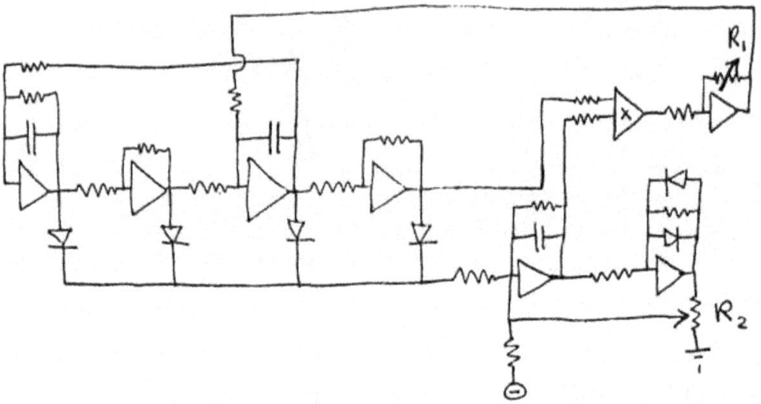

Figure 4.7.6.1: Electronic circuit exhibiting invariant tori and their breakup

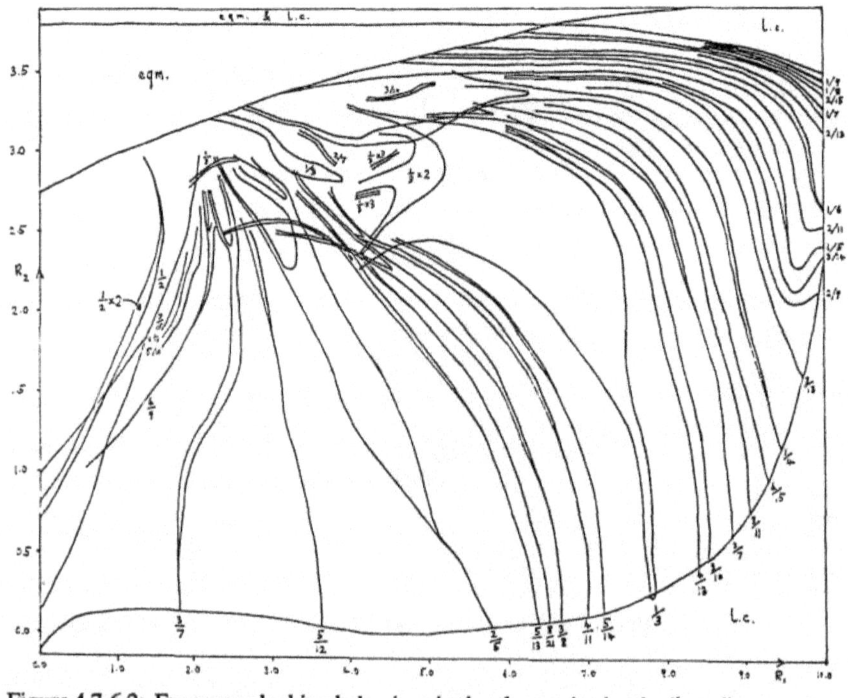

Figure 4.7.6.2: Frequency locking behaviour in the electronic circuit. "eqm" means there is an attracting equilibrium, and "l.c." means there is an attracting limit cycle

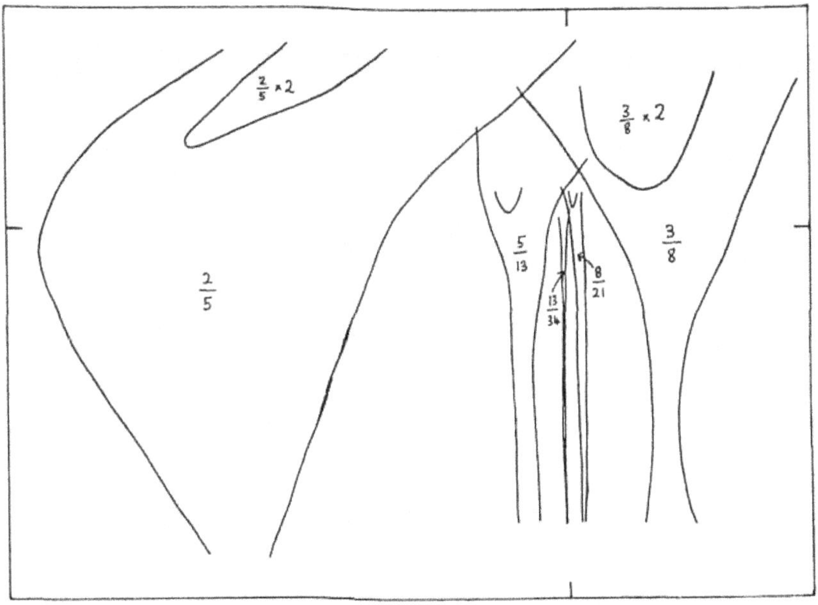

Figure 4.7.6.3: Part of figure 4.7.6.2 in scaling parameters

Other orbits are possible, however, giving complicated behaviour.

Finally, one might ask what happens in a family with more than two parameters. If it contains a critical case, then typically it lies in a codimension two set of parameter values also attracted to the fixed point. The behviour in the other two parameter directions is given by the universal 2pf. A family with more than two parameters, however, is likely to contain special cases corresponding to other fixed points of the renormalisation, giving different behaviour near them.

§4.7.7 **Poincaré's theorem**. In this section I show that Poincaré's theorem generalises from periodic maps to commuting pairs, as claimed in §4.1.2. I require that there exist $K > 0$, $m, n \in \mathbf{Z}$, such that defining:

$$V = U^m T^n \tag{4.7.7.1}$$

then

$$V(x) \geq x + K \text{ for all } x. \tag{4.7.7.2}$$

Note that it follows that:

$$V^{-1}(x) \leq x - K \text{ for all } x. \tag{4.7.7.3}$$

Compare $R^{-1}(x) = x + 1$ in the periodic case.

Theorem: If U, T: $\mathbb{R} \to \mathbb{R}$ are orientation preserving homemorphisms, commute, and satisfy the above condition, then $\exists\, \mu{:}\nu \in \mathbb{R}P$, independent of x, such that:

$$\frac{U^q T^p x}{r} \to 0 \text{ as } r = \max(|p|, |q|) \to \infty \text{ iff } \frac{q\mu - p\nu}{r} \to 0. \quad (4.7.7.4)$$

It is clearly unique from the "only if" part.

Proof:

Case (i): Suppose $\exists\,(i, j)$, coprime, s.t. $U^iT^jx - x$ changes sign or takes the value 0 somewhere. If it changes sign then it has to take the vlaue 0 at some intermediate point x_0, by continuity. So in either case there is periodic point x_0, of type (i, j). Write:

$$W = U^iT^j \quad (4.7.7.5)$$

and

$$N = in - jm. \quad (4.7.7.6)$$

Then $N \neq 0$, because $N = 0$ would imply that (m, n) is a multiple of (i, j), and so $Vx_0 = x_0$ (contradiction). Any pair $(p, q) \in \mathbb{Z}^2$ is within (N, N) of some point (Na, Nb). But:

$$U^{Na}T^{Nb} = W^cV^d \quad (4.7.7.7)$$

where

$$\begin{vmatrix} c \\ d \end{vmatrix} = \begin{vmatrix} n & -m \\ -j & i \end{vmatrix} \begin{vmatrix} a \\ b \end{vmatrix}. \quad (4.7.7.8)$$

Thus

$$\frac{|U^p T^q x_0|}{r} \geq \frac{Kd}{r} \sim \frac{K}{N} \frac{ip - jq}{r}. \quad (4.7.7.9)$$

Thus x_0 has rotation number $i{:}j$. By commutation, the same is true for V^kx_0. Also every point lies in some interval $[V^kx_0, V^{k+1}x_0]$, so by order preservation, the result holds for all x.

Case (ii): Next suppose $\forall\, p, q$ coprime, $U^qT^px - x$ has constant sign and is never zero. Then there exists a (unique) ray in $\mathbb{Z}^2$, passing through (0,0) but through no other points of $\mathbb{Z}^2$, and dividing $\mathbb{Z}^2\backslash(0,0)$ into pairs (p, q) for which $U^qT^px \gtrless x$ for all x. In other words,

$$\exists\,(\mu, \nu) \in \mathbb{R}^2\backslash(0,0) \text{ s.t. } U^qT^px \gtrless x\ \forall x \text{ according as } q\mu - p\nu \gtrless 0. \tag{4.7.7.10}$$

To show this, first note that $U^qT^px > x\ \forall x$ iff $U^{-q}T^{-p}x < x\ \forall x$. Secondly, if:

$$U^{q_i}T^{p_i}x > x\ \ \forall x,\ i = 0,1 \tag{4.7.7.11}$$

with (p_i, q_i), $i = 0, 1$ not on the same ray (wlog, $q_0p_1 > p_0q_1$), then $U^qT^px > x$ for all x, for any (p, q) in the smaller angle between (p_0, q_0) and (p_1, q_1), i.e. satisfying:

$$\begin{aligned} p_1q &\geq q_1p \\ p_0q &\leq q_0p \end{aligned} \tag{4.7.7.12}$$

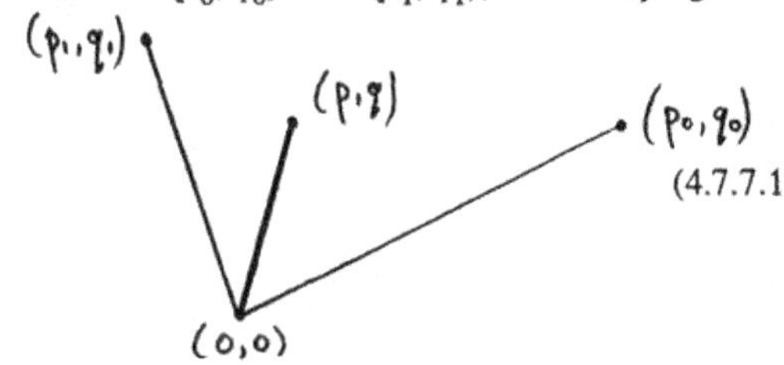

because there exist integers $a \geq 0, b \geq 0, N > 0$ such that:

$$\begin{aligned} aq_0 + bq_1 &= Nq \\ ap_0 + bp_1 &= Np \end{aligned} \tag{4.7.7.13}$$

viz.:

$$N = q_0p_1 - p_0q_1 \tag{4.7.7.14}$$

$$\begin{vmatrix} a \\ b \end{vmatrix} = \begin{vmatrix} p_1 & -q_1 \\ -p_0 & q_0 \end{vmatrix} \begin{vmatrix} q \\ p \end{vmatrix}. \tag{4.7.7.15}$$

Thus

$$(U^qT^p)^Nx > x,\ \ \forall x \tag{4.7.7.16}$$

and so $U^qT^px > x$, $\forall x$ (otherwise U^qT^px would be less than x, for all x, and (4.7.7.16) would not hold).

This implies the existence of the above ray. It is unique because $U^qT^px - x$ has constant sign for all (p, q). Now V moves all points to the right, so $A \equiv m\mu - n\nu > 0$. Given (p, q), with $q\mu - p\nu > 0$, say, write it as:

$$(i+cm, j+cn) \tag{4.7.7.17}$$

with c the largest integer such that $i\mu - j\nu > 0$. Then:

$$\frac{|U^p T^q x - x|}{r} > \frac{cK}{r} \sim \frac{K}{A} \frac{p\mu - q\nu}{r}. \tag{4.7.7.18}$$

So the pair has rotation number $\mu{:}\nu$.

§4.7.8 **Types of irrational number**. In this section I give a summary of various types of irrational number. A number ω is said to have *Liouville exponent* τ if:

$$\exists C > 0 \text{ s.t. } |\omega - \frac{q}{p}| \geq \frac{C}{q^\tau} \; \forall p, q \in \mathbb{Z}, q > 0. \tag{4.7.8.1}$$

I define LE_τ to be the set of numbers with Liouville exponent τ. Clearly

$$LE_\tau \subset LE_{\tau'} \text{ for } \tau \leq \tau' \tag{4.7.8.2}$$

so they form an increasing sequence. The "pigeon-hole principle" (Niven, 1963) shows that:

$$LE_\tau = \emptyset \text{ for } \tau < 2. \tag{4.7.8.3}$$

For $\tau > 2$, LE_τ can be shown to have full measure. On the other hand, LE_2 is precisely those numbers whose continued fraction expansion (1.3.1.5.10) has bounded coefficients. This is a set of measure zero. They are also known as numbers of *constant type*.

The transition at $\tau = 2$ can be broken up as follows:

Nobles $\subset QI \subset LE_2 \subset DB \subset$ condn. $A \subset$ Roth $\subset LE_\tau, \tau > 2 \subset$ Diophantines $\subset$ Irrationals

measure zero | full measure (4.7.8.4)

QI is the *quadratic irrationals*, those irrationals which satisfy a quadratic equation with integer coefficients. Equivalently, they are the numbers whose continued fraction expansion is eventually periodic. DB (*bounded density*) is the set of numbers whose continued fraction expansion $[a_0,...]$ has bounded coefficients in an average sense:

$$\sup \frac{1}{n} \sum_{i=0}^{n} a_i < \infty. \tag{4.7.8.5}$$

It has measure zero. $[a_0,...]$ satisfies *condition A* (Herman, 1977) if:

$$\lim_{B \to \infty} \limsup_{N \to \infty} \frac{\sum_{\substack{1 \le i < N \\ a_i \ge B}} \log(a_i + 1)}{\sum_{1 \le i \le N} \log(a_i + 1)} = 0. \qquad (4.7.8.6)$$

Equivalently, if given $\varepsilon > 0$, $\exists B$ s.t.

$$\prod_{\substack{i=1 \\ a_i \ge B}}^{N} (1 + a_i) = 0(q_{N-1}^{\varepsilon}) \text{ as } N \to \infty \qquad (4.7.8.7)$$

where q_n is the denominator of the nth convergent. The *Roth* numbers are

$$\bigcap_{\tau > 2} LE_{\tau}. \qquad (4.7.8.8)$$

Equivalently, they are the numbers for which:

$$a_i = 0(q_i^{\varepsilon}) \ \forall \varepsilon > 0. \qquad (4.7.8.9)$$

Note that the algebraics (irrationals which satisfy a polynomial equation with integer coefficients) are all Roth numbers (Stolarsky, 1974). Finally, the *Diophantines* are

$$\bigcup_{\tau} LE_{\tau}. \qquad (4.7.8.10)$$

Each of the inclusions in (4.7.8.4) is strict. For example, e is a Roth number, but does not satisfy condition A (continued fraction expansion $[2, 1, 2, 1, 1, 4, 1, 1, 6, 1, 1, ...]$) (Knuth, 1968). The number:

$$\sum_{n} 2^{-n!} \qquad (4.7.8.11)$$

is irrational but not Diophantine (Niven, 1963). The complement of the Diophantines in the irrationals is called the *Liouville numbers*. The above example can be generalised to show that there are uncountably many.

Summary

In summary, the main result of this thesis is that there are universal aspects of self-similarity in area preserving maps, connected with fixed points of certain renormalisation operators. I studied two particular renormalisations. The first is relevant to the disintegration of the stable region associated with an initially stable periodic orbit, as a parameter is varied. The disintegration is accompanied by an infinite sequence of successive period doublings, starting from the initial periodic orbit. The accumulation point of the sequence corresponds to a fixed point of the renormalisation. Beyond the accumulation point, there are locally only very small regions of stability.

The second is relevant to the breakup of invariant circles. An important quantity for an invariant circle is its rotation number. I concentrated on noble rotation numbers, which I showed to be the most significant in a certain sense. Invariant circles break up, as a parameter is varied, by developing thin spots which are visited very infrequently. These thin spots turn into gaps which are not visited at all by the original orbit. The critical case is given by a fixed point of a renormalisation. I discussed the relevance of this renormalisation to problems involving rotation number, and I developed a scheme which finds this fixed point to high accuracy. Properties of the fixed point lead to an approximate criterion for existence of invariant circles, which should have significant practical value.

The renormalisation for invariant circles also has a simple fixed point, corresponding to a linear twist map. Analysis of the stability of this fixed point under the renormalisation shows that it is attracting, which (modulo some details to be tidied up)[57] gives a new proof of persistence of noble circles for small enough perturbations of a linear twist map.

PROBLEMS FOR THE FUTURE

In this section I collect together a list of problems for further work, raised by this thesis.

(1) Find the generality of reversibility in area preserving maps (§1.1.4.5).
(2) Investigate further Greene's result on lack of symmetry of perturbations of the standard map (§1.1.4.5).
(3) Prove/disprove the Jacobian conjecture (§1.1.4.5).
(4) Examine the Krein crunch (§1.2.1.6).[58]
(5) Write actions for a general symmetric periodic orbit in terms of halfway round the orbit only (§1.2.3.3).

---

[57] Which have been done.

[58] This has been done by Bridges, Cushman and MacKay (1993).

(6) Explain the dominant symmetry for Birkhoff orbits (§1.2.3.4).
(7) Find the generic bifurcations from symmetric orbits of reversible maps, not necessarily area preserving (§1.2.4.6).[59]
(8) Use Rimmer bifurcation to construct a non-reversible area preserving map (§1.2.4.6).[60]
(9) Find the generic bifurcations from strongly symmetric periodic orbits (§1.2.4.7).
(10) Derive the fine structure constant as a property of a fixed point of an appropriate renormalisation operator (§2.1.3).[61]
(11) Do renormalisation for turbulence theory, and get universal equations for a turbulent fluid (§2.1.4).[62]
(12) Find universal accelerator modes after breakup of invariant circles (§2.2.4).
(13) Do renormalisation for horseshoes or homoclinic tangencies (§2.2.4).
(14) Prove existence of the fixed point for 1-D period doubling in C^3 (§2.3.10).
(15) Explain why β is so close to α^2 (§3.1.2).
(16) Determine α, β in Rannou's map (§3.1.2).
(17) Find universal horseshoe in the universal lpf for period doubling in area preserving maps (§3.3.4).
(18) Look again for infinite period doubling sequences in 4-D maps (§3.4.3).[63]
(19) Define the Calabi invariant for commuting pairs without using generating functions (§4.1.2).[64]
(20) Generalise Denjoy, Poincaré-Birkhoff, Moser twist, and Mather's theorems to commuting pairs (§4.1.2).
(21) Fix up the use of Mather's theorem for existence of a golden circle (§4.2.2).
(22) Show that all the Jordan blocks for DN_1 at the simple fixed point are simple (§4.3.3).
(23) Prove existence of a golden curve for pairs converging to the simple fixed point without use of Moser twist or Mather's theorems. Find how smooth they will be (§4.3.5).[65]
(24) Construct a C^4 neighbourhood of attraction for the area preserving simple fixed point (§4.3.5).

---

59 Has been done by Roberts and Quispel (1992).

60 This has been done by Roberts & Capel (1992).

61 It seems this is still open e.g. Feynman, 1985.

62 There are some theories in this direction, e.g. Yakhot & Orszag (1986), Avellaneda & Majda (1991). A good problem, suggested to me by Saffman, would be to prove the "log-law of the wall", in particular deriving the Karman constant (e.g. Tennekes and Lumley, 1972).

63 This has been done by Mao & Helleman (1986).

64 This has been done in a footnote to that section.

65 This has been done by Khanin & Sinai (1986), and Haydn (1990).

(25) Explain why δ is so close to γ (§4.4.1).
(26) Explain the triples of eigenvalues of DN_{s1} (§4.4.2).
(27) Relate my approach to finding the critical fixed point to that of Kadanoff and Shenker (§4.4.2).
(28) Use Tchebyshev projection to reduce truncation error in composition.[66] Try Padé approximants (equivalently, continued fraction expansion for functions). Try other bases than powers (§4.4.2).
(29) Measure the fraction of area occupied by invariant circles for $\mu < 0$, in the universal 1pf, and some sort of diffusion rate for $\mu > 0$ (§4.5.1).
(30) Try rotation number $[(2,\ )^{\infty}]$ and check that there are lots of noble circles left at criticality(§4.5.2).[67]
(31) Formulate the fixed point at infinity (§4.5.3).
(32) Integrate W^u into the neighbourhood of attraction for the simple fixed point (§4.6.1).
(33) Figure out the behaviour under renormalisation for other rotation numbers. In particular, show convergence to a linear twist for close enough maps (§4.6.3).[68]
(34) Do continued fraction expansions in higher dimensions.[69] Find the analogue of nobles. Investigate breakup of invariant tori in higher dimensional symplectic maps (§4.6.3).
(35) Optimise the neighbourhood of attraction for the 1-D simple fixed point, and write a calculator program that does interval arithmetic (§4.7.4).
(36) Check the essential convergence rate for the critical fixed point in 1-D (§4.7.6).[70]
(37) Obtain the universal frequency locking picture for the neighbourhood of nobles (§4.7.6).[71]

Additional problems

(38) Determine the non-area preserving spectrum at an area-preserving fixed point of the renormalisation for breakup of invariant circles (§4.2.4(v)).[72]
(39) Understand the breakup of invariant circles in non-reversible maps, where there is no obvious dominant line (§4.4.1).
(40) Prove existence of the critical golden fixed point and its hyperbolicity (§4.4.2).

---

66 Lanford does this routinely now.

67 This has been done by Stark (MacKay & Stark, 1992).

68 This has been done; see discussion in footnotes to §4.6.3.

69 There are lots of possibilities e.g. Jacobi-Perron, Kim-Ostlund, Poincaré, all essentially equivalent and none ideal.

70 Done; see footnote in §4.7.6.

71 Done; see footnote in §4.7.6.

72 This has now been done: MacKay (1992d).

(41) Prove non-existence of golden invariant circles for systems above the stable manifold of the critical fixed point (§4.6.1).
(42) Determine the universality class for the critical golden fixed point (§4.6.1).
(43) Apply the results of this thesis to practical problems, such as the ones which initiated this work: confinement in magnetic fusion reactors.[73]

[73] The residue criterion has been applied to the design of stellarators: see Cary and Hanson (1986). Most applications require treatment of systems with more degrees of freedom, but I believe the results given here could also be useful for problems like confinement/disspersion of short waves in axisymmetric plasmas and for the design of steady fluid flows with good mixing properties

References

Volumes frequently referred to

[1] Topics in Nonlinear Dynamics, ed. S. Jorna, Am. Inst. Phys. Conf. Proc. vol. 46 (AIP, New York, 1978).

[2] Nonlinear dynamics and the beam-beam interaction, eds. M. Month and J.C. Herrera, AIP Conf. Proc. **57** (1979).

[3] "Longtime prediction in Dynamics", Nonequilibrium Problems in Statistical Mechanics, vol. 2, eds. W. Horton, L. Reichl and V. Szebehely (Wiley, New York, 1983).

[4] Chaotic behaviour in deterministic systems, Les Houches 1981, eds. G. Iooss, R.H.G. Helleman & R. Stora (N. Holland, 1983).

[5] Proceedings of the conference on Order in Chaos, Los Alamos (May, 1982), Physica **7D** (1983).

Papers and books

Abraham, R. and Marsden, J., 1978, Foundations of mechanics (Benjamin, New York).

Amit, D.J., 1978, Field theory, the Renormalisation Group, and Critical Phenomena (McGraw-Hill) (& 2nd ed. World Sci., 1984).

Arnold, V.I., 1963, Proof of a theorem of A.N. Kolmogorov on the invariance of quasiperiodic motions under small perturbations of the Hamiltonian, Russ. Math. Surveys **18:5**, 9-36; and Small denominators and problems of stability of motion in classical and celestial mechanics, Russ. Math. Surveys **18:6**, 85-91.

Arnold, V.I., 1978, Mathematical Methods of Classical Mechanics (Springer, New York).

Arnold, V.I., 1980, Chapitres supplémentaires de la théorie des équations différentielles ordinaires (Editions de Moscou) (roughly equivalent English version: Geometric Methods in the Theory of Ordinary Differential Equations, Springer, 1983).

Arnold, V.I. and Avez, A., 1968, Ergodic problems of Classical Mechanics (Benjamin; reprinted by Addison Wesley, 1989).

Aubry, S., 1983, The twist map, the extended Frenkel-Kontorova model and the devil's staircase, in [5], 240-258.

Aubry, S., Le Daeron, P.Y. and André, G., 1982, "Classical ground-states of a one dimensional model for incommensurate structures", preprint, Saclay (never published, but see Aubry & Le Daeron, 1983 for most of it).

Bak, P. and Jensen, M.H., 1982, "Bifurcations and chaos in the φ^4 theory on a lattice", J. Phys A**15**, 1893.

Balescu, R., 1975, Equilibrium and non-equilibrium statistical mechanics (Wiley, New York).

Belitskii, G.R., 1978, Russ. Math. Surveys **33**, 107.

Benettin, G., Cercignani, C., Galgani, L. and Giorgilli, A., 1980a, Universal properties in conservative dynamical systems, Lettere Nuovo Cimento **28**, 1-4.

Benettin, G., Galgani, L. and Giorgilli, A., 1980b, Further results on universal properties in conservative dynamical systems, Lettere Nuovo Cimento **29**, 163-6.

Birkhoff, G.D., 1932, Sur quelques courbes fermées remarquables, Bull. Soc. Math. de France **60**, 1-26, reprinted in Collected Mathematical Papers, Vol. II, AMS (New York, 1950), 111.

Birkhoff, G.D., 1927, Dynamical Systems (AMS Colloq. Publ., vol. 9, revised 1966).

Bountis, T.C., 1981, Period doubling bifurcations and universality in conservative systems, Physica **3D**, 577-589.

Bredon, G.E., 1972, Introduction to compact groups (Academic Press).

Campanino, M., Epstein, H. and Ruelle, D., 1982, "On Feigenbaum's functional equation", Topology **21**, 125.

Campbell, L.A., 1973, Math. Ann. **205**, 243.

Cary, J.R., 1981, "Lie transform perturbation theory for Hamiltonian systems", Phys. Rept **79**, 129-159.

Chang, S-J., Wortis, M. and Wright, J.A., 1981 Phys. Rev. **24A**, 2669.

Chenciner, A., 1983, in [4].

Chirikov, B.V., 1979, A universal instability of many-dimensional oscillator systems, Phys. Repts **52**, 263-379.

Collet, P. and Eckmann, J-P., 1980, Iterated maps on the interval as dynamical systems (Birkhäuser, Boston).

Collet, P., Eckmann, J-P. and Koch, H., 1981a, J. Stat. Phys. **25**, 1.

Collet, P., Eckmann, J-P. and Koch, H., 1981b, On universality for area-preserving maps of the plane, Physica **3D**, 457-467.

Coullet, P. and Tresser, C., 1978, C.R. Acad. Sc. Paris **287**, 577, and J. de Physique **39**, Colloq. C5-25.

Curry, J.H. and Yorke, J.A., 1978, Springer Lect. Note Math. **668**, 48.

Dahlquist, G. and Björck, A., 1974, Numerical Methods (Prentice-Hall Inc., Englewood Cliffs, NJ)

Derrida, B. and Pomeau, Y., 1980, Feigenbaum's ratio of two-dimensional area-preserving maps, Phys. Lett. **80A**, 217-9.

Devaney, R., 1976, Reversible diffeomorphisms and flows, Trans. AMS **218**, 89-113.

Devaney, R. and Nitecki, Z., 1979, Shift automorhisms in the Hénon mapping, Commun. Math. Phys. **67**, 137-146.

DeVogelaere, R., 1958, On the structure of symmetric periodic solutions of conservative systems, in Contributions to the Theory of Nonlinear Oscillations, vol. IV, ed. S. Lefschetz, p. 53-84.

DeVogelaere, R., 1962, Tech.rep. 62-1, Dept. Math., U.C. Berkeley.

Dewar, R.L., 1976, J. Phys. **9A**, 2043.

Dragt, A.J. and Finn, J.M., 1976, Lie series and invariant functions for analytic symplectic maps, J. Math. Phys. **17**, 2215.

Eckmann, J-P., 1983, in [4].

Eckmann, J-P., Koch, H. and Wittwer, P., 1984, "A computer-assisted proof of universality for area preserving maps", Mem. AMS **47:289**, 1-122.

Engel, W., 1955, Math. Ann. **130**, 11.

Engel, W., 1958, Math. Ann. **136**, 319.

Escande, D.F. and Doveil, F., 1981a, Renormalisation method for the onset of stochasticity in Hamiltonian systems, Phys. Lett. **83A**, 307-310.

Escande, D.F. and Doveil, F., 1981b, Renormalisation method for computing the threshold of the large scale stochastic instability in two degree of freedom Hamiltonian systems, J. Stat. Phys. **26**, 257-284.

Escande, D.F., 1982a, "Renormalisation approach to non-integrable Hamiltonians", in [3].

Escande, D.F., 1982b, "Large scale stochasticity in Hamiltonian systems", Physica Scripta **T2:1**, 126-141.

Escande, D.F. and Mehr, A., 1982, "Link between KAM tori and nearby cycles", Phys. Lett. **91A**, 327.

Feigenbaum, M.J., 1978, J. Stat. Phys. **19**, 25.

Feigenbaum, M.J., 1979, J. Stat. Phys. **21**, 669.

Feigenbaum, M.J., Kadanoff, L.P. and Shenker, S.J., 1982, "Quasiperiodicity in dissipative systems: a renormalisation analysis", Physica **5D**, 370.

Finn, J.M., 1974, Doctoral Thesis (Univ. of Maryland).

Gallavotti, G., 1982, "Perturbation theory of classical Hamiltonian systems", Lecture notes in Physics, ed. J. Frolich (Springer).

Gollub, J.P., 1982, in [4].

Grebogi, C. and Kaufman, A.N., 1981, Phys. Rev. **24A**, 2829.

Greene, J.M., 1968, Two-dimensional area preserving mappings, J. Math. Phys. **9**, 760-8.

Greene, J.M., 1979a, A method for computing the stochastic transition, J. Math. Phys. **20**, 1183-1201.

Greene, J.M., 1979b, KAM surfaces computed from the Hénon-Heiles Hamiltonian, in [2], p. 257-271.

Greene, J.M., 1980, The calculation of KAM surfaces, Annals of New York Acad. Sci. **357**, 80-89.

Greene, J.M., 1981, "Noncanonical Hamiltonian mechanics", in Mathematical methods in Hydrodynamics, La Jolla (Dec. 1981).

Greene, J.M., MacKay, R.S., Vivaldi, F. and Feigenbaum, M.J., 1980, paper 7T5, Bull. Am. Phys. Soc. **25**, 987.

Greene, J.M., MacKay, R.S., Vivaldi, F. and Feigenbaum, M.J., 1981, Universal behaviour in families of area-preserving maps, Physica **3D**, 468–486.

Grossmann, S., and Thomae, S., 1977, Z. Naturforsch **32a**, 1353.

Halmos, P.R., 1956, Lectures on Ergodic Theory (Math. soc. of Japan, Tokyo).

Helleman, R.H.G., 1980, in Fundamental problems in Statistical Mechanics, vol. 5, ed. E.G.D. Cohen, p. 165 (North Holland, Amsterdam).

Helleman, R.H.G. 1983, in [3].

Hénon, M., 1969, Numerical study of quadratic area-preserving mappings, Quart. Appl. Math. **27**, 291–312.

Hénon, M. and Heiles, C., 1964, The applicability of the third integral of motion: some numerical experiments, Astron. J. **69**, 73–79.

Herman, M.R., 1976, C.R. Acad. Sc. Paris série A **282**, 503 and **283**, 579.

Herman, M.R., 1977, Springer Lecture notes in Math. **597**, 271.

Herman, M.R., 1979, Publ. Math. IHES **49**, 5.

Herman, M.R., 1981, "Demonstration du théorème des courbes translatées de nombres de rotation de type constant", manuscript, Paris, and other notes (see Herman, 1983).

Herman, M.R., 1982,unpublished lectures given at [4].

Janssen, T. and Tjon, J.A., 1982, Phys. Lett. **87A**, 139.

Kadanoff, L.P., 1966, Physics **2**, 263.

Kadanoff, L.P., 1981a, Scaling for a critical Kolmogorov-Arnol'd-Moser trajectory, Phys. Rev. Lett. **47**, 1641–3.

Kadanoff, L.P., 1981b, in Proceedings of the 9th Midwestern Solid State Theory Seminar (Argonne).

Kappraff, J.M. and Marzec, C., 1982, "Plant phyllotaxis", submitted to J. Theo. Biol.

Karney, C.F.F., Rechester, A.B. and White, R.B., 1982, Effect of noise on the standard mapping, Physica **4D**, 425–438.

Katok, A., 1979, Bernoulli diffeomorphisms on surfaces, Annals of Math. **110**, 529–547.

Katok, A., 1982, "Some remarks on Birkhoff and Mather twist map theorems", Erg. Th. Dyn. Sys. **2**, 185.

Khovanskii, A.N., 1963, The application of continued fractions and their generalisations to problems in approximation theory (Noordhoff, Groningen).

Knuth, D.E., 1968, The art of computer programming, vol. 2, Seminumerical algorithms (Addison Wesley, Reading, Mass.)

Kolmogorov, A.N., 1954, Preservation of conditionally periodic movements with small change in the Hamiltonian function, Dokl. Akad. Nauk **98**, 527–531 (reprinted in English, in Stochastic behaviour in classical and quantum Hamiltonian systems, eds. G. Casati and J. Ford, Lect. notes in Phys., vol. 93, Springer, 1979).

Krasnosel'skii, M.A., Vainikko, G.M., Zabreiko, P.P., Rutitskii, Ya. B. and Stetsenko, V. Ya., 1972, Approximate solution of Operator equations (Wolters-Noordhoff).

Lanford III, O.E., 1980, Springer Lecture notes in Physics **116**, 340.

Lanford III, O.E., 1981, "A computer-assisted proof of the Feigenbaum conjectures",

Bull. AMS **6** (1982) 427.

Lanford III, O.E., 1983, in [4].

Laplace, in Essai philosophique sur les probabilités, Mecanique Céleste, Livre XV, Ch. 1, p. 324.

Laslett, L.J., 1978, in [1], p. 221.

Lichtenberg, A.J., 1979, in Intrinsic Stochasticity in Plasmas, eds. G. Laval and D. Gresillon (Editions de Physique, Orsay).

Lichtenberg, A.J., and Liebermann, M.A. 1983, "Regular and Stochastic Motion" (Applied Mathematical Sciences **38**, Springer, New York).

Littlejohn, R.G., 1981, "Efficient, higher order guiding centre calculations via the action form *p.dq - Hdt*", preprint, UCLA, PPG-582, and "Hamiltonian formulation of guiding centre motion", Phys. Fl 24 (1981) 1730-49 (see also Littlejohn, 1983 and Cary and Littlejohn, 1983)

MacKay, R.S., 1983, in [3].

MacKay, R.S., 1982, Islets of stability beyond period-doubling, Phys. Lett. **87A**, 321-4.

MacKay, R.S., 1983, A renormalisation approach to invariant circles in area-preserving maps, in [5], 283-300.

Mandelbrot, B.B., 1977, Fractals: Form, Chance and Dimension (W.H. Freeman, San Francisco).

Markus, L. and Meyer, K.R., 1974, Generic Hamiltonian dynamical systems are neither integrable nor ergodic, Mem.Amer. Math. Soc. **144**.

Markus, L. and Meyer, K.R., 1980, Periodic orbits and solenoids in generic Hamiltonian dynamical systems, Am. J. Math. **102**, 25-92.

Mather, J.N., 1984, "Non-existence of invariant circles", Erg. Th. Dyn. Sys. **4**, 301-309.

Mather, J.N., 1982b, "Existence of quasiperiodic orbits for twist homeomorphisms of the annulus", Topology **21**, 457-467; also Comm. Math. Helv. **57** (1982) 356, Non-uniqueness of solutions of Percival's Euler-Lagrange equation, Comm. Math. Phys. **86** (1982) 465-473.

Mather, J.N., 1986, "A criterion for the non-existence of invariant circles", Publ. Math. IHES **63**, 153-204.

Mather, J.N., 1982d, "Glancing billiards", Erg. Th. Dyn. Sys. **2**, 397-403.

May, R.M., 1974, Science **186**, 645.

May, R.M., 1976, Nature **261**, 259.

May, R.M. and Oster, G.F., 1976, Amer. Naturalist **110**, 573.

McLaughlin, J.B., 1981, J. Stat. Phys. **24**, 375.

McMillan, E.M., 1971, in Topics in Modern Physics (Colo. Ass.U.P., Boulder, CO), p. 219.

Metropolis, M., Stein, M.L. and Stein, P.R., 1973, J. Comb. Theory **15**, 25.

Meyer, K.R., 1970, Generic bifurcation of periodic points, Trans AMS **149**, 95-107.

Meyer, K.R., 1971, Generic stability of elliptic points, Trans AMS **154**, 273-277.

Meyer, K.R., 1975, Generic bifurcations in Hamiltonian systems, Springer Lecture notes in

Mathematics **468**, 62-70.
Milton, Paradise lost, Poetical works, ed. D. Bush (OUP, 1966).
Montgomery, D. and Zippin, L., 1955, Topological transformation groups (Interscience, New York).
Morrison,P.J. and Greene, J.M., 1980, Non-canonical Hamiltonian density formulation of hydrodynamics and ideal magnetohydro-dynamics, Phys. Rev. Lett. **45**, 790-4.
Moser, J.K., 1956, The analytic invariants of an area-preserving mapping near a hyperbolic fixed point, Comm.PAM **9**, 673-692.
Moser, J.K., 1958, New aspects in the theory of stability of Hamiltonian systems, Comm. PAM **11**, 81 (see also Gelfand & Lidskii, 1955).
Moser, J.K., 1960, Bol. Soc. Mat. Mex. **5**, p. 176.
Moser, J.K., 1962, On invariant curves of area-preserving mappings for an annulus, Nachr. Akad. Wiss. Göttingen, Math. Phys. **IIa**, 1.
Moser, J.K., 1973, Stable and Random Motions (Princeton Univ. Press).
Nelkin, M., 1974, Phys. Rev. **9A**, 388.
Nelson, E., 1969, Topics in dynamics I: flows (Princeton Univ. Press).
Newhouse, S.E., 1979, Publ. Math. IHES **50**, 101.
Newhouse, S.E., 1980, in Dynamical Systems: Bressanoné, CIME, 1978, ed. J. Guckenheimer (Birkhäuser, Boston).
Newman, R.A.P.C. and Percival, I.C., 1983, "Definite paths and upper bounds on regular regions of velocity phase space", Physica **6D**, 249-259.
Newton, I., Principia, A revision of Motte's translation, F. Cajori (UCP Berkeley, 1946), p. 547.
Nitecki, Z., 1971, Differentiable Dynamics (MIT press, Cambridge, Mass.).
Niven, I., 1963, Irrational Numbers (Carus Math. Monographs, no. 11, Wiley).
Oseledec, V.I., 1968, A multiplicative ergodic theorem: Lyapunov characteristic numbers for dynamical systems, Trans. Mosc. Math. Soc. **19**, 197-231.
Percival, I.C., 1979, A variational principle for invariant tori of fixed frequency, J. Phys. **12A**, L57-60.
Percival, I.C., 1979, Variational principles for invariant tori and cantori in [2], p. 302-310.
Percival, I.C., 1982, "Chaotic boundary of a Hamiltonian map", Physica **6D**, 67-77.
Pesin, Ya. B., 1977, Lyapunov characteristic exponents and the smooth ergodic theory, Russ. Math. Surveys **32:4**, 55-114.
Poincaré, H., Les Méthodes Nouvelles de la Méchanique Céleste (NASA Translation TTF-450/452, US Fed. Clearinghouse, Springfield VA, 1967).
Prasad, A.V., 1948, J. London Math. Soc. **23**, 169.
Rand, D., Ostlund, S., Sethna, J. and Siggia, E., 1982, Phys. Rev. Lett. **49**, 132 (see also Ostlund et al, 1983).
Rannou, F., 1974, Numerical study of discrete plane area-preserving mappings, Astron. & Astrop. **31**, 289-301.

Rimmer, R., 1978, Symmetry and bifurcation of fixed points of area-preserving maps, J. Diff. Eqns **29**, 329-344.

Rimmer, R., 1983, "Generic bifurcations from fixed points of involutory area preserving maps", Memoirs of AMS **41** (1983).

Robinson, R.C., 1970, Generic properties of conservative systems I & II, Am. J. Math. **92**, 562-603 and 897-906.

Rössler, O.E., 1979, Ann. N.Y. Ac. Sci. **316**, 376.

Rüssmann, H., 1983, "On the existence of invariant curves of twist mappings of an annulus", Springer Lecture Notes in Math **1007**, 677-718.

Schmidt, G., 1980, Stochasticity and fixed point transitions, Phys. Rev. **22A**, 2849-54.

Schmidt, G. and Bialek, J., 1982, Fractal diagrams for Hamiltonian stochasticity, Physica **5D**, 397.

Shenker, S.J., 1982, "Scaling behaviour in a map of a circle to itself: empirical results", Physica **5D**, 405.

Shenker, S.J., 1982, in [5].

Shenker, S.J. and Kadanoff, L.P., 1982, Critical behaviour of a KAM surface: I Empirical results, J. Stat. Phys. **27**, 631-656.

Siegel, C.L. and Moser, J.K., 1971, Lectures on Celestial Mechanics (Springer).

Siggia, E., 1982, in [5].

Simo, C., 1979, J. Stat. Phys. **21**, 465.

Sinclair, R.M., Hosea, J.C. and Sheffield, G.V., 1970, A method for mapping a toroidal magnetic field by storage of phase stabilised electrons, Rev. Sci. Instruments **41**, 1552-9.

Spivak, M., 1965, Calculus on manifolds (Benjamin, New York).

Stolarsky, K.B., 1974, Algebraic numbers and Diophantine approximation (Marcel Dekker Inc.).

Stueckelberg, E.C.G. and Petermann, A., 1953, Helv. Phys. Acta **26**, 499.

Sulem, P.L., Fournier, J.D. and Pouquet, A., 1979, Springer Lecture notes in Phys. **104**, 321.

Tabor, M., 1981, The onset of chaotic motion in dynamical systems, Adv. Chem. Phys. vol. 46, eds. S.A. Rice and I. Prigogine (Wiley, New York), 73-152.

Tennyson, J.L., Liebermann, M.A. and Lichtenberg, A.J., 1979, Diffusion in near-integrable Hamiltonian systems with three degrees of freedom, in [2], p. 272-301.

Vivaldi, F. and Ford, J., 1981, "Symmetries in systems with $1\frac{1}{2}$ degrees of freedom", preprint, Atlanta (never published).

White, R.B. et al., 1982, IAEA conference proceedings, IAEA-CN-41/T-3.

Whiteman, K.J., 1977, Invariants and stability in classical mechanics, Rep. Prog. Phys. **40**, 1033-99.

Widom, M. and Kadanoff, L.P., 1982, Renormalisation group analysis of bifurcations in area preserving maps, Physica **5D**, 287-292.

Wightman, A.S., 1981, "The mechanisms of stochasticity in classical dynamical systems",

in Perspectives in Statistical Physics (N. Holland), ed. H.J. Raveché, 343-363.
Wilson, K.G., 1971, Phys. Rev. **3D**, 1818.
Wilson, K. and Kogut, J., 1974, Phys. Reports **12C**, 75.
Wright, D., 1981, Illinois J. Math. **25**, 423; H. Bass, E.H. Cornell & D. Wright, Bull AMS **7** (1982) 287.
Zehnder, E., 1973, Homoclinic points near elliptic fixed points, Commun. PAM **26**, 131-182.
Zisook, A.B., 1981, Phys. Rev. **A24**, 1640.
Zisook, A.B., 1982, talk given at [5].
Zisook, A.B. and Shenker, S.J., 1982, Renormalisation group for intermittency in area-preserving mappings, Phys. Rev. **25A**, 2824-6.

Supplementary reference list

Anderson, P.W., 1984, Basic notions of condensed matter physics (Addison Wesley).
Arnol'd, V.SI. (ed.) 1988, Dynamical systems III (Springer).
Aubry, S, Le Daeron, P.Y., 1983, The discrete Frenkel-Kontorova model and its extensions, Physica **8D**, 381-422.
Aubry, S. and Abramovici, G., 1990, Chaotic trajectories in the standard map: the concept of anti-integrability, Physica D **47**, 461-497.
Avellaneda, M. & Majda, A.J., 1991, Approximate and exact renormalisation theories for a model for turbulent transport, subm. to Phys Fl A.
Bridges, T., Cushman, R. and MacKay, R.S., 1993,Dynamics near irrational collision of eigenvalues for symplectic maps, preprint.
Bullet, S., 1986, Invariant circles for the piecewise linear standard maps, Comm. Math. Phys. **107**, 241-262.
Cary, J.R, Hanson, J.D., 1986, Stochasticity reduction, Phys. Fl **29**, 2464-2473.
Cary, J.R, Littlejohn, R.G., 1983, Noncanonical Hamiltonian mechanics and its application to magnetic field line flow, Ann. Phys. **151**, 1-34.
Celletti, A. and Chierchia, L., 1988, Construction of analytic KAM surfaces and effective stability bounds, Comm. Math. Phys. **118**, 119-161.
Davie, A.M. and Dutta, T.K., 1992, Period-doubling in two-parameter families, Physica D, to appear.
Davis, M.J., MacKay, R.S., and Sannami, A., 1991, Markov shifts in the Hénon family, Physica D **52**, 171-8.
Epstein, H. & Lascoux, J., 1981, Analyticity properties of the Feigenbaum function, Commum. Math. Phys. **81**, 437-53.
Falconer, K., 1985, The geometry of fractal sets (Cambridge).
Feigenbaum, M.J., 1988, Presentation fucntions and scaling function theory for circle maps, Nonlin. **1**, 577-602.
Feynman, R.P., 1985 QED (Penguin).

Fisher, M.E., 1983, Springer Lecture notes in physics **186**.

Gelfand, I.M. and Lidskii, V.B., 1955, On the structure of regions of stability of linear canonical systems of differential equations with periodic coefficients, Usp. Mat. Nauk **10**, 3-40 [AMS Transl. series 2, **8** (1958) 143-182].

Greene, J.M., MacKay, R.S., Stark, J, 1986, Boundary circles for area-preserving maps, Physica D **21**, 267-295.

Haydn, N.T.A., 1990, On invariant curves under renormalisation, Nonlinearity **3**, 887-912.

Herman, M.R., 1983, Sur les courbes invariantes par les difféomorphismes de l'anneau préservant les aires pt. I, Astérisque **103-104**.

Herman, M.R., 1986, Sur les courbes invariantes par les difféomorphismes de l'anneau préservant les aires pt. II, Astérisque **144**.

Kadanoff, L.P., 1983, Supercritical behaviour of an ordered trajectory, J. Stat. Phys. **31**, 1.

Ketoja, J. and MacKay, R.S., 1989, Fractal boundary for the existence of invariant circles for area-preserving maps, Physica D **35**, 318-334.

Khanin, K. and Sinai, Ya. G, 1986, The renormalisation group method and KAM-theory, in Nonlinear Phenomena in Plasma Physics and Hydrodynamics, ed. R.Z. Sagdeev (Mir) 93-118.

Kosygin, D. 1991, Multidimensional KAM theory from the renormalisation group viewpoint, in Adv. Sov. Math. vol. 3, Dyn. Sys. and Stat. Mech., Sinai Ya. G. (ed). (AMS).

Littlejohn, R.G., 1983, Variational principles of guiding centre motion, J. Plasma Phys. **29**, 111-125.

Llave R. de la, Rana, D., 1990, Accurate strategies for small divisor problems, Bull. AMS **22**, 85-90.

MacKay, R.S., 1984, Equivariant universality classes, Phys Lett A **106**, 99-100.

MacKay, R.S. Meiss, J.D., Percival, I.C., 1984, Transport in Hamiltonian systems, Physica D **13**, 55-81.

MacKay, R.S. and Percival, I.C., 1985, Converse KAM: theory and practice, Commun. Math. Phys. **98**, 469-512.

MacKay, R.S. and Percival, I.C., 1987, Universal small-scale structure near the boundary of Siegel disks of arbitrary rotation number, Physica **26D**, 193-202.

MacKay, R.S., 1986, Transition to chaos for area-preserving maps, in Nonlinear dynamics aspects of particle accelerators, eds. J.M. Jarrett, M. Month and S. Turner, Springer Lecture Notes in Phys. **247**, 390-454.

MacKay, R.S. and Meiss, J.D., 1983, Linear Stability of periodic orbits in Lagrangian systems, Phys. Lett **A98**, 92-94.

MacKay, R.S. and Tresser, C., 1986, Transition to topological chaos for circle maps, Physica D **19**, 223; Erratum, Physica D **29** (1988) 427.

MacKay, R.S. and Meiss, J.D., 1987, (eds) Hamiltonian dynamical systems: a reprint collection (Adam Hilger, Bristol).

MacKay, R.S., 1987, Introduction to the dynamics of area-preserving maps, in: Physics of Particle accelerators, eds. Month, M. & Dienes, M, Am. Inst. Phys. Conf. Proc. **153** vol. 1, 534-602.

MacKay, R.S. 1988, A simple proof of Denjoy's theorem, Math. Proc. Camb. Phil. Soc. **103**, 299-303.

MacKay, R.S., Zeijts, J.B.J van, 1988, Period doubling for bimodal maps: a horseshoe for a renormalisation operator, Nonlinearity **1**, 253-277.

MacKay, R.S. 1991, Scaling exponents at the transition by breaking of analyticity for incommensurate structures, Physica D **50**, 71-79.

MacKay, R.S., 1992b, Greene's residue criterion, Nonlinearity **5**, 161-187.

MacKay, R.S. 1992c, Hyperbolic structure in classical chaos, in Quantum Chaos, eds. Smilansky, Guarneri I and Casati G (Ital. Phys. Soc.).

MacKay, R.S. 1992a, Some aspects of the dynamics and numerics of Hamiltonian systems, in The Dynamics of numerics and numerics of dynamics, eds. D.S. Broomhead and A. Iserles (Oxford) 137-193.

MacKay, R.S. and Shardlow, T., 1992, Multiplicity of bifurcations in area-preserving maps, subm. to Bull LMS.

MacKay, R.S., 1992d, Non-area-preserving directions from area-preserving fixed points of golden circle renormalisation, subm. to Nonlinearity.

MacKay, R.S. and Stark, J. 1992, Locally most robust and boundary circles for area-preserving maps, Nonlinearity **5**, 867-888.

MacKay, R.S. and Baesens, C. 1993, The one-to-two-hole transition for cantori, in preparation.

Manton, N.S. and Nauenberg, M., 1983, Universal scaling behaviour for iterated maps in the complex plane, Commun. Math. Phys. **89**, 555-570.

Mao, J-M. and Helleman, R.H.G., 1987, New Feigenbaum constants for four-dimensional volume-preserving symmetric maps, Phys. Rev. A **35**, 1847.

Mather, J.N. 1990, Variational construction of orbits of twist diffeomorphisms, preprint ETH.

Meiss, J.D., 1986, Class renormalisation: islands around islands, Phys. Rev. A**34**, 2375-2383.

Meyer, K.R., 1981, Hamiltonian systems with a discrete symmetry, J. Diff. Eq. **41**, 228-238.

Meyer, K.R., and Hall, G.R., 1992, Introduction to Hamiltonian dynamical systems and the N-body problem (Springer).

Nehari, Z., 1952, Conformal mapping (Dover).

Ostlund, S. Rand, D.A. Sethna, J. and Siggia, E., 1983, Universal properties of the transition from quasi-periodicity to chaos in dissipative systems, Physica D8, 303-342.

Peyrard, M. and Aubry, S., 1983, Critical behaviour at the transition by breaking of analyticity in the discrete Frenkel-Kontorova model, J. Phys. C **16**, 1593-1608.

Rand, D.A. 1987, Fractal bifurcation sets, renormalisation strange sets and their universal invariants, Proc. Roy. Soc. A **413**, 45-61.

Rand, D.A. 1988, Universality and renormalisation in dynamical systems, in New directions in dynamical systems, eds. Bedford, T., Swift J.W. (CUP) 1-56.

Rand, D.A., 1988, Global phase space universality, smooth conjugacies and renormalisation: 1. The $C^{1+\alpha}$ case, Nonlinearity **1**, 181-202.

Rand, D.A. 1992, Existence, non-existence and universal breakdown of dissipative golden invariant tori, I, II & III, Nonlinearity **5**, 639-662, 663-680, 681-706.

Roberts, J.A.G. and Quispel, G.R.W. 1992, Chaos and time-reversal symmetry, Phys. Repts, to appear.

Roberts, J.A.G. and Capel, H.W. 1992, Area-preserving mappings that are not reversible, Phys Lett A **162**, 243-248.

Sinai, Ya. G 1989, (ed.) Dynamical Systems II (Springer).

Sinai, Ya. G, and Khanin, K, 1988, Int. J. Mod. Phys B**2**, 147-165.

Stirnemann, A., 1992, Renormalisation for golden circles, preprint ETH, Zurich.

Sullivan, D. 1991, Bounds, quadratic differentials and renormalisation conjectures, in Mathematics into the 21st century (AMS).

Takens, F. 1974, Forced oscillations and bifurcations, Comm. Math. Inst. Rijksuniv Utrecht **3**, 1-59.

Tennekes, H. and Lumley, J.L., 1972, A first course in turbulence (MIT).

Widom, M. 1983, Renormalisation group analysis of quasiperiodicity in analytic maps, Comm. Math. Phys. **92**, 121-136.

Veerman, J.J.P. and Tangerman, F.M. 1991, Intersection properties of invariant manifolds in certain twist maps, Comm. Math. Phys. **139**, 245-265.

Vul, E.B. and Khanin, K.M. 1982, On the unstable separatrix of Feigenbaum's fixed point, Russ. Math. Surv. 37:5, 200-201.

Yakhot, V, Orszag, S.A. 1986, Renormalisation group analysis of turbulence, I. Basic theory, J. Sci. Comp. **1**, 3-51.

www.ingramcontent.com/pod-product-compliance
Lightning Source LLC
Chambersburg PA
CBHW070322060426
42445CB00001BB/1
* 9 7 8 9 8 1 0 2 1 3 7 1 8 *